Advanced Cold Spraying Technology

Advanced Cold Spraying Technology

Guest Editors

Wen Sun
Xin Chu
Adrian Wei Yee Tan

Basel • Beijing • Wuhan • Barcelona • Belgrade • Novi Sad • Cluj • Manchester

Guest Editors

Wen Sun
Institute of New Materials
Guangdong Academy of
Sciences
Guangzhou
China

Xin Chu
Institute of New Materials
Guangdong Academy of
Sciences
Guangzhou
China

Adrian Wei Yee Tan
University of Southampton
(Malaysia)
Johor
Malaysia

Editorial Office
MDPI AG
Grosspeteranlage 5
4052 Basel, Switzerland

This is a reprint of the Special Issue, published open access by the journal *Coatings* (ISSN 2079-6412), freely accessible at: https://www.mdpi.com/journal/coatings/special_issues/R9X14V488I.

For citation purposes, cite each article independently as indicated on the article page online and as indicated below:

Lastname, A.A.; Lastname, B.B. Article Title. *Journal Name* **Year**, *Volume Number*, Page Range.

ISBN 978-3-7258-2437-3 (Hbk)
ISBN 978-3-7258-2438-0 (PDF)
https://doi.org/10.3390/books978-3-7258-2438-0

Cover image courtesy of Wen Sun

Contents

Article

Balanced Anti-Corrosion Action of Reduced Graphene Oxide in Zn-Al Coating during Medium-Term Exposure to NaCl Solution

Qifeng Shi [1,2,3], Huishu Wu [1,3,4,*], Peipei Zhang [1,3], Dongsheng Wang [1,2,3,5,*], Jingwen Wang [1,2,3,5] and Xiaohua Jie [4]

1   College of Mechanical Engineering, Tongling University, No. 4 Cuihu Road, Tongling 244000, China; 13431093689@163.com (Q.S.); wangjingwen@tlu.edu.cn (J.W.)
2   Advanced Copper-Based Materials Industry Generic Technology Research Center of Anhui Province, Tongling 244000, China
3   Key Laboratory of Additive Manufacturing, Anhui Higher Education Institutes, Tongling University, No. 4 Cuihu Road, Tongling 244000, China
4   School of Materials and Energy, Guangdong University of Technology, Guangzhou 510006, China; cnxyyz3@gdut.edu.cn
5   Key Laboratory of Construction Hydraulic Robots, Anhui Higher Education Institutes, Tongling 244000, China
*   Correspondence: ggwuhuishu@163.com (H.W.); wangdongsheng@tlu.edu.cn (D.W.)

Abstract: Considering the electronegativity and shielding anti-sepsis characteristic of reduced graphene oxide (G), we design a Zn-Al coating with embedded G (Zn-G/Al) on low-carbon steel using the low-pressure cold spray (LPCS) method. In this method, G-coated Al powders (G/Al) prepared using in situ reduction and Zn powders were mixed as a raw material for spraying. Embedding G could boost the cathodic protection performance of Zn-Al (70 wt.% zinc and 30 wt.% aluminum) coating, as has been confirmed in previous work. In this work, the microstructure, composition and electrochemical parameters of Zn-G/Al coating during full immersion were measured to investigate G's effect on the corrosion protection properties of the Zn-Al coating. The test results showed that embedded G could facilitate the generation of many corrosion products and pile on the coating surface to form a corrosion product film during full immersion. The corrosion product film on the Zn-0.2 wt.%G/Al coating surface demonstrated an excellent protective property, which reflects the fact that the E_{corr} and i_{corr} values for Zn-0.2 wt.%G/Al after 20d immersion ($E_{corr} = -1.143$ V$_{vs.SCE}$, $i_{corr} = 49.96$ µA/cm^2) were lower than the initial value ($E_{corr} = -1.299$ V$_{vs.SCE}$, $i_{corr} = 82.16$ µA/cm^2). It can be concluded that adding an appropriate amount of G to the coating can balance the cathodic protection and shielding property of the coating. The equilibrium mechanism was also analyzed in this work.

Keywords: Zn-Al coating; reduced graphene oxide; shielding protection; electrical conductivity; low-pressure cold sprayed; low-carbon steel

Citation: Shi, Q.; Wu, H.; Zhang, P.; Wang, D.; Wang, J.; Jie, X. Balanced Anti-Corrosion Action of Reduced Graphene Oxide in Zn-Al Coating during Medium-Term Exposure to NaCl Solution. *Coatings* **2023**, *13*, 1570. https://doi.org/10.3390/coatings13091570

Academic Editor: Alina Vladescu

Received: 7 August 2023
Revised: 4 September 2023
Accepted: 5 September 2023
Published: 8 September 2023

1. Introduction

Coating technology is an economical and effective surface treatment technology, which is widely used to prevent the corrosion and destruction of metal constructions in a marine environment [1,2]. A cathodic protection coating is one of many protective coatings [3,4]. The protection mechanism of cathodic protection coating is reflected in multiple ways. On the one hand, cathodic protection coatings act as a sacrificial anode to guard the substrate by making use of the potential difference between it and the substrate. On the other hand, the coating is also an excellent physical shielding coating, involving the shielding of the coating itself and the blocking of the corrosion products forming a film via the degradation of the coating [5].

Zn-based coatings are a promising cathodic protective coating that can prevent the corrosion of carbon steel due to their lower self-corrosion potential (around -1.1 V$_{SCE}$~-1.3 V$_{SCE}$)

compared to that of low-carbon steel (-0.72 V_{SCE}) [6–11]. For instance, the self-corrosion potential for a hot-dipped Zn-Al coating is -1.05 V_{SCE} [6] and that for cold-sprayed Zn-xNi is around -0.98 $V_{SCE} \sim -1.01$ V_{SCE} [7]. Among the Zn-based composite coatings, the Zn-Al coating has certain advantages. For example, the potential difference between Zn and Al is tiny, meaning that the galvanic corrosion tendency in the coating is not obvious. Moreover, the corrosion products of Zn-Al are mainly insoluble layered double hydroxides (LDHs), which could delay the penetration of the corrosive medium [10]. Therefore, Zn-Al coatings often act as an anode to protect low-carbon steel.

Many researchers have focused on the fabrication and anti-corrosion performance of Zn and Zn-Al coatings as a protective coating for steel in marine environments [9–14]. Tachiban et al. [9] studied the corrosion resistance of hot-dipped Zn and Zn-Al coatings in a coastal area. Xue et al. [11] investigated the microstructure and corrosion resistance behavior of a Zn-Al coating co-deposited on low-carbon steel via pack cementation. How the Al content affects the corrosion behavior of arc-sprayed Zn-Al coatings was further studied in the work of Zhu et al. [12]. The results indicated that the Zn-Al coating has a higher corrosion resistance to seawater than the pure Zn coating, and Zn-30Al exhibited the highest corrosion resistance. The experiments of Kim et al. [13] made clear that Zn-Al coatings improve the corrosion resistance of the base material in marine environments by forming ZnO and $Zn_5(OH)_8Cl_2 \cdot H_2O$ corrosion products. However, the pure Zn-30Al coating represented a limited cathodic protection efficiency in a harsh corrosive environment [15].

Usually, the potential difference between the coating and the substrate is increased to improve the efficiency of the cathodic protection of the coating. Liu [16] and Zhu et al. [17] added active elements Mg and Cu to a Zn-Al coating, respectively. Their experiments show that the obtained coatings have a lower electrode potential and express an enhanced cathodic protection efficiency. As a rule, a lower electrode potential means a higher active dissolution rate, which greatly reduces the service life of the coating. Hence, the key point when improving the comprehensive protective performance of Zn-Al coatings is to attain a balance between sacrificial behavior and dissolution resistance. Huang [15] increased the cathodic protection without causing too fast a coating dissolution by matching the ratio of Zn and Al in a Zn-Al coating. Arrighi [4] added Ce(III) as a potential corrosion inhibitor of sacrificial Zn-Fe coatings that were electrodeposited on steel. Liu [16] added an active Mg element to Al-Zn-Si, which decreased the electrode potential and enhanced the cathodic protection efficiency of an Al-Zn-Si alloy. Moreover, an appropriate amount of magnesium in the alloy can promote the formation of a MgZn phase, which could inhibit the corrosion of the alloy. This means that the appropriate alloying element Mg has a balancing effect on the service life and cathodic protection property of Al-Zn-Si-Mg alloys. Is there an adequate additive that could enhance both the service life and cathodic protection property of Zn-Al coatings?

Reduced graphene oxide (rGO; in this work, rGO is denoted as G) is a form of graphene. It is a special corrosion protection material with a large surface area and strong conductivity and a promising and versatile additive [18]. The anti-sepsis characteristics of graphene are related to its distribution state and the properties of the composite materials. When a complete graphene coating is distributed on the substrate surface [19], or a graphene composite is made with organic and inorganic coatings [20], graphene mainly plays a shielding role, using its large surface area to extend the propagation path of corrosive media. If graphene is composited with metals, the graphene in these composites may promote the dissolution of metal by accelerating electron transfer when in a corrosive environment, by which passive metals could form a passive film [21] and active metal would rapidly be degraded [22,23]. Based on the abovementioned points, it can be speculated that G is a suitable additive, which can enhance both the service life and cathodic protection properties of a Zn-Al coating, and the key is to distribute G evenly in the coating.

We deposited a Zn-G/Al composite coating on low-carbon steel using LPCS [24,25]. Using G-coated Al powders prepared by chemical reduction as a feedstock, G was successfully embedded in the interface of Zn and Al. According to experiments [24], G lowered the potential of the Zn-G/Al coating and thus enhanced its potential difference with the substrate. In this way, the boosting potential difference facilitates the cathode polarization behavior and induces the formation of corrosion products when in a harsh environment. In this work, Zn-G/Al composite coatings were immersed in NaCl solution, and the medium-term immersion behavior was observed to study how the embedded G affects the medium-term protection performance of Zn-G/Al coatings.

2. Experiments

2.1. Preparation and Characteristics of Zn-x wt.%G/Al Coatings

Zn-x wt.%G/Al coatings were deposited with 70 wt.% zinc (Zn) powders and 30 wt.% G-coated aluminum (G/Al) powders; x represents the mass G-to-Al particle ratio, which was 0, 0.1, 0.2 wt.% and 0.3 wt.%, respectively. G-coated Al powders were prepared by the chemical reduced method, as shown in ref. [24]. Taking the preparation method of 0.1 wt.%G/Al powders as an example: 20 mg graphene oxide (GO) was sonicated in 100 mL deionized (DI) water until it was well-dispersed in solvent. Afterward, 20 g pure Al powder was added to the dispersed GO solution and stirred until the solution became colorless and transparent. Then, the mixture was filtered and cleaned with ethanol, and finally dried under vacuum at 50 °C for 8 h to obtain the 0.1 wt.%G/Al powders. Due to the reduction in Al, GO was reduced to rGO and coated on the surface of the Al powder to form reduced graphene-oxide-coated Al powder (rGO/Al, namely G/Al). The preparation process of 0.2 wt.%G/Al and 0.3 wt.%G/Al powders is the same as for 0.1 wt.%G/Al; the only difference is the dispersed GO solution, which is 40 mg/200 mL for 0.2 wt.%G/Al and 60 mg/300 mL for 0.3 wt.%G/Al.

Zn-x wt.%G/Al coatings were deposited on a low-carbon steel plate with the size of 2 mm × 10 mm × 10 mm (low-carbon steel is composed of 0.16% C, 0.53% Mn, 0.30% Si, 0.055% S. 0.045% P and balanced Fe) by low-pressure cold spray from the Obninsk Center for Powder Spraying (Russia). Powders blended with 70 wt.% Zn and 30 wt.% G/Al were used as raw spraying materials via mechanical vibration. Using compressed air as the driving gas for the particles, the temperature and pressure of the air were set as 400 °C and 0.8 MPa, respectively. The surface optical images and cross-section SEM images of the obtained Zn-x wt.%G/Al coatings are depicted in our previous work [25].

2.2. Testing Barrier Properties of Zn-x wt.%G/Al Coatings

The medium-term shielding performance for the coating was evaluated by a full-immersion test. The full-immersion test was carried out in 3.5 wt.% NaCl solution at 25 °C; immersion time was 20 d. Before testing, the coating surface was sanded with emery papers up to a grit size of 1200# and mechanically polished; after testing, the samples were washed and dried. After that, the coating sample was molded in epoxy resin, with a 1.0 cm^2 area exposed to the solution. The surface corrosion morphology of the coatings after 10 and 20 days' immersion was first investigated by macro-observations and then analyzed by a field emission scanning electronic microscope (FE-SEM, Hitachi S-4800, Tokyo, Japan). The component characteristics of corrosion products after 10 and 20 days' immersion were detected on an X-ray diffraction diffractometer (XRD, D/Max 2500 PC) with a Cu target (λ = 0.154 nm) at a scanning rate of 4/min ranging from 10° to 80°.

The electrochemical corrosion behavior of the coating was investigated by testing the potentiodynamic polarization (PDP) and carrying out electrochemical impedance spectroscopy (EIS). An electrochemical test was performed in a conventional three-electrode system at room temperature. The working electrode was the exposed surface of the coating sample. A saturated calomel electrode (SCE) and platinum foil were used as the reference and auxiliary electrode, respectively. The PDP experiments of the samples were conducted at the beginning of immersion (2 h) and after 20 days' immersion. The polarization data were acquired at a sweep rate of 1 mV/s, from 100 mV below the open-circuit potential (OCP) to 300 mV above the OCP. The self-corrosion potential (E_{corr}) and self-corrosion current density (i_{corr}) values of the tested coatings were obtained using Tafel extrapolation.

EIS for the samples was acquired at 1, 5, 10 and 20 days' immersion, respectively. The EIS test was conducted at OCP; a single-amplitude perturbation of 5 mV was applied with a frequency range from 10^5 Hz to 10^{-2} Hz. ZsimpWin was used for the data fitting of impedance spectra. For accuracy, three parallel samples were tested in all electrochemical experimental tests and the average value was reported.

3. Results and Discussion

The SEM morphologies of Zn powders, Al powders and 0.2 wt.%G/Al are presented in Figure 1. The size for Zn and Al is around 15–20 μm; some wrinkles appeared on the 0.2 wt.%G/Al powders' surface, as shown in Figure 1c. A Raman test was conducted to measure G; D (1310 cm^{-1}) and G (1595 cm^{-1}) peaks were observed on the Raman spectrum acquired from the surface of 0.2 wt.%G/Al powder (Figure 1d). As depicted, the D/G intensity ratio of G (1.13) was higher than that of GO (0.84), which confirmed that GO was reduced and coated on the Al powder surface.

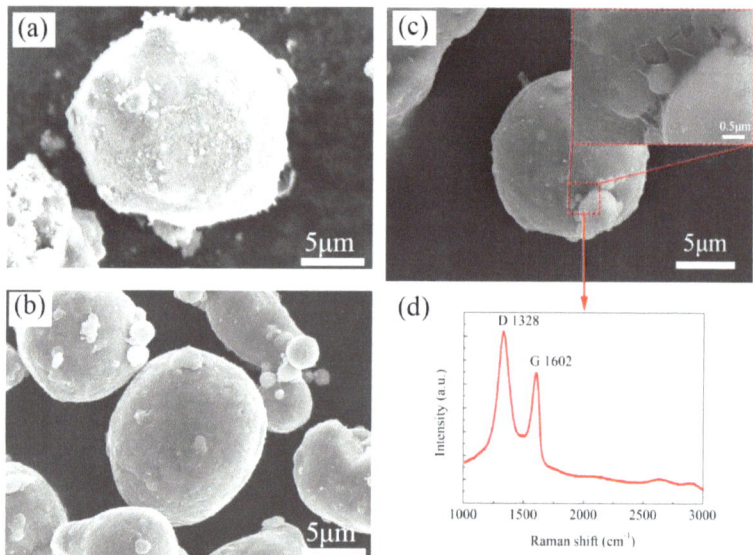

Figure 1. The morphology of Zn powders (**a**), Al powders (**b**) and 0.2 wt.%G/Al powders (**c**), and the Raman spectrum acquired from the surface of 0.2 wt.%G/Al powder (**d**).

Figure 2 shows the macroscopic corrosion appearance of the Zn-x wt.%G/Al coating after immersion in 3.5 wt.% NaCl solution for 5 d, 10 d and 20 d. As presented, the white corrosion products gradually covered the coating surface during the immersion time. During the early and middle immersion time (5 d and 10 d), the area of the coating surface covered by corrosion products differed. The cover area is arranged in the order of Zn-0.2 wt.%G/Al, Zn-0.1 wt.%G/Al, Zn-0.3 wt.%G/Al, Zn-Al, from large to small. The results indicate that the presence of G in the coating could accelerate the dissolution of the coating when in a corrosive solution, and the Zn-0.2 wt.%G/Al coating experienced the fastest corrosion rate at the initial immersion time. With prolonged immersion, the corrosion products on Zn-x wt.%G/Al coatings become more compact and evenly distributed.

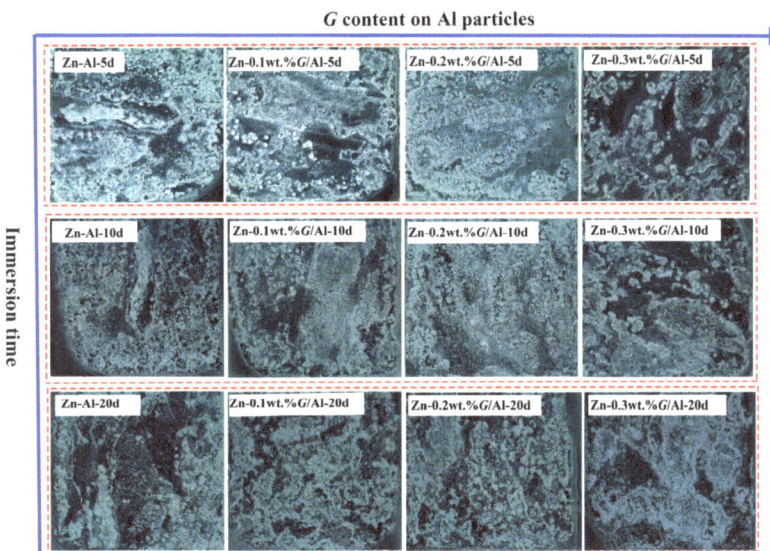

Figure 2. Surface macro-morphology of Zn-x wt.%G/Al coatings after immersion in NaCl solution.

Figure 3 shows the SEM morphology of corrosion products generated on Zn-x wt.%G/Al coatings' surface after immersion in 3.5 wt.% NaCl for 10 d. As demonstrated in Figure 3a, some loose corrosion products accumulated on the Zn-Al coating surface, and part of the bare coating substrate was still exposed after 10 d of immersion. The Zn-0.1 wt.%G/Al and Zn-0.3 wt.%G/Al coatings' surfaces were covered by a mass of nanoplatelets with a size of 2–3 μm and the nanoplatelets were vertically cross-linked, as presented in Figure 3b,d. Corrosion products on the Zn-0.2 wt.%G/Al coating were the most compact, and nanosheets for the products were very tiny, with a size of 2–3 μm, as shown in Figure 3c.

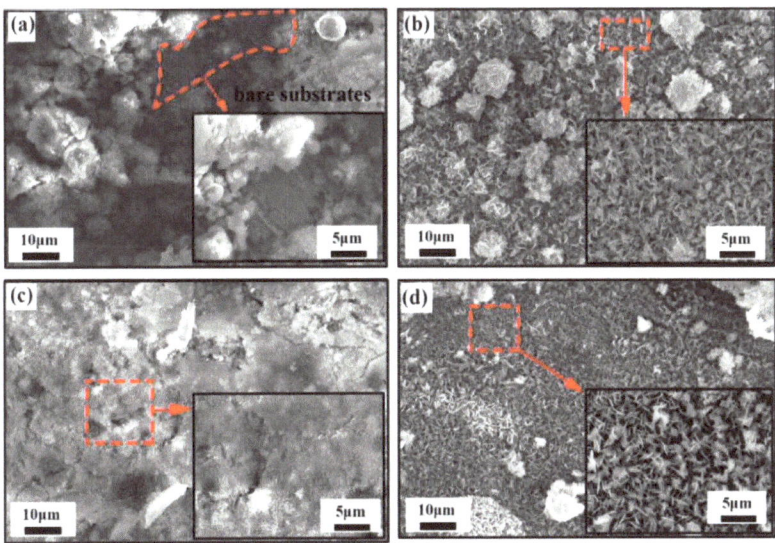

Figure 3. Surface SEM morphology for Zn-x wt.%G/Al coatings after immersion in NaCl solution for 10 d. (**a**) Zn-Al coating. (**b**) Zn-0.1 wt.%G/Al. (**c**) Zn-0.2 wt.%G/Al. (**d**) Zn-0.3 wt.%G/Al.

Figure 4 shows the SEM morphology of corrosion products generated on Zn-x wt.%G/Al coatings' surface after immersion in 3.5 wt.% NaCl solution for 20 d. After 20 d immersion, clusters of corrosion products accumulated loosely on the Zn-Al coating surface (Figure 4a), with apparent gaps between the cluster layers. Rod-shaped nanoparticles clustered and densely covered the Zn-0.1 wt.%G/Al coating surface to form a film of corrosion products, as shown in Figure 4b. For the Zn-0.2 wt.%G/Al coating after 20 d immersion, as presented in Figure 4c, the corrosion product layer is compact and composed of nanosheets. Flake-like corrosion products were formed on the Zn-0.3 wt.%G/Al coating surface, and some cracks and holes occurred on the corrosion product layer's surface.

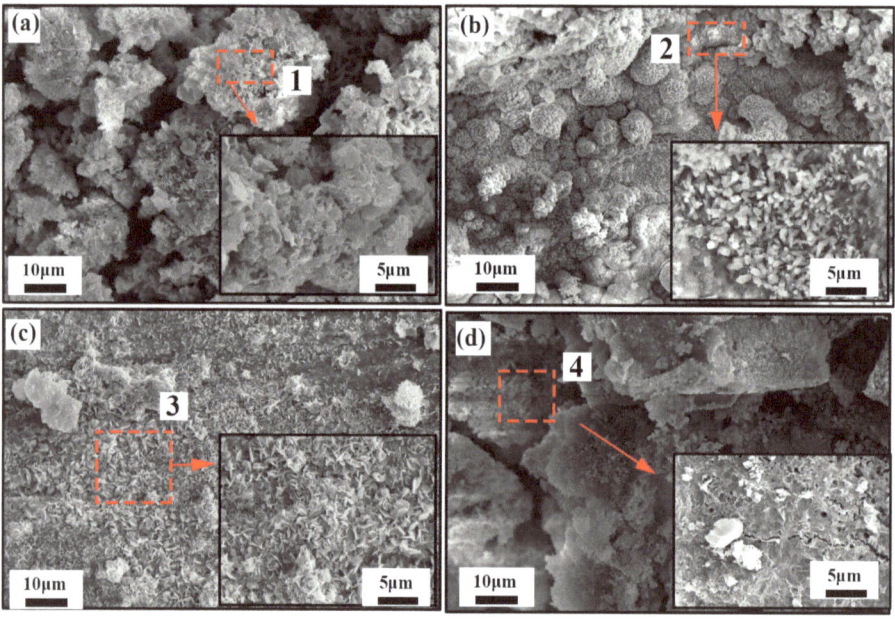

Figure 4. Surface SEM morphology for Zn-x wt.%G/Al coatings after immersion in 3.5 wt.% NaCl solution for 20 d. (**a**) Zn-Al coating. (**b**) Zn-0.1 wt.%G/Al. (**c**) Zn-0.2 wt.%G/Al. (**d**) Zn-0.3 wt.%G/Al.

Figure 5 is EDX results acquired from regions 1, 2, 3 and 4 shown in Figure 4. After 20 d immersion, the corrosion products are mainly composed of elements Zn, Al, Cl, C and O. The element Cl came from corrosive media, and C is derived from CO_2 that dissolved in solution. As shown in EDX results, it can be seen that O content in corrosion products formed on the Zn-0.2 wt.%G/Al surface is highest, which may be related to the formation of $ZnAl_2CO_3(OH)_{16}\cdot 4H_2O$.

Figure 6 shows the XRD results for corrosion products on the Zn-x wt.%G/Al coating after 10 d and 20 d immersion in NaCl solution. After 10 d immersion, an outstanding peak clearly appeared, centered around 11.3°, 43° and associated with hydrozincite [22]. Peaks caused by $Zn_4CO_3(OH)_6\cdot H_2O$ were present at 37°, 54° and 72° in the XRD pattern [20]. The peak of zinc oxide (ZnO) at 39.5° [26] was relatively higher than the peaks of simonkolleite ($Zn_5(OH)_8Cl_2$) at 11.3°, $ZnAl_2CO_3(OH)_{16}\cdot 4H_2O$ around 11.3° and 43° and $Zn_4CO_3(OH)_6\cdot H_2O$ at 54° and 72° [10]. There were some changes in the intensity of the diffraction peak of the corrosion products when immersion time was prolonged to 20 d, as shown in Figure 6. The diffraction peak intensity of ZnO at 39.5° decreased, and the peak intensity of $Zn_4CO_3(OH)_6\cdot H_2O$ at 54°, 72° and $ZnAl_2CO_3(OH)_{16}\cdot 4H_2O$ at 43° was strengthened. Some mini peaks around 33.5°, ascribed to $Zn_5(OH)_8Cl_2$ and $Zn_4CO_3(OH)_6\cdot H_2O$, successively appeared, which may be due to the transformation of ZnO. ZnO is unstable, which means that it readily reacts with water and gradually turns into simonkolleite with low dissolving properties in water. After immersion in NaCl solution, simonkolleite could provide protection for the coating. For Zn-0.2 wt.%*G*/Al coating, the diffraction peak intensity of $ZnAl_2CO_3(OH)_{16} \cdot 4H_2O$ at 43° strengthened remarkably, which may be related to the translation of Al.

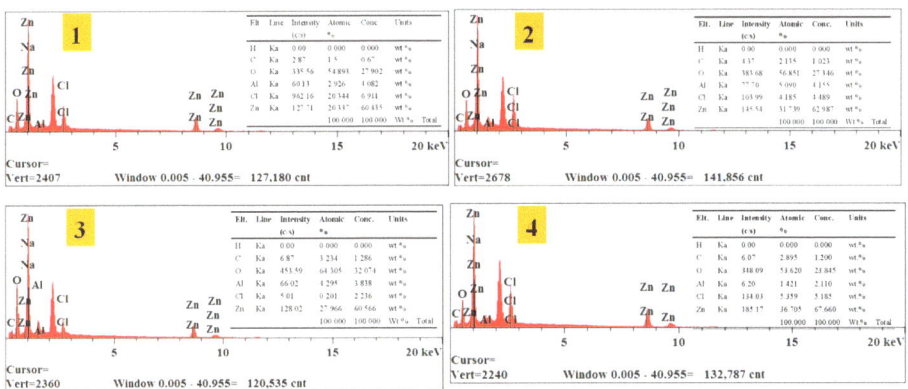

Figure 5. EDX analysis of corrosion products generated on Zn-x wt.%*G*/Al coatings after immersion for 20 d in NaCl solution, 1, 2, 3, 4 is corresponding with point 1, 2, 3, 4 in Figure 4.

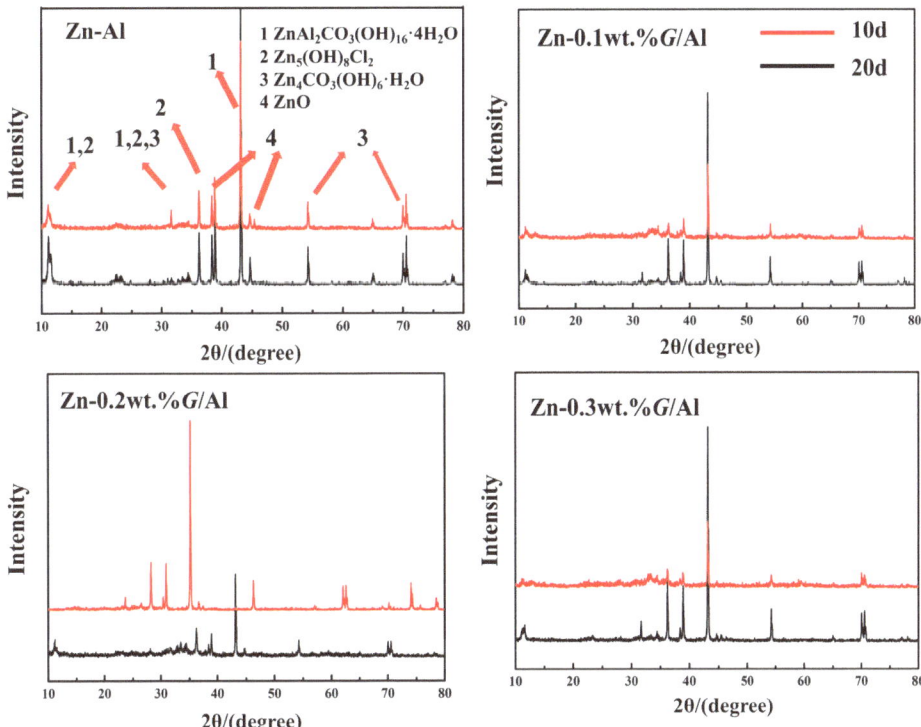

Figure 6. XRD results for corrosion products on Zn-x wt.%*G*/Al coatings after 10 d and 20 d immersion in 3.5 wt.% NaCl, respectively.

Figure 7 presents the potentiodynamic polarization curves (PPCs) of Zn-x wt.%G/Al coatings after immersion for 2 h and 20 d. The specific E_{corr} and i_{corr} values fitted from the PPC are displayed in Table 1. For the Zn-Al coating, the E_{corr} value of the Zn-Al coating after immersion for 2 h is -1.055 V_{SCE}; the value for the coating after 20 d immersion decreased to -1.083 V_{SCE}. After 20 d immersion, the i_{corr} value for the Zn-Al coating is 15.8 $\mu A/cm^2$, which is two and a half times higher than the value of the Zn-Al coating after immersion for 2 h (6.508 $\mu A/cm^2$). The E_{corr} values for Zn-0.1 wt.%G/Al, Zn-0.2 wt.%G/Al and Zn-0.3 wt.%G/Al coatings are -1.180 V_{SCE}, -1.143 V_{SCE} and -1.152 V_{SCE} after 20 d immersion, which are elevated compared with the coating values acquired at 2 h immersion time. The variation trend of the i_{corr} value for the Zn-G/Al coating is the opposite. The i_{corr} value for Zn-0.1 wt.%G/Al and Zn-0.3 wt.%G/Al coatings increased, but that for the Zn-0.2 wt.%G/Al coating decreased.

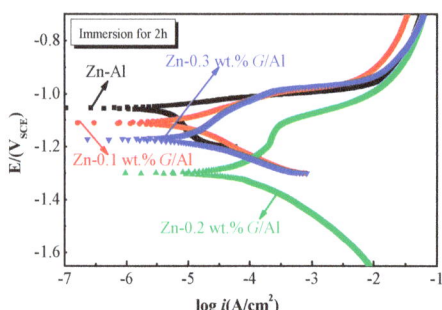 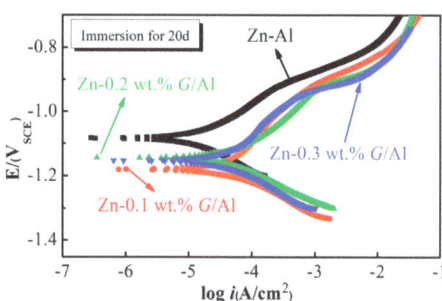

Figure 7. Polarization curves for Zn-x wt.%G/Al coatings after immersion for 2 h and 20 d.

Table 1. E_{corr} and I_{corr} values fitted from PPS for Zn-x G/Al composite coatings after immersion in 3.5 wt. % NaCl solution for 2 h and 20 d.

Coatings	E_{corr} ($V_{vs.SCE}$)		I_{corr} ($\mu A/cm^2$)	
	2 h	20 d	2 h	20 d
Zn-Al	-1.055	-1.083	6.508	15.8
Zn-0.1 wt.%G/Al	-1.187	-1.180	55.59	58.07
Zn-0.2 wt.%G/Al	-1.299	-1.143	82.16	49.96
Zn-0.3 wt.%G/Al	-1.172	-1.152	14.58	52.06

In PPCs, the E_{corr} value indicates the corrosion tendency of the coating and a higher E_{corr} value means a lower corrosion tendency; the i_{corr} value represents the corrosion rate of the coating and the value of i_{corr} is inversely proportional to its anti-corrosion properties. For the steel with a protective coating, the higher the potential difference between the coating and steel matrix, the stronger the coating's propensity for cathodic protection. A higher corrosion current density indicated a better cathodic protection efficiency. Compared with the E_{corr} value for steel (-0.72 V_{SCE}), Zn-G/Al coating values were all below -1V_{SCE}, which indicates that all Zn-G/Al coatings are cathodic protection coatings. Increasing with immersion time, the E_{corr} values for Zn-0.1 wt.%G/Al, Zn-0.2 wt.%G/Al and Zn-0.3 wt.%G/Al coatings at 20 d increased by 0.5%, 12% and 1.7%, respectively, in comparison with the value at the initial immersion time but were still lower than that of steel (-0.72 V_{SCE}). Therefore, the coating could protect steel through its sacrificial anode action during full immersion.

Cathodic protection is the method of sacrificing the anode to avoid the corrosion of the cathode. The sacrificial ability of the anode increases with the increase in dissolution rate of the coating layer, while the large dissolution rate of the coating layer decreases service life for the coating [8]. The key point is the balance between sacrificial behavior and dissolution resistance. In the work of Jonathan Elvins et al. [27], the effect of magnesium additions on cut edge corrosion resistance of zinc aluminum alloy galvanized steel was studied. They measured individual anodic activity against anode number for the cut edge by SVET. Results indicated that the addition of Mg resulted in an increase in zinc loss, an increase in active anode numbers and an increase in the number of long-lived anodes. Ding et al. [23] prepared graphene/zinc-containing coatings on steel, and electrochemical test results indicated that 0.3%graphene-70%Zn coating has the highest current corrosion identity, which presented that adding graphene prolonged the duration and increased the values of the stable cathodic protection currents of the coating. Hence, a higher cathodic protection efficiency calls for a higher corrosion current density during the process of cathodic protection. As shown in Table 1, the i_{corr} value for the Zn-G/Al coating is higher than that of Zn-Al, which means that adding G to the coating could enhance the activity and strengthen the cathodic protection effect of the Zn-Al composite coating during the full-immersion process.

During the anodic protection of metals, the anodic protection coating is also required to have a low corrosion rate to extend its service life. However, the higher the anodic protection efficiency of the coating, the higher the corrosion rate. Therefore, if the coating can generate a protective corrosion product layer to reduce the degradation rate during the anodic protection process, then this contradiction may be alleviated. It is surprising that the i_{corr} value for Zn-0.2 wt.%G/Al went down after 20 d immersion compared with the value at the initial immersion time, which demonstrated that the corrosion products generated on the Zn-0.2 wt.%G/Al coating could delay the degradation and prolong the service life of the coating. Therefore, we conclude that inserting G into the coating endowed the coating with an enduring activity and equipped it with a persistent cathodic protection ability during full immersion. When the G content in powders of the Al surface is 0.2 wt.%, the obtained composite coating surface generated a protective corrosion product film after 20d immersion, which could retard the degradation of coating.

Compared with casted Zn-Al alloy [28] or hot-dipped Zn-Al composite coating (E_{corr} is around -1.0) [16], Zn-G/Al coating has an excellent cathodic protection efficiency. However, the i_{corr} value for Zn-G/Al coating is still higher than that of traditional cathodic protection coating.

To illustrate the equilibrium effect, the impedance change for Zn-Al and Zn-0.2 wt.%G/Al coatings during the full immersion is presented in Figure 8. Actually, Zn-G/Al coating during immersion experienced two processes: firstly, the generation of corrosion products; secondly, the compaction of corrosion products. The densification of corrosion products is related to activity of the coating. For the Zn-Al coating, the electrochemical impedance spectrum (EIS) after 5–10 d immersion was fitted by the equivalent circuit $R_s(Q_1(R_1Q_2(R_2Z_w)))$ [29]. Rs, corresponding to a high-frequency region, represents the resistance of the electrolyte; Q_1 and R_1 are linked to the capacitance and resistance of the corrosion products stacked on the coating surface, respectively. The reactions of the coating are described by the charge transfer resistance R_2 and the double-layer capacitance Q_2 which correspond to the low-frequency region. As demonstrated in EIS curves, a 45° near-line appears in the low-frequency region, which is caused by the diffusion of the porous corrosion product layer, for which a corresponding semi-infinite-length Warburg element Z_W is added in series with R_2. The EIS after 20 d immersion was fitted by the equivalent circuit $R_s(Q_1(R_1(R_2Q_2)))$ [30]; there was no apparent diffusion phenomenon due to the corrosion product layer has basically formed. For the Zn-0.2 wt.%G/Al coating, before 5 d immersion, EIS curves were fitted by an equivalent circuit of $R_s(Q_1(R_1Q_2(R_2Z_w)))$. After 10 d immersion, corrosion products were generated and stacked on the Zn-0.2 wt.%G/Al coating surface to form a layer. Therefore, the equivalent circuit $R_s(Q_1(R_1(R_2Q_2)))$ can be used to fit the EIS curve of the coating after 10 d immersion.

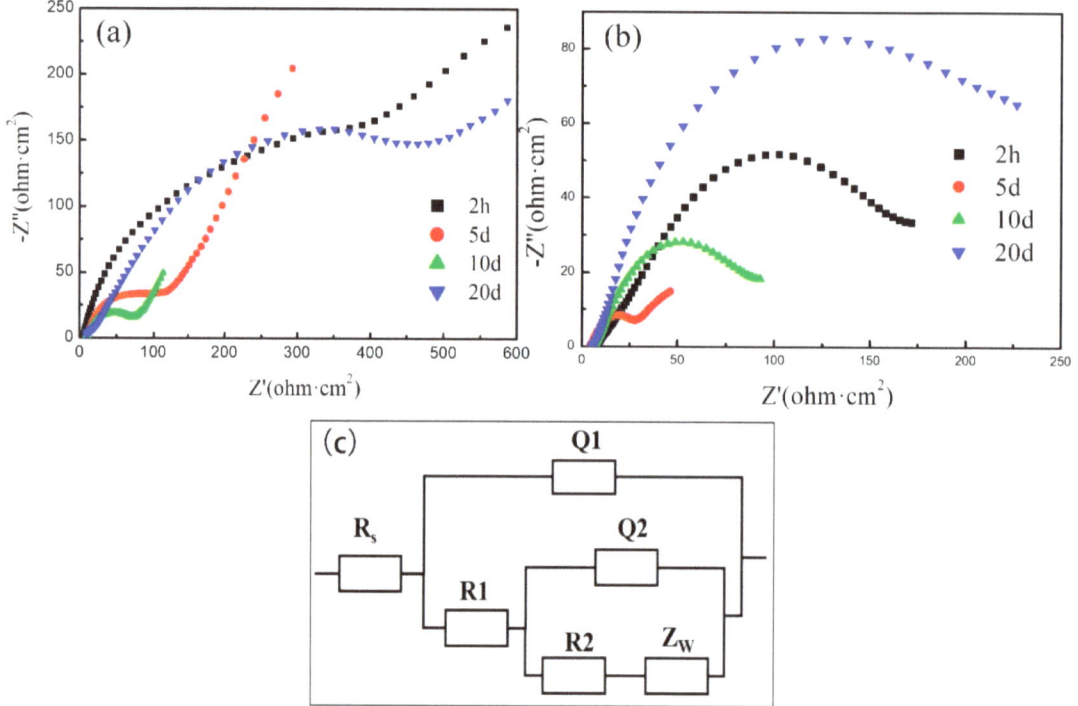

Figure 8. Impedance change for Zn-Al and Zn-0.2 wt.%G/Al coatings during full immersion. (**a**) Zn-Al. (**b**) Zn-0.2 wt.%G/Al coating. (**c**) equivalent circuit model.

The variation in the EIS radius for the coating during mid-term immersion showed that the Zn-x wt.%G/Al coating experienced three phases: the first is the shielding of the coating, the second is the erosion of the coating, which provides cathodic protection, and the third is the barrier of the formed erosion product layer. For the Zn-Al coating, the cathodic protection stage occurred in the first 10 d of immersion. The arc radius increased with immersion time due to the corrosion product layer formed on the Zn-Al coating. Although the service life of the coating can be extended due to the shielding protection of the corrosion product layer, the cathodic protection efficiency is still low, as shown in Table 1 (E_{corr} value of Zn-Al at 20d immersion time is -1.083 V_{SCE}, which is higher than that of Zn-x wt.%G/Al coating). For Zn-0.2 wt.%G/Al, the cathodic protection stage occurred in the first 5 d of immersion, and the R_1 value (arc radius value) dramatically decreased. After 5 d immersion, the arc radius value of the EIS curve gradually increased. The increased R_1 value shows that the formed corrosion product layer has a better shielding protection effect. With a further increase in immersion time (after 20 d), the shielding protection effects of the corrosion product layer significantly increased beyond the shielding protection of the Zn-0.2 wt.%G/Al coating at the initial immersion time, as shown by the largest arc radius value of the EIS curve. The excellent protection of the corrosion products on the Zn-0.2 wt.%G/Al coating surface could be the result of the shielding effect of G in the corrosion products. The Zn-0.2 wt.%G/Al coating has a superior cathodic protection efficiency after 20 d immersion (with a lower E_{corr} value of -1.152 V_{SCE}), which can be ascribed to the bridge connection with the conductivity of G.

4. Mechanism

According to the results of PPS and EIS, we deduced that the Zn-0.2 wt.%G/Al coating possesses a continuous cathodic protection ability compared with the other three coatings after 20 d immersion, and the corrosion product on Zn-0.2 wt.%*G*/Al coating could delay the dissolution of the coating, which could prolong the service life of the coating during the full-immersion process. The special protective properties of the Zn-0.2 wt.%*G*/Al coating produced during medium-term immersion can be attributed to the equilibrium effect of *G*. Actually, graphene is a double-edged sword when applied in the corrosion protection of a coating [22]. When *G* completely coats metal surfaces or is used as an additive in organic or inorganic coatings, *G* prolongs the propagation path of the corrosive medium through its large specific surface area; thus, the use of *G* in the coating mainly plays a role in shielding protection [31,32]. However, the conductivity of *G* is inclined to expedite the metal degradation when *G* metal matrix composites are used [18]. Due to the potential difference between Zn, Al and *G*, the reaction process of the Zn-*G*/Al coating in an aggressive medium involves galvanic reactions. Figure 9 shows the microbattery reaction process of the two coatings soaked in an aggressive NaCl medium.

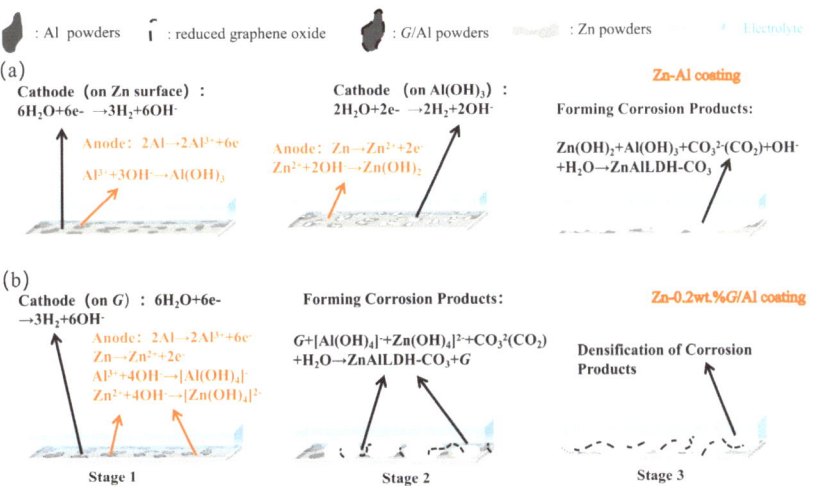

Figure 9. Model for microbattery reaction process of Zn-Al and Zn-0.2 wt.%*G*/Al coatings during immersion in 3.5 wt.% NaCl solution. (**a**) Zn-Al coating. (**b**) Zn-0.2 wt.%*G*/Al coating.

For the Zn-Al coating (Figure 9a), Al with a low self-corrosion potential is preferentially dissolved and promotes the release of OH^-, by which the 3.5 wt.% NaCl solution gradually becomes alkaline and promotes the dissolution of Zn. The dissolved zinc and aluminum first formed oxides that were easily hydrolyzed. The oxides hydrolyzed to hydroxides $Zn(OH)_4^{2-}$ and $Al(OH)_4^-$, which continuously promote the dissolution of the coating. When the hydroxide content in solution is high enough, hydroxides coordinate to form stable bimetallic hydroxides and deposit on the coating surface. In the Zn-*G*/Al coating (Figure 9b), the addition of conductive *G* transforms Zn and Al into anode, so that Zn and Al dissolve simultaneously and promote the formation of corrosion products, as demonstrated in Figure 2. During the corrosion product layer formation, the solvent-less *G* in the Zn-Al coating was exposed by the dissolution of Zn and Al and entered into the corrosion product layer, accompanied by the stacking of corrosion products. According to the XRD results exhibited in Figure 4, corrosion products mainly consist of LDH. LDH is an inorganic compound with an impotent electron conductivity, which would weaken the cathodic protection property of the Zn-Al coating. *G* in the corrosion product layer acts as both a barrier and means of electron conduction. Firstly, the barrier property of *G* prolongs the permeation path of the erosion media, which enhanced the shielding ability of the corrosion product film. Secondly, *G* in the corrosion product layer bridged the coating and the corrosive medium and facilitated electron propagation, which provided the coating with continuous cathodic protection.

Among the four coatings, we deduced that the Zn-0.2 wt.%*G*/Al coating possesses a continuous cathodic protection ability, and the corrosion product film generated on Zn-0.2 wt.%*G*/Al coating after immersion in the medium has the best shielding ability. This phenomenon is related to the distribution and concentration of *G* in the coating. In our previous work, we proved that the deposition effect of Al powders is strongly associated with the *G* concentration when coated on Al. During cold spraying, properly increasing the difference in hardness between the sprayed particles can enhance the particle deposition efficiency; however, an exorbitant particle hardness will cause particles to violently rebound. *G* possesses an excellent mechanical strength; metal materials combined with *G* have a significantly enhanced hardness. In addition, a mechanically bonded heterogeneous interface is formed between *G* and Zn particles during cold spraying; therefore, with an increase in *G* concentration on the Al surface, the interface energy between *G* and Zn will increase and the *G*/Al in the coating will be distributed more evenly. In the Zn-Al coating, poorly dispersed Al agglomerates are found. When the *G*-coated Al is applied, the distribution of Al becomes uniform; moreover, *G*/Al in the Zn-0.2 wt.%*G*/Al composite coating is significantly decreased and uniformly scattered. When the concentration of *G* coated on Al is increased to 0.3 wt%, the main component of the coating is Zn, due to the rebound of 0.3 wt.%*G*/Al. Therefore, among the four coatings, the Zn-0.2 wt.%*G*/Al coating is equipped with a continuous cathodic protection ability and the corrosion products on the Zn-*G*/Al coating surface have better shielding properties than the initial Zn-*G*/Al coating due to the shielding of *G*, which could delay the dissolution of that coating when dipped in aggressive solution.

5. Conclusions

In this work, the corrosion behavior of a Zn-x wt.%*G*/Al coating in an aggressive NaCl solution was explored. Results indicated the Zn-0.2 wt.%*G*/Al coating is equipped with a continuous cathodic protection ability and formed a protective layer of corrosion products on the Zn-*G*/Al coating surface with better shielding properties than the initial Zn-*G*/Al coating, which could delay the dissolution of that coating when dipped in aggressive solution.

The special protective properties of the Zn-0.2 wt.%*G*/Al coating can be attributed to the equilibrium effect of *G*. Firstly, *G* in Zn-0.2 wt.%*G*/Al coating could facilitate the dissolution of the initial coating and, at the same time, the conductivity of *G* in the corrosion product layer bridged the coating and the corrosive medium and facilitated electron propagation, which provided the coating with continuous cathodic protection. Secondly, the barrier *G* in the corrosion product layer could prolong the permeation path of the aggressive medium, which enhanced the shielding ability of the corrosion product film.

Author Contributions: Methodology, Q.S.; Resources, D.W. and J.W.; Writing—original draft, Q.S.; Writing—review & editing, H.W.; Supervision, P.Z. and X.J. All authors have read and agreed to the published version of the manuscript.

Funding: This research was funded by Guangdong Social Science and Technology Development Key Project (No. 2021A1515110154), Natural Science Foundation of Anhui Province (No. 2008085ME149), Natural Science Foundation of Anhui Province (No. 2308085QE132, No. 2308065ME171), Anhui University Scientific Research Project (No. 2023AH051660, No. 2022AH040247).

Institutional Review Board Statement: Not applicable.

Informed Consent Statement: Not applicable.

Data Availability Statement: Not applicable.

Conflicts of Interest: The authors declare no conflict of interest.

References

1. Pourhashem, S.; Seif, A.; Saba, F.; Nezhad, E.G.; Ji, X.; Zhou, Z.; Zhai, X.; Mirzaee, M.; Duan, J.; Rashidi, A.; et al. Antifouling nanocomposite polymer coatings for marine applications: A review on experiments, mechanisms, and theoretical studies. *J. Mater. Sci. Technol.* **2022**, *118*, 73–113. [CrossRef]
2. Ou, Y.; Wang, H.; Hua, Q.; Liao, B.; Ouyang, X. Tribocorrosion behaviors of superhard yet tough Ti-C-N ceramic coatings. *Surf. Coat. Technol.* **2022**, *439*, 128448. [CrossRef]
3. Xu, M.; Lam, C.C.; Wong, D.; Asselin, E. Evaluation of the cathodic disbondment resistance of pipeline coatings—A review. *Prog. Org. Coat.* **2020**, *146*, 105728. [CrossRef]
4. Arrighi, C.; Nguyen, T.T.; Paint, Y.; Coelho, C.S.; Coelho, L.B.; Creus, J.; Olivier, M.G. Study of Ce(III) as a potential corrosion inhibitor of Zn-Fe sacrificial coatings electrodeposited on steel. *Corros. Sci.* **2022**, *200*, 110249. [CrossRef]
5. Luo, Q.; Li, Q.; Zhang, J.Y.; Lu, H.S.; Li, L.; Chou, K.C. Microstructural evolution and oxidation behavior of hot-dip 55 wt.% Al-Zn-Si coated steels. *J. Alloys Compd.* **2015**, *646*, 843–851. [CrossRef]
6. Malihe, Z.; Hossein, A.; Parvin, G.; Singh, C.N.P.; Sargazi, G. Application of Artificial Neural Networks for Corrosion Behavior of Ni-Zn Electrophosphate Coating on Galvanized Steel and Gene Expression Programming Models. *Front. Mater.* **2022**, *9*, 823155.
7. Wu, P.P.; Song, G.L.; Zhu, Y.X.; Zheng, D.J. Intelligentization of traditional sacrificial anode Zn by Mg-alloying for reinforcing steel. *Corros. Sci.* **2022**, *194*, 109943. [CrossRef]
8. Shimpei, T.; Izumi, M.; Yu, S.; Takahashi, M.; Matsumoto, M.; Nobuyoshi, H. Micro-electrochemical investigation on the role of Mg in sacrificial corrosion protection of 55mass%Al-Zn-Mg coated steel. *Corros. Sci.* **2017**, *129*, 126–135.
9. Tachiban, K.; Morina, Y.; Mayuzumi, M. Hot dip fine Zn and Zn–Al alloy double coating for corrosion resistance at coastal area. *Corros. Sci.* **2007**, *49*, 149–157. [CrossRef]
10. Azevedo, M.S.; Allély, C.; Ogle, K.; Volovitch, P. Corrosion mechanisms of Zn (Mg, Al) coated steel: 2. The effect of Mg and Al alloying on the formation and properties of corrosion products in different electrolytes. *Corros. Sci.* **2015**, *90*, 482–490. [CrossRef]
11. Xue, Q.; Sun, C.Y.; Yu, J.Y.; Huang, L.; Wei, J.; Zhang, J. Microstructure evolution of a Zn-Al coating co-deposited on low carbon steel by pack cementation. *J. Alloys Compd.* **2017**, *699*, 1012–1021. [CrossRef]
12. Zhu, Z.X.; Xu, B.X.; Chen, Y.X. Effect of Al Content on Electrochemical Corrosion Behavior of Arc Sprayed Zn-Al Coating. *China Surf. Eng.* **2011**, *24*, 58–61.
13. Kim, J.H.; Lee, M.H. A Study on Cavitation Erosion and Corrosion Behavior of Al-, Zn-, Cu-, and Fe-based Coatings Prepared by Arc Spraying. *J. Therm. Spray Technol.* **2010**, *19*, 1224–1230. [CrossRef]
14. Darabi, A.; Azarmi, F. Investigation on relationship between microstructural characteristics and mechanical properties of wire-arc-sprayed Zn-Al coating. *J. Therm. Spray Technol.* **2020**, *29*, 297–307. [CrossRef]
15. Huang, G.S.; Lou, X.D.; Wang, H.R.; Li, X.B.; Xing, L.K. Investigation on the Cathodic Protection Effect of Low Pressure Cold Sprayed AlZn Coating in Seawater via Numerical Simulation. *Coatings* **2017**, *7*, 93. [CrossRef]
16. Liu, W.; Li, M.C.; Luo, Q.; Fan, H.Q.; Zhang, J.Y.; Lu, H.S.; Chou, K.C.; Wang, X.L.; Li, Q. Influence of alloyed magnesium on the microstructure and long-term corrosion behavior of hot-dip Al-Zn-Si coating in NaCl solution. *Corros. Sci.* **2016**, *104*, 217–226. [CrossRef]
17. Zhu, L.Q.; Xu, B.P.; Zhou, M.X.; Hu, S.J.; Zhang, G.D. Fabrication and characterization of Al-Zn-Cu composite coatings with different Cu contents by cold spraying. *Surf. Eng.* **2020**, *36*, 1090–1096. [CrossRef]
18. Sun, J.L.; Chen, Y.; Zhai, P.; Zhou, Y.H.; Zhao, J.; Huang, Z.F. Tribological performance of binderless tungsten carbide reinforced by multilayer graphene and SiC whisker. *J. Eur. Ceram. Soc.* **2022**, *42*, 4817–4824. [CrossRef]
19. Zhang, Y.M.; Sun, J.L.; Xiao, X.Z.; Wang, N.; Meng, G.Z.; Gu, L. Graphene-like two-dimensional nanosheets-based anticorrosive coatings: A review. *J. Mater. Sci. Technol.* **2022**, *129*, 139–162. [CrossRef]
20. Chen, Y.N.; Wu, L.; Yao, W.H.; Wu, J.H.; Xiang, J.P.; Dai, X.W.; Wu, T.; Yuan, Y.; Wang, J.F.; Jiang, B.; et al. Development of metal-organic framework (MOF) decorated graphene oxide/MgAl-layered double hydroxide coating via microstructural optimization for anti-corrosion micro-arc oxidation coatings of magnesium alloy. *J. Mater. Sci. Technol.* **2022**, *130*, 12–26. [CrossRef]
21. Wang, X.J.; Zhang, L.Y.; Zhou, X.L.; Wu, W.; Jie, X.H. Corrosion behavior of Al$_2$O$_3$-reinforced graphene encapsulated Al composite coating fabricated by low pressure cold spraying. *Surf. Coat. Technol.* **2020**, *386*, 125486. [CrossRef]
22. Muhammad, R.; Pan, F.S.; Muhammad, A.; Chen, X.H. Corrosion behavior of magnesium-graphene composites in sodium chloride solutions. *J. Magnes. Alloys* **2017**, *5*, 271–276.
23. Ding, R.; Zheng, Y.; Yu, H.B.; Li, W.H.; Wang, X.; Gui, T.J. Study of water permeation dynamics and anti-corrosion mechanism of graphene/zinc coatings. *J. Alloys Compd.* **2018**, *748*, 481–495. [CrossRef]
24. Wu, H.S.; Zhang, L.Y.; Liu, C.S.; Mai, Y.J.; Zhang, Y.C.; Jie, X.H. Deposition of Zn-G/Al composite coating with excellent cathodic protection on low carbon steel by low-pressure cold spraying. *J. Alloys Compd.* **2020**, *821*, 153483. [CrossRef]
25. Wu, H.S.; Shen, G.Z.; Li, R.X.; Zhang, L.Y.; Jie, X.H.; Liu, G. Influence of embedded reduced graphene oxide on the corrosion-wear performance of cold sprayed Zn-rGO/Al coating in NaCl solution. *Surf. Coat. Technol.* **2022**, *429*, 127856. [CrossRef]
26. Chen, M.; Wang, Y.; Song, L.; Gunawan, P.; Zhong, Z.; She, X.; Su, F. Urchin-like ZnO microspheres synthesized by thermal decomposition of hydrozincite as a copper catalyst promoter for the Rochow reaction. *RSC Adv.* **2012**, *2*, 4164–4168. [CrossRef]
27. Elvins, J.; Spittle, J.A.; Sullivan, J.H.; Worsley, D.A. The effect of magnesium additions on the microstructure and cut edge corrosion resistance of zinc aluminium alloy galvanised steel. *Corros. Sci.* **2008**, *50*, 1650–1658. [CrossRef]

28. Prosek, T.; Hagström, J.; Persson, D.; Fuertes, N.; Lindberg, F.; Chocholatý, O.; Taxén, C.; Šerák, J.; Thierry, D. Effect of the microstructure of Zn-Al and Zn-Al-Mg model alloys on corrosion stability. *Corros. Sci.* **2016**, *110*, 71–81. [CrossRef]

29. Oliveira, C.G.; Ferreira, M. Ranking high-quality paint systems using EIS. Part II: Defective coatings. *Corros. Sci.* **2003**, *45*, 139–147. [CrossRef]

30. Mouanga, M.; Bercot, P. Comparison of corrosion behaviour of zinc in NaCl and in NaOH solutions; Part II: Electrochemical analyses. *Corros. Sci.* **2010**, *52*, 3993–4000. [CrossRef]

31. Ding, R.; Li, W.; Wang, X.; Gui, T.; Li, B.; Han, P.; Tian, H.; Liu, A.; Wang, X.; Liu, X.; et al. A brief review of corrosion protective films and coatings based on graphene and graphene oxide. *J. Alloys Compd.* **2018**, *764*, 1039–1055. [CrossRef]

32. Sun, J.; Martinsen, K.H.; Klement, U.; Kovtun, A.; Xia, Z.; Silva, P.F.B.; Hryha, E.; Nyborg, L.; Palermo, V. Controllable Coating Graphene Oxide and Silanes on Cu Particles as Dual Protection for Anticorrosion. *ACS Appl. Mater. Interfaces* **2023**, *15*, 38857–38866. [CrossRef] [PubMed]

Article

Mechanical and Tribological Study on Aluminum Coatings with High-Pressure and Low-Pressure Cold-Spray Processes

Abreeza Manap [1,2,*], NF Afandi [1], Savisha Mahalingam [2], Siti Nurul Akmal Yusof [3] and Zulkifli Mohd. Rosli [4]

1 College of Engineering, Universiti Tenaga Nasional, Jalan IKRAM-UNITEN, Kajang 43000, Selangor, Malaysia
2 Institute of Sustainable Energy, Universiti Tenaga Nasional, Jalan IKRAM-UNITEN, Kajang 43000, Selangor, Malaysia
3 Department of Mechanical Precision Engineering, Malaysia-Japan International Institute of Technology, Universiti Teknologi Malaysia, Jalan Sultan Yahya Petra, Kuala Lumpur 54100, Selangor, Malaysia
4 Fakulti Teknologi Kejuruteraan Mekanikal Dan Pembuatan, Universiti Teknikal Malaysia Melaka, Hang Tuah Jaya, Durian Tunggal 76100, Melaka, Malaysia
* Correspondence: abreeza@uniten.edu.my; Tel.: +60-3-89212020 (ext. 7310)

Abstract: Cold spray is a promising approach to repair all damages and defects in aluminum (Al) constituent elements. The study aims to investigate the mechanical and tribological properties of Al coatings deposited using high-pressure cold-spray (HPCS) and low-pressure cold-spray (LPCS) techniques. Al powder was sprayed on a cold-rolled plate of aluminum 1100, which was used as the substrate. The results showed that the micro-hardness of the LPCS Al coating reached up to 196.6 HV before the wear test compared to that of HPCS (174.3 HV). Moreover, more low friction coefficients obtained by LPCS (0.798) than HPCS (0.807) indicated good tribological properties with a high amount of oxide composition. Meanwhile, the wear studies reveal that the specific wear rate of the Al coating of LPCS (0.008) was lower than the HPCS (0.009) as the load increased from 3 N to 5 N, thus providing excellent wear resistance. Therefore, the results exhibited greater mechanical and tribological characteristics for Al coatings produced by the LPCS process than by the HPCS process.

Keywords: energy; friction coefficient; high-pressure cold spray; low-pressure cold spray; specific wear rate

Citation: Manap, A.; Afandi, N.; Mahalingam, S.; Yusof, S.N.A.; Mohd. Rosli, Z. Mechanical and Tribological Study on Aluminum Coatings with High-Pressure and Low-Pressure Cold-Spray Processes. *Coatings* **2022**, *12*, 1792. https://doi.org/10.3390/coatings12111792

Academic Editor: Jose Ygnacio Pastor

Received: 21 October 2022
Accepted: 17 November 2022
Published: 21 November 2022

Publisher's Note: MDPI stays neutral with regard to jurisdictional claims in published maps and institutional affiliations.

1. Introduction

Aluminum (Al) alloys are lightweight, non-ferrous metal with high corrosion resistance [1,2] and ductility. The high st rength of Al provides reduces fuel consumption in transportation systems such as aircraft, railcars, and light vehicles. Furthermore, it is known as an excellent electric conductor [3]; Al wire is used for transmitting electrical power over long distances. It also has the potential to be used in heat exchangers, refrigerators, and air conditioners because of its good thermal-conductor properties. However, damages may occur in Al components that cannot be repaired with conventional technologies [1]. For instance, bores in the Al castings are subject to damages such as corrosion and wear. Corrosion takes place as Al components are readily oxidized, including when Al is present either in solid solution or intermetallic particles. Welding Al components is difficult because of its high specific thermal conductivity and high coefficient of thermal expansion [1,4].

Within the last decade, numerous studies have been carried out to investigate the wear performance of metal coatings produced by thermal spray, including high-velocity oxy-fuel, combustion flame, vacuum plasma, and two-wire electric arc [5]. Even though these thermal-spray processes use strong and clean spraying methods, the heat dissipation may change the microstructure of the coatings and their mechanical behavior, leading to coating particle oxidation, decomposition, and grain growth, which is a major threat in coating technology [6,7]. In order to overcome these damages, cold spray (CS) can be utilized as a new approach in the coating technique. The main benefit of CS over thermal spray techniques is that the coating material is not thermally altered. This event minimizes any possible phase transformation and sustains the particles in their unmodified solid state.

There are two types of CS methods: high-pressure cold spray (HPCS) and low-pressure cold spray (LPCS). The HPCS process used helium or nitrogen as working gas, at high pressure (2.0–4.0 MPa), and preheated it (up to 1000 °C) inside the de Laval type nozzle [8–11]. The particle velocities for this process range from 200 to 1200 m/s [12], which allows for the deposition of various materials ranging from pure metal to alloys. A study on Al coatings deposited using HPCS has been reported extensively by several researchers [9–15]. Efficient bonding for Al coating occurs between 600 and 900 m/s, within which the highest deposition efficiency can be achieved [9–13]. However, the mechanical properties of Al coating deposited using HPCS were unsatisfactory. Some particles did not severely deform, which resulted in lower hardness and high coating porosity [16–18]. A similar outcome was also reported for other light materials. The examination of these light-metal coatings revealed a decrease in hardness with increasing distance from the substrate, despite displaying excellent impact behaviors such as severe plastic deformation in the substrate, intimate metallurgical (atomic) bond-forming along the interface, and an increase in hardness at the interface region as a result of grain refinement [19–21]. Thus, HPCS is deemed as not suitable for depositing soft and light materials [22].

The LPCS process used compressed gas as a working gas at relatively low pressure (below 0.6 MPa) and preheated it (up to 550 °C) inside the de Laval type nozzle [8–10]. The particle velocities for LPCS range from 350 to 700 m/s [23]. The velocity for depositing Al using LPCS ranges from 300 to 500 ms^{-1} [11–14]. Due to the relatively lower velocity of LPCS, it is believed that erosion on the Al substrate surface can be avoided. Despite the potential exhibited by the LPCS method, the impact behavior of light metals, including Al, has not been sufficiently investigated using this technique. In our previous work, computer simulation was used to investigate Al particle impact on the Al substrate during HPCS and LPCS coating processes [16]. It was found that Al deposited using LPCS resulted in lower porosity than Al deposited using HPCS. Therefore, it is of great importance to extend the study and investigate the coating properties as well as the tribological properties.

Moreover, there is no study comparing HPCS and LPCS methods with Al deposition in a single work. Therefore, the present work aims to investigate the mechanical and tribological properties of Al coatings deposited using HPCS and LPCS. Several analyses were conducted by evaluating the different coating spray processes' microstructure, elemental composition, hardness, wear rate, friction of coefficient, and wear resistance.

2. Materials and Methods

2.1. Cold-Spray Deposition Process

The HPCS and LPCS coatings were produced by PCS-203 (Plasma Giken Kogyo Co., Ltd., Saitama, Japan) and DYMET403J (Obninsk Center for Powder Spraying, Kaluga Oblast, Russia), respectively. Figure 1 shows the schematic diagram of the HPCS and LPCS system with the spraying parameters, as listed in Table 1. The spray material (pure Al powder of AL G-AT; particle diameter: 25 μm) was purchased from Fukuda Metal Foil and Powder Corporation (Kyoto, Japan). A cold-rolled plate of aluminum 1100 was used as the substrate. Helium and compressed air were used as the carrier gas for HPCS (2.0 MPa) and LPCS (0.6 MPa), respectively. The temperature was set up at 300 K with a nozzle distance of 15 mm.

2.2. Mechanical and Wear Test

The hardness of the different coating processes was measured using the Vickers hardness test machine (Hitachi, Universiti Tenaga Nasional, Selangor, Malaysia). The test was conducted at room temperature (30 °C), and the measurement of hardness was taken at three different places on each sample to obtain the average value of hardness. The pore size was analyzed using the image analysis method. Friction and wear tests were performed with micro pin-on-disc tribotester (model CM-9109, Ducom, Bangalore, India) according to the G99 ASTM standard test method. The pin-on-disc tribotester schematic is illustrated in Figure 2. Wear test samples were placed inside the steel pin holder, which was 15 mm in diameter. The test surface and wear-disc holder were cleaned with soft cotton after each run on the machine to remove any wear debris. The sliding tests were performed on a wear track diameter of 30 mm for a constant sliding distance of 400 m under ambient conditions. The tests were conducted for different loads (2 N, 3 N, and 5 N) and a sliding velocity constant at 0.1 m/s. The wear-test parameters are given in Table 2.

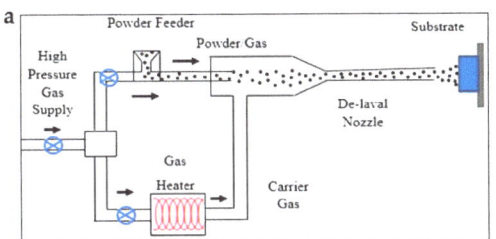 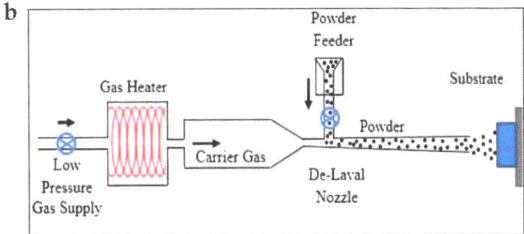

Figure 1. Schematic diagram of (**a**) HPCS and (**b**) LPCS [24].

Table 1. Spraying parameters of HPCS and LPCS.

Parameters	HPCS	LPCS
Size (μm)	25	25
Working gas	Helium	Air
Gas pressure (MPa)	2.0	0.6
Initial temperature (K)	300	300
Nozzle distance (mm)	15	15
Number of layers	20	20

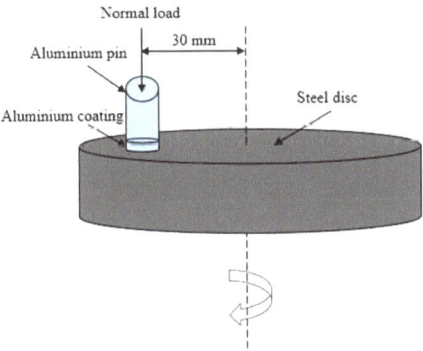

Figure 2. Schematic diagram of micro pin-on-disc tribotester.

Table 2. Wear-test parameters.

Parameters	Selected Value
Applied load (N)	2, 3 and 5
Velocity (m/s)	0.1
Environment	Air
Temperature (°C)	25 ± 2
Humidity (%)	55 ± 5
Speed of motor (rpm)	200
Sliding distance (m)	400
Track diameter (mm)	30
Duration (s)	960

2.3. Characterization

The samples were cut using the Buehler Isomet Precision Saw. The cutter speed was set to a low-speed setting, which is 100 rpm. Then, the samples were mounted using the cold mounting technique. The mounting materials used were resin from Struers Epoxy kit that included epoxy resin and epoxy hardener. In order to obtain a highly reflective surface that was free from scratches and deformation, the samples had to be be carefully ground and polished before they could be examined under the microscope. The samples were ground using silicon carbide papers of 800 and 1000 grit sizes and followed by a polishing process using a diamond solution as the polisher. The cross-sectional microstructures, surface morphologies, and wear tracks were examined using scanning electron microscopy (SEM) (Hitachi SU1510, Tokyo, Japan). The elemental composition of the coating and the substrate were determined using energy-dispersive X-ray spectroscopy (EDX) (Hitachi SU1510, Tokyo, Japan).

3. Results

3.1. Microstructure and Mechanical

The microstructure images of Al coatings deposited using HPCS and LPCS are shown in Figure 3. According to visual examination from Figure 3a, pores indicated by the arrow show that HPCS coating exhibited larger pores between the particles compared to that of LPCS in Figure 3b. To confirm this observation, the pore-size distribution was plotted and quantified using a histogram, as shown in Figure 4. The average pore size of the HPCS coating is approximately 0.35 μm, and the average pore size of the LPCS coating is approximately 0.18 μm.

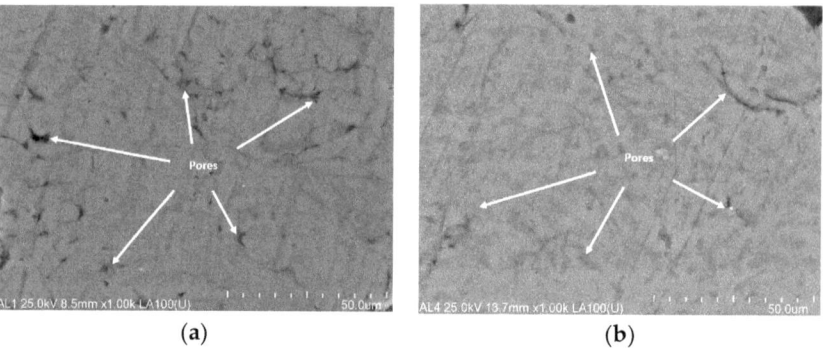

(a)   (b)

Figure 3. SEM images of Al coatings using (**a**) HPCS and (**b**) LPCS.

Smaller average pore size is obtained in the LPCS coating due to the full deformation of the sprayed Al particles that leads to good bonding between the sprayed particles and the substrate. Furthermore, the smaller pore-size distribution in the LPCS coating contributes to denser coating with lower porosity. The porosity of the Al coatings deposited using HPCS and LPCS on Al substrates is given in Table 3. It is the measure of pore area over total area. The LPCS coating has lower porosity (3.48%) than the HPCS coating (7.72%). A previous study in [17] claimed that high porosity in Al coatings reduces the mechanical properties where porosity is one of the key factors that influences the mechanical and wear performance in coatings. Hence, it is inferred that the lower porosity observed in LPCS coating could improve the mechanical and wear performances, which can be assessed through microhardness, friction, and wear tests.

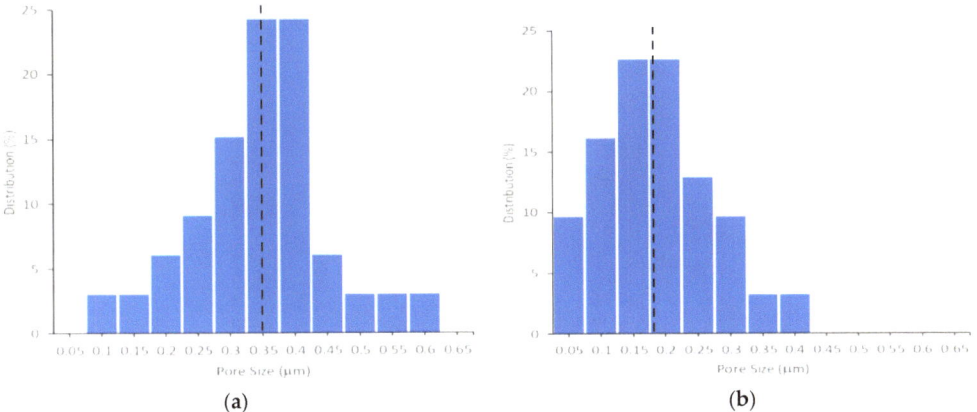

Figure 4. Pore size distribution of Al coatings using (**a**) HPCS and (**b**) LPCS. (Dotted line is average pore size, for (**a**) 0.35 micrometer and (**b**) 0.18 micrometer).

Table 3. Vickers hardness and porosity of HPCS and LPCS coatings.

Process	Vickers Hardness (HV)	Porosity (%)
HPCS	174.3	7.72
LPCS	196.6	3.48

The microhardness obtained through the Vickers Hardness test of Al coatings deposited using HPCS and LPCS on Al substrates is also summarized in Table 3. The LPCS coating has a higher hardness of 196.6 HV than the HPCS coating, which is 174.3 HV. This is attributed to the lower porosity of the LPCS coating. According to Lee et al., the Al coating deposited using LPCS has higher hardness due to the peening effect from the low coating porosity [18].

Higher hardness indicates work-hardening resulting from plastic deformation and strain-hardening during particle impacts [18], which can be easily achieved using HPCS through the severe deformation of the particles on the substrate. However, the presence of larger pores in the HPCS-sprayed Al coating is due to the peculiarities of the HPCS process on lightweight materials. Although the first layer of particles was fully deformed, the continuous bombardment of subsequent particles at high velocity resulted in large pores between the particles. These large pores prevent the formation of compact and coherent layers. On the other hand, the smaller pores formed between the Al particles through the LPCS result in a dense coating. Therefore, the LPCS induces a high level of plastic deformation on the Al coating, as well as strain hardening, which in turn increases hardness in the coating.

3.2. Friction and Wear

Wear, and friction, tests were performed on a pin-on-disc wear-test machine. Figure 5 shows the wear load and frictional force of HPCS and LPCS coatings as a function of time under 2 N, which are used to calculate the wear rate and the specific wear rate. The wear rate and specific wear rate are given as in the following equation:

$$\dot{W} = \frac{v}{F_n d} \tag{1}$$

$$\dot{V} = K\dot{W} \tag{2}$$

where $\dot{W}$ is the specific wear rate, which is simply the wear volume v divided by the product of the normal load F_n and the sliding distance d. Moreover, $\dot{V}$ is the wear rate and K is the wear constant.

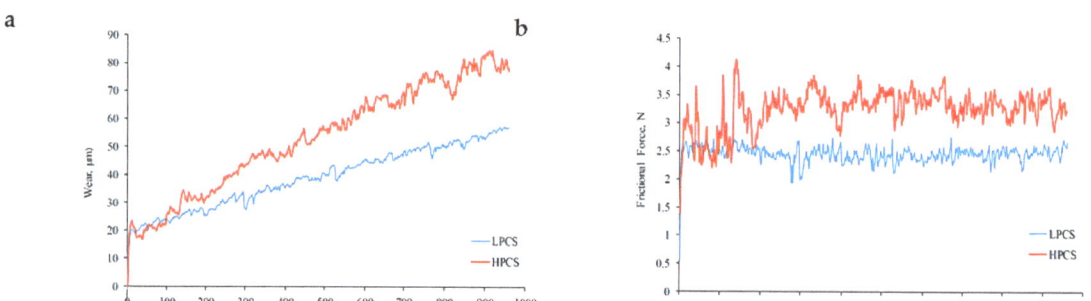

Figure 5. (**a**) Wear load; (**b**) frictional force of Al coatings using HPCS and LPCS.

Figure 6 shows the variation of the friction coefficient of Al coating deposited using HPCS and LPCS as a function of time subjected to 2 N. The wear tests conducted on the HPCS and LPCS coatings demonstrated that the friction coefficient varied between 0.001 and 1.8 μ and between 0.3 and 1.4 μ for HPCS and LPCS, respectively, as seen in Figure 6. These results show that the LPCS coating has a smaller range of friction coefficient than the HPCS coating.

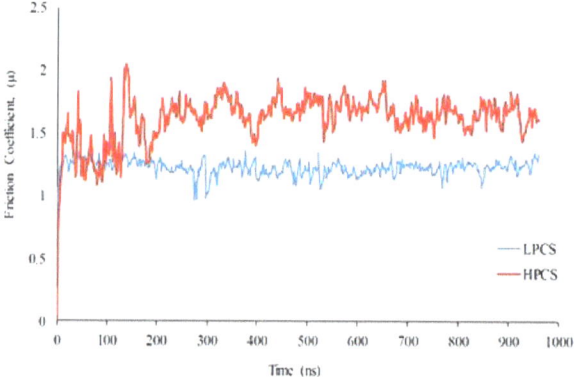

Figure 6. Friction coefficient at 2 N load of Al coatings using HPCS and LPCS.

Furthermore, Figure 7a,b demonstrate the variation of friction coefficients and the wear rate of Al coatings as a function of applied load (2 N, 3 N, and 5 N) deposited using HPCS and LPCS, respectively. The friction coefficient for both the HPCS and LPCS coatings decreases as the load increases from 2 N to 5 N. This shows that CS technology decreases the friction coefficient [22]. Comparing the HPCS and LPCS coatings, the friction coefficient of the LPCS coating decreased gradually from 1.343 to 1.297 and then to 0.798 with small gaps in the values as the load was increased, whereas the HPCS coating decreased from 1.609 to 1.067 and then to 0.807 with larger gaps. As stated in [22], dense coatings with low porosity generate a low friction coefficient that improves wear resistance.

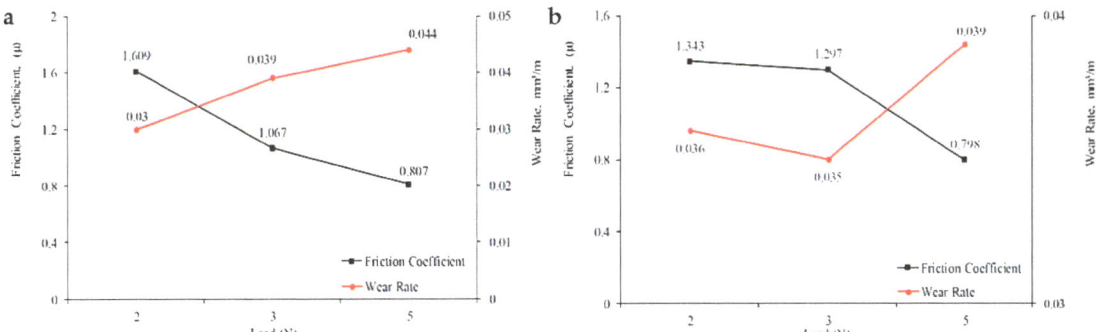

Figure 7. Friction coefficient and wear rate at 2, 3, and 5 N loads of Al coatings using (**a**) HPCS and (**b**) LPCS.

However, the wear rate increased for both HPCS and LPCS coatings as the load increased from 2 N to 5 N, as seen in Figure 7. This is because the wear rate is independent of the applied load, as observed in Equation (2). Figure 8a,b show the specific wear rate and wear resistance of Al coatings as a function of applied load (2 N, 3 N, and 5 N) deposited using HPCS and LPCS, respectively. The specific wear rate decreased for both HPCS and LPCS coatings as the applied load increased. Even though the specific wear rate of LPCS coating (0.016 mm^3/Nm) was higher than the HPCS coating (0.015 mm^3/Nm) at the low load of 2 N, the specific wear rate of the LPCS coating decreased as the load was increased to 5 N. The lower specific wear rate of the LPCS coating at a higher applied load led to a higher value of wear resistance (98.5 Nm/mm^3) than the HPCS coating (98.2 Nm/mm^3).

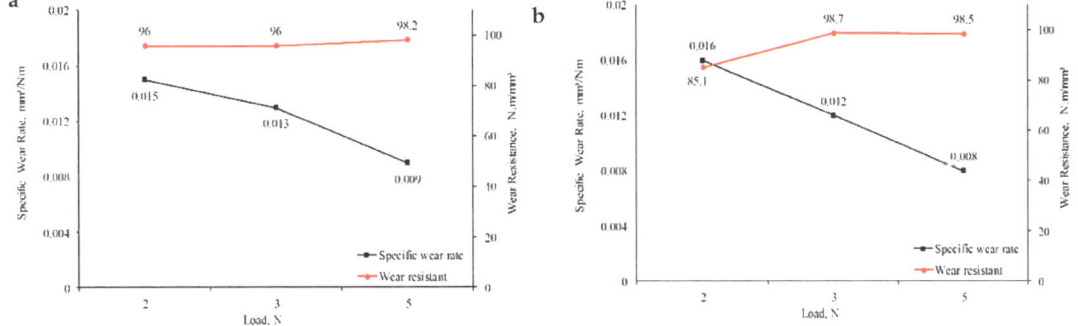

Figure 8. Specific wear rate and wear resistance at 2, 3, and 5 N loads of Al coatings using (**a**) HPCS and (**b**) LPCS.

Figures 9 and 10 show the SEM images of wear tracks subjected to different loads of 2 N, 3 N, and 5 N, formed on HPCS and LPCS coatings, respectively. The wear track on the HPCS coating shows a smoother surface with only small patches and grooves in the 2 N load compared to the 3 N and 5 N loads. The wear tracks on higher loads (3 N and 5 N) have more patches with deep grooves. On the other hand, the wear track on the LPCS coating with a low load of 2 N provides a regular surface profile with a small number of cracks and cavities (Figure 10a). The wear track with 3 N of the applied load demonstrates grooves on the surface (Figure 10b).

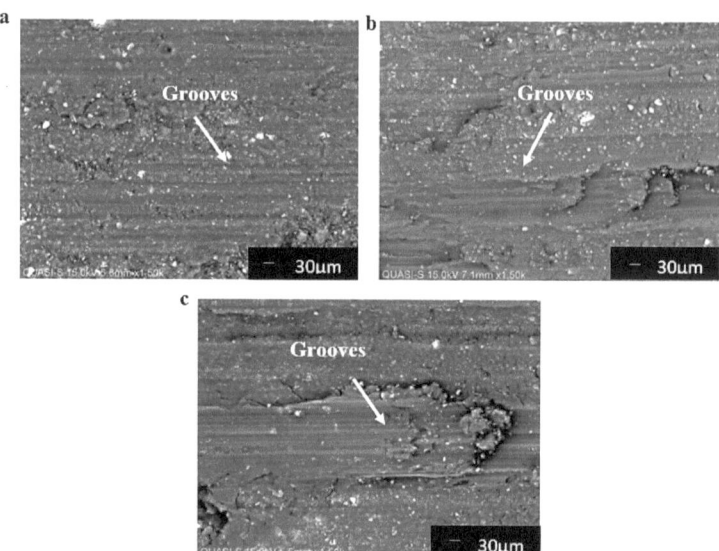

Figure 9. SEM image of HPCS Al coatings at an applied loads of (**a**) 2 N; (**b**) 3 N; and (**c**) 5 N.

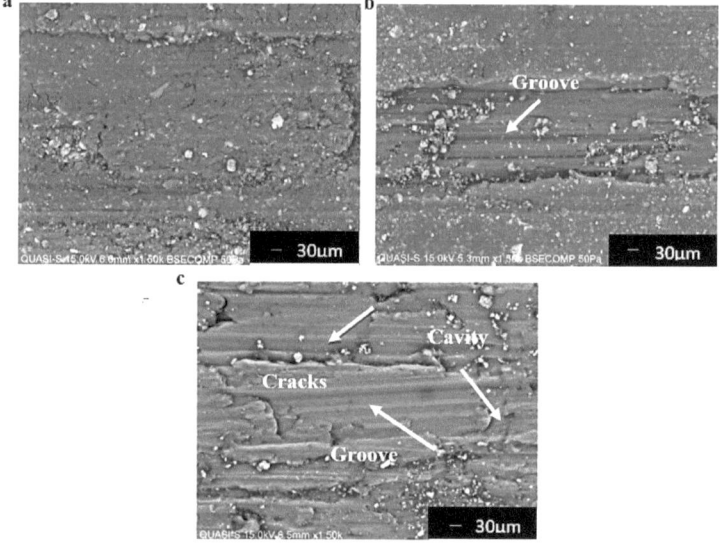

Figure 10. SEM image of LPCS Al coatings at an applied load of (**a**) 2 N; (**b**) 3 N; and (**c**) 5 N.

In contrast, the wear track with the 5 N load shows the formation of larger cracks, small cavities, and grooves on the surface (Figure 10c). The composition of oxygen in the HPCS and LPCS coatings is tabulated in Table 4. The results show that the composition of oxygen increases as the load increases for both the HPCS and the LPCS coating. The LPCS coating, however, exhibits a higher amount of oxygen with increasing load. This may contribute to its low friction coefficient observed in Figure 7b.

Table 4. EDX result of HPCS and LPCS.

Process	Oxygen Composition (%)			
	Initial	2 N	3 N	5 N
HPCS	16.49	17.02	31.20	45.38
LPCS	18.85	16.59	35.23	47.18

According to Asuke et al. [25], the wear rate increases with increasing applied load because the amount of wear loss is small when a small load is applied. Besides that, the drastic reduction in specific wear rate for HPCS and LPCS coatings from 0.015 mm^3/Nm to 0.009 mm^3/Nm (Figure 8a) and from 0.016 mm^3/Nm to 0.008 mm^3/Nm (Figure 8b), respectively, is due to the effectiveness of the oxide layer formed on the surface. Stott mentioned that the high oxide layer formed on the surface of the coating contributes to a low specific wear rate [26,27]. The lower specific wear rate exhibited by the LPCS coating at higher loads (3 N and 5 N) than the HPCS coating indicates that the oxide layer inhibits contact between the surfaces, which may improve the wear resistance in the coating. Accordingly, the LPCS coating showed more excellent wear resistance at higher applied loads of 3 N and 5 N, with a high amount of oxygen composition of 35.23% and 47.18%, respectively, whereby high wear resistance indicates improved wear resistance and, therefore, good tribological properties [25].

The obtained results depict that the worn surfaces on the wear track of the HPCS and LPCS coatings denote the presence of an oxide layer resulting from the formation of cracks and cavities. This oxide layer, which is formed by oxidative wear, protects the Al coating from wear when it is in contact with a counter-face and induces low friction. One reason for the low friction is the formation of low-strength microfilms, in this case the growth of aluminum oxide (Al_2O_3) between the contacting surfaces. Therefore, the high amount of oxygen (47.18%) in the LPCS coating subjected to the 5 N load contributes to the large cracks and small cavities formed on the worn surfaces, as observed in Figure 10c. Moreover, the formation of an oxide layer leads to a low friction coefficient, as observed in Figure 7. Therefore, due to the higher amount of oxygen composition in the LPCS coating, the Al coating deposited using LPCS exhibits better wear properties than that of HPCS.

4. Conclusions

In summary, the mechanical and tribological properties of Al coatings using the HPCS and LPCS processes were successfully studied. The micro-hardness of the Al coating using the LPCS process before the wear test reached up to 196.6 HV, which is higher than the Al coating using the HPCS process (174.3 HV). Meanwhile, the pore size increases with increasing porosity, which influences the mechanical and wear performance. Al coating using the LPCS process has a lower percentage of pore-size diameter than the HPCS process. The lower percentage of pore diameter contributes to a dense coating with less porosity. Moreover, the low friction coefficients obtained by the LPCS process indicate the good tribological properties of the Al coating, with a high amount of oxide composition. The wear studies reveal that the specific wear rate of Al coating using the LPCS process is lower as the load increases from 3 N to 5 N, thus providing greater wear resistance than that of the HPCS coating.

Author Contributions: Conceptualization, A.M. and S.N.A.Y.; methodology, A.M. and S.N.A.Y.; software, A.M., N.A., and S.N.A.Y.; validation, A.M. and S.N.A.Y.; formal analysis, S.N.A.Y.; investigation, S.N.A.Y.; resources, A.M.; data curation, S.N.A.Y.; writing—original draft preparation, S.M. and S.N.A.Y.; writing—review and editing, A.M., N.A., S.M., S.N.A.Y. and Z.M.R.; visualization, N.A. and Z.M.R.; supervision, A.M.; project administration, A.M.; and funding acquisition, A.M. All authors have read and agreed to the published version of the manuscript.

Funding: This research was funded by the Ministry of Higher Education Malaysia (MoHE) and the Universiti Tenaga Nasional (UNITEN) for funding this study under the fundamental research-grant scheme (FRGS) of Grant No. 20130108FRGS, and the APC was funded by UNITEN, Grant No. Grant No. J510050002—IC-6 BOLDREFRESH2025—CENTRE OF EXCELLENCE.

Institutional Review Board Statement: Not applicable.

Informed Consent Statement: Not applicable.

Data Availability Statement: Not applicable.

Acknowledgments: The authors would like to thank the Ministry of Higher Education Malaysia (MoHE) and Universiti Tenaga Nasional (UNITEN) for funding this study under the fundamental research-grant scheme (FRGS) of Grant No. 20130108FRGS.

Conflicts of Interest: The authors declare no conflict of interest.

References

1. Ogawa, K.; Ito, K.; Ichimura, K.; Ichikawa, Y.; Ohno, S.; Onda, N. Characterization of low-pressure cold-sprayed aluminum coatings. *J. Therm. Spray Technol.* **2000**, *17*, 728–735. [CrossRef]
2. Nordheim, E.; Barrasso, G. Sustainable development indicators of the European aluminium industry. *J. Clean. Prod.* **2007**, *15*, 275–279. [CrossRef]
3. Wang, G.; Zhang, L.; Zhang, J. A review of electrode materials for electrochemical supercapacitors. *Chem. Soc. Rev.* **2012**, *41*, 797–828. [CrossRef] [PubMed]
4. Attia, H.; Meshreki, M.; Korashy, A.; Thomson, V.; Chung, V. Fretting wear characteristics of cold gas-dynamic sprayed aluminum alloys. *Tribol. Int.* **2011**, *44*, 1407–1416. [CrossRef]
5. Singh, J.; Lal, H.; Bala, N. Study of wear behaviour of HVOF and cold sprayed coatings on boiler steels. *Int. J. Mech. Eng. Robot. Res.* **2013**, *2*, 134–138.
6. Bolelli, G.; Lusvarghi, L.; Barletta, M. Heat treatment effects on the corrosion resistance of some HVOF-sprayed metal alloy coatings. *Surf. Coat. Technol.* **2008**, *202*, 4839–4847. [CrossRef]
7. Papyrin, A.; Kosarev, V.; Klinkov, S.; Alkhimov, A.; Fomin, V.M. *Cold Spray Technology*, 1st ed.; Elsevier: Amsterdam, The Netherlands, 2007.
8. Manap, A.; Seo, D.; Ogawa, K. Characterization of thermally grown oxide on cold sprayed CoNiCrAlY bond coat in thermal barrier coating. *Mater. Sci. Forum* **2011**, *696*, 324–329. [CrossRef]
9. Manap, A.; Nooririnah, O.; Misran, H.; Okabe, T.; Ogawa, K. Experimental and SPH study of cold spray impact between similar and dissimilar metals. *Surf. Eng.* **2014**, *30*, 335–341. [CrossRef]
10. Schmidt, T.; Gärtner, F.; Assadi, H.; Kreye, H. Development of a generalized parameter window for cold spray deposition. *Acta Mater.* **2006**, *54*, 729–742. [CrossRef]
11. Manap, A.; Mahalingam, S.; Yusof, S.N.A.; Afandi, N.; Abdullah, H. Impact Behaviour of Aluminum Particles upon Aluminum, Magnesium, and Titanium Substrates using High Pressure and Low-Pressure Cold Spray. *Sains Malays.* **2022**, *51*, 585–597. [CrossRef]
12. Schmidt, T.; Assadi, H.; Gärtner, F.; Richter, H.; Stoltenhoff, T.; Kreye, H.; Klassen, T. From particle acceleration to impact and bonding in cold spraying. *J. Therm. Spray Technol.* **2009**, *18*, 794–808. [CrossRef]
13. Manap, A.; Ogawa, K.; Okabe, T. Numerical analysis of interfacial bonding of aluminum powder particle and aluminum substrate by cold spray technique using the SPH method. *J. Solid Mech. Mater. Eng.* **2012**, *6*, 241–250. [CrossRef]
14. Singh, H.; Sidhu, T.S.; Kalsi, S.B. Cold spray technology: Future of coating deposition processes. *Frat. Integrità Strutt.* **2012**, *22*, 69–84. [CrossRef]
15. Wang, Q.; Birbilis, N.; Zhang, M.X. Interfacial structure between particles in an aluminum deposit produced by cold spray. *Mater. Lett.* **2011**, *65*, 1576–1578. [CrossRef]
16. Yusof, S.N.A.; Manap, A.; Misran, H.; Othman, S.Z. Computational analysis of single and multiple impacts of low pressure and high pressure cold sprayed aluminum particles using SPH. *Adv. Mater. Res.* **2014**, *974*, 147–151. [CrossRef]
17. Yalcin, B. Effect of porosity on the mechanical properties and wear performance of 2% copper reinforced sintered steel used in shock absorber piston production. *J. Mater. Sci. Technol.* **2009**, *25*, 577–582.
18. Lee, H.; Shin, H.; Lee, S.; Ko, K. Effect of gas pressure on Al coatings by cold gas dynamic spray. *Mater. Lett.* **2008**, *62*, 1579–1581. [CrossRef]
19. Wang, Q.; Qiu, D.; Xiong, Y.; Birbilis, N.; Zhang, M.X. High resolution microstructure characterization of the interface between cold sprayed Al coating and Mg alloy substrate. *Appl. Surf. Sci.* **2014**, *289*, 366–369. [CrossRef]
20. Rokni, M.R.; Widener, C.A.; Crawford, G.A.; West, M.K. An investigation into microstructure and mechanical properties of cold sprayed 7075 Al deposition. *Mater. Sci. Eng. A* **2015**, *625*, 19–27. [CrossRef]
21. Marzbanrad, B.; Jahed, H.; Toyserkani, E. On the evolution of substrate's residual stress during cold spray process: A parametric study. *Mater. Des.* **2018**, *138*, 90–102. [CrossRef]

22. Koivuluoto, H.; Coleman, A.; Murray, K.; Kearns, M.; Vuoristo, P. High pressure cold sprayed (HPCS) and low pressure cold sprayed (LPCS) coatings prepared from OFHC Cu feedstock: Overview from powder characteristics to coating properties. *J. Therm. Spray Technol.* **2012**, *21*, 1065–1075. [CrossRef]

23. Ning, X.J.; Jang, J.H.; Kim, H.J.; Li, C.J.; Lee, C. Cold spraying of Al-Sn binary alloy: Coating characteristics and particle bonding features. *Surf. Coat. Technol.* **2008**, *202*, 1681–1687. [CrossRef]

24. Julio, V. Current and future applications of cold spray technology. *Met. Finish.* **2010**, *108*, 37–39.

25. Asuke, F.; Abdulwahab, M.; Aigbodion, V.S.; Fayomi, O.S.; Aponbiede, O. Effect of load on the wear behaviour of polypropylene/carbonized bone ash particulate composite. *Egypt. J. Basic Appl. Sci.* **2014**, *1*, 67–70. [CrossRef]

26. Stott, F.H. The role of oxidation in the wear of alloys. *Tribol. Int.* **1998**, *31*, 61–71. [CrossRef]

27. Manap, A.; Okabe, T.; Ogawa, K.; Mahalingam, S.; Abdullah, H. Experimental and smoothed particle hydrodynamics analysis of interfacial bonding between aluminum powder particles and aluminum substrate by cold spray technique. *Int. J. Adv. Manuf. Technol.* **2019**, *103*, 4519–4527. [CrossRef]

Article

Influence of Remaining Oxide on the Adhesion Strength of Supersonic Particle Deposition TiO$_2$ Coatings on Annealed Stainless Steel

Noor irinah Omar [1,*], Yusliza Yusuf [1], Syahrul Azwan bin Sundi [1], Ilyani Akmar Abu Bakar [2], Verry Andre Fabiani [3], Toibah Abdul Rahim [4] and Motohiro Yamada [5]

[1] Faculty of Mechanical Engineering and Manufacturing Technology, Universiti Teknikal Malaysia Melaka, Durian Tunggal, Melaka 76100, Malaysia
[2] School of Civil Engineering, College of Engineering, Universiti Teknologi MARA, Shah Alam 40450, Malaysia
[3] Department of Chemistry, Faculty of Engineering, Universitas Bangka Belitung, Pangkalpinang 33172, Indonesia
[4] Faculty of Manufacturing Engineering, Universiti Teknikal Malaysia Melaka, Durian Tunggal, Melaka 76100, Malaysia
[5] Department of Mechanical Engineering, Toyohashi University of Technology, 1-1 Tempaku-Cho, Toyohashi 441-8580, Japan
* Correspondence: nooririnah@utem.edu.my; Tel.: +60-1-2366-1044

Abstract: The cold spray or Supersonic Particle Deposition technique has great potential for producing ceramic nanostructured coatings. This technique operates at a processing temperature that is low enough to preserve the initial feedstock materials' microstructure. Nevertheless, depositing ceramic powders using a cold spray can be challenging because of the materials' brittle nature. The interaction between substrate and particles is significantly influenced by substrate attributes, including hardness, material nature, degree of oxidation and temperature. In this study, the effect of the substrate's remaining oxide composition on the adhesion strength of an agglomerated nano-TiO$_2$ coating was investigated. The results showed that the coating adhesion strength increased for hard materials such as stainless steel and pure chromium as the annealed substrate temperature also increased from room temperature to 700 °C, indicating thicker oxide on the substrate surface. TiO$_2$ particles mainly bond with SUS304 substrates through oxide bonding, which results from a chemical reaction involving TiO$_2$-OH$^-$. Chromium oxide (Cr$_2$O$_3$) is thermodynamically preferred in SUS304 and provides the OH$^-$ component required for the reaction. SUS304 shows a thermodynamic preference for chromium oxide (Cr$_2$O$_3$), and this enables Cr$_2$O$_3$ to provide the necessary OH$^-$ component for the reaction.

Keywords: supersonic particle deposition; titanium dioxide; adhesion bonding; hard material; stainless steel SUS304; pure chromium

Citation: Omar, N.i.; Yusuf, Y.; Sundi, S.A.b.; Abu Bakar, I.A.; Andre Fabiani, V.; Abdul Rahim, T.; Yamada, M. Influence of Remaining Oxide on the Adhesion Strength of Supersonic Particle Deposition TiO$_2$ Coatings on Annealed Stainless Steel. *Coatings* **2023**, *13*, 1086. https://doi.org/10.3390/coatings13061086

Academic Editor: Lech Pawlowski

Received: 31 March 2023
Revised: 29 May 2023
Accepted: 29 May 2023
Published: 12 June 2023

1. Introduction

Cold spray is a deposition technique that operates in a solid state without melting the feedstock. High-velocity particles transfer their kinetic energy to the substrate, resulting in localized heat and interfacial deformation [1–3]. This in turn leads to mechanical interlocking and metallurgical bonding [4]. Although the bonding process is not fully understood, optimal temperature, critical velocity and energy levels are needed for effective bonding to take place [5,6]. High strain rate formation and localized heating at the particle-substrate interface result in microscopic protrusions, which can cause metallurgical bonding through material deformation and atomic-level interactions [7,8].

Severe plastic deformation of materials determines interface bonding, resulting in metal-jet formation and adiabatic shear instability (ASI) at the interface [9–12]. The native oxide layer is broken due to the impact of the high-velocity particles, thus creating contact between particles and substrates and resulting in jet formation via ASI [13–15]. As opposed to the results of Hassani-Gangaraj et al. [16], Grujicic et al. [17] and Assadi et al. [5], presented evidence that contradicted the notion that ASI is not a mandatory factor for adhesion in the cold spray process. In response to the statements made by Assadi et al. [18], Hassani-Gangaraj et al. [19] upheld their simulation-driven study, which indicated that ASI was unnecessary for bonding.

The inadequate adhesion and delamination issues of soft and hard or hard and hard interfaces pose significant challenges to industries. Therefore, comprehending the bonding mechanism helps to address these concerns and benefit the advanced manufacturing sector. There is a significant demand for thick copper coatings on steel (SS316L) plates, which are similar to those of bulk copper and show strong adhesion, particularly for application in Tokamaks' vacuum vessels [20,21]. Singh et al. [7] conducted a study to analyze the bonding process between copper particles and steel substrates (soft-on-hard interface) by modifying substrate conditions and cold spray parameters. Drehmann et al. [22], Wustefled et al. [23] and Dietrich et al. [24] discovered that deformation-induced recrystallization near the particle-substrate interface was responsible for the bonding between aluminum particles and a super-finished monocrystalline sapphire substrate. Metallurgical bonding was facilitated by the development of nanoparticles at the interface, leading to increased adhesion strength between the Al_2O_3 monocrystalline ceramic substrate and the malleable Al particles.

The substrate surface plays a crucial role in achieving strong adhesion in cold-sprayed coatings as it determines the bonding strength between the first layer of TiO_2 particles and the outermost surface. When a metal substrate is used, an active oxide layer with a thickness of several micrometers is formed on the outer surface. Upon contact with the substrate surface, the agglomerated TiO_2 particles disintegrate, leading to particle-substrate contact a few nanometers beneath the oxide layer.

Cold spraying is a common method for creating TiO_2 coatings on metallic objects. Combining a powdered photocatalytic oxide metal with a ductile metal powder allows for plastic deformation upon particle impact. However, the presence of metal oxide particles covering 30–80% of the top surface can hinder the performance of the ceramic TiO_2 coating. Nevertheless, studies have shown that thick, pure agglomerated TiO_2 coatings without additional substances exhibit photocatalytic activity comparable to the raw powder [25]. However, these pure agglomerated coatings have weak interfacial adhesion strength, and the bonding mechanism is not fully understood.

This research aimed to investigate how substrate oxidation levels affect the adhesion of coatings to surfaces. Stainless steel (SUS304) was chosen as the substrate to prevent the development of substrate metal jets caused by particle impact. The study considered factors including substrate oxide thickness, chemical composition, and atomic composition.

2. Equipment and Procedures

2.1. Spray Process

Throughout all coating experiments, the spray parameters used are shown in Table 1 below.

Table 1. Spray parameters.

Cold Spray Machine	De-Laval 24TC Nozzle
Gas carrier	N_2
Operating temperature	500 °C
Operating pressure	3 MPa
Traverse speed	10 mm/s
Spray distance	20 mm
Powder feed rate	3 g/min
Pass number	1

2.2. Materials

Agglomerated TiO_2 powder was used as a feedstock (TAYCA Corporation, Osaka, Japan). It had an anatase crystalline form; its average particle size was 7.55 μm (Figure 1).

Figure 1. SEM image of Titanium Dioxide powder.

The study utilized two substrates: pure chromium (Cr) and stainless steel (SUS304). To explore the influence of surface oxide at varying temperatures, the substrates were initially subjected to grit-blasting, which was then followed by annealing. The annealing process was conducted under atmospheric conditions using an electric furnace that was set to two different temperatures, specifically 400 °C and 700 °C. The substrates were heated gradually at a rate of 15 °C per 5 min and then soaked for 5 min before being cooled down to room temperature inside the furnace.

2.3. Testings

2.3.1. Testing Tensile-Strength

The study adhered to JIS H 8402 standards and utilized specimens with dimensions of Ø 25 mm × 10 mm to test the adhesion strength. The fracture load value, indicating adhesion strength, was measured using a universal testing machine (Autograph AGS-J, Shimadzu, Kyoto, Japan). For every spraying condition, five specimens were tested, and the average fracture load value was recorded. Furthermore, the fractured coating surface underwent EDX analysis to examine its composition.

2.3.2. Coatings Characterizations

To analyze the TiO_2 coating cross-sectional microstructures on annealed substrates, a scanning electron microscope (SEM: JSM-6390, JEOL, Tokyo, Japan) was employed. A sample measuring 25 mm × 10 mm was embedded in a hardenable resin for observation. The embedded sample underwent a grinding process using silica papers until a #3000 grit size was achieved. Subsequently, it was polished using 1 μm and 0.3 μm alumina suspension.

2.3.3. Micro-Vickers Hardness

The correlation between the adhesion strength of the TiO_2 coating on the annealed substrate and the hardness of the substrate's surface was investigated using a micro-Vickers hardness tester (HMV-G, Shimadzu, Kyoto, Japan). The substrate's hardness was measured with a 98.07 mN test load and a 10 s dwell time applied to the cross-sectional sample. The measurement revealed a hardness value of HV 0.1.

2.3.4. Substrate Oxide Evaluations

The substrate oxide thickness was measured using X-ray photoelectron spectroscopy (XPS) with the ULVAC-PHI Quantera SXM-CI, Kanagawa, Japan. XPS analysis involved using a monochromatic Al Kα source with a current of 15 mA and a voltage of 10 kV. Narrow scans ranging from 0 to 1000 eV were performed for Fe 2p, Cr 2p, and O 1s in order to analyze the different annealed substrates. The measured binding energies were corrected using C 1s at 285.0 eV. Pre-sputtering, which could potentially alter the sample surface and affect the measurements, was not conducted. Detailed XPS analysis conditions for substrate oxide analysis can be found in Table 2.

Table 2. XPS parameter for substrate oxide layer analysis [26].

Regions Measured	Fe 2p, Cr 2p, O 1s
X-ray output	10
Probe diameter	50
Time per step	30
Pass energy	140
Cycle	30

2.3.5. Wipe Test

To investigate the deformation of a single TiO$_2$ particle on different substrates, a wipe test was performed. Prior to deposition, the substrates were prepared by grinding and polishing to achieve a smooth, mirror-like surface. The spraying process was carried out using nitrogen as the process gas at a temperature of 500 °C and a pressure of 3 MPa. The distance between the nozzle and substrate was maintained at 20 mm, and the process was conducted at a traverse speed of 2000 mm/s. Before spraying, the substrates were cleaned with acetone. The deposition of a single TiO$_2$ particle on a mirror-polished annealed substrate was observed using the FEI Helios Dual Beam 650 field emission scanning electron microscope (FESEM, FEI, Hilsboro, OR, USA) and focused ion beam (FIB, FEI, Hilsboro, OR, USA) microscope from FEI, based in Oregon, USA.

2.3.6. TEM Testing

To examine the oxide layer post high-velocity cold spraying with various pressures impacting the substrate surface, TEM (transmission electron microscopy) testing was performed. The sample was prepared by creating a thin film, which was then analyzed using field emission gun (FEG) electron microscopy. The analysis was conducted on an EOL JEM-2100F, Tokyo, Japan field emission transmission electron microscope instrument in scanning mode at 200 kV.

3. Result

3.1. Adhesion Strength Testing and Fracture Surface Analysis on Annealed Substrates

Figure 2 illustrates the adhesion strength of TiO$_2$ coating on pure chromium (Cr) and annealed stainless steel (SUS304). The adhesion strength of both substrates increased as the annealing temperature increased, with values ranging from 0.51 to 2.63 MPa for SUS304 and 0.71 to 1.44 MPa for pure Cr. Figure 3 displays the fracture surface of SUS304 and TiO$_2$ coating after tensile strength testing, confirming that the fracture occurred at the interface between the substrate and coating and indicating a strong cohesive bond between TiO$_2$ particles during the coating formation process.

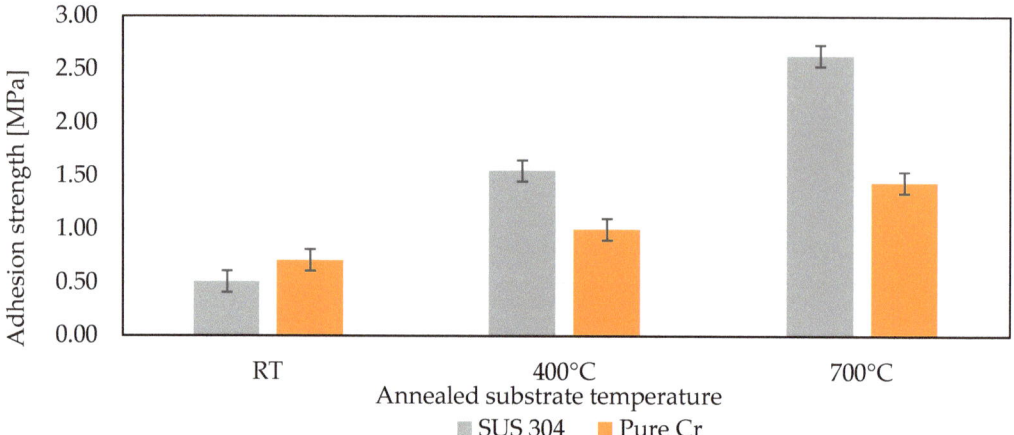

Figure 2. TiO$_2$ coating adhesion strength on SUS304 and pure Cr from room temperature to 700 °C annealed.

Figure 3. SUS304 fractured surface and TiO$_2$ coating post tensile strength testing at (**a**) room temperature; (**b**) annealed 400 °C; and (**c**) annealed 700 °C.

Figure 4 depicts the cross-sectional microstructure of SUS304 and TiO$_2$ coatings at room temperature and 700 °C, representing the conditions with the lowest and highest adhesion strengths, respectively. The coating exhibited a thickness of 200 to 300 μm, indicating the achievement of the critical velocity of particles during the spraying process. Figures 5 and 6 present the results of EDS analysis and the spectrum of the fractured TiO$_2$ coating on pure chromium substrates annealed at room temperature and 700 °C, respectively. Figure 6 demonstrates a notable presence of chromium in the fractured coating of the substrate annealed at 700 °C compared to the room-temperature substrate. This suggests that chromium oxide may influence the bonding mechanism, potentially contributing to the observed increase in adhesion strength in Figure 2. However, further investigation is required to fully comprehend the role of chromium oxide in the bonding process.

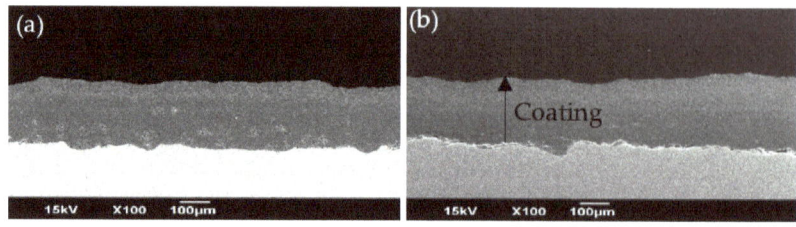

Figure 4. SUS 304 cross-sectional microstructure and TiO$_2$ coating at (**a**) room temperature; and (**b**) annealed 700 °C.

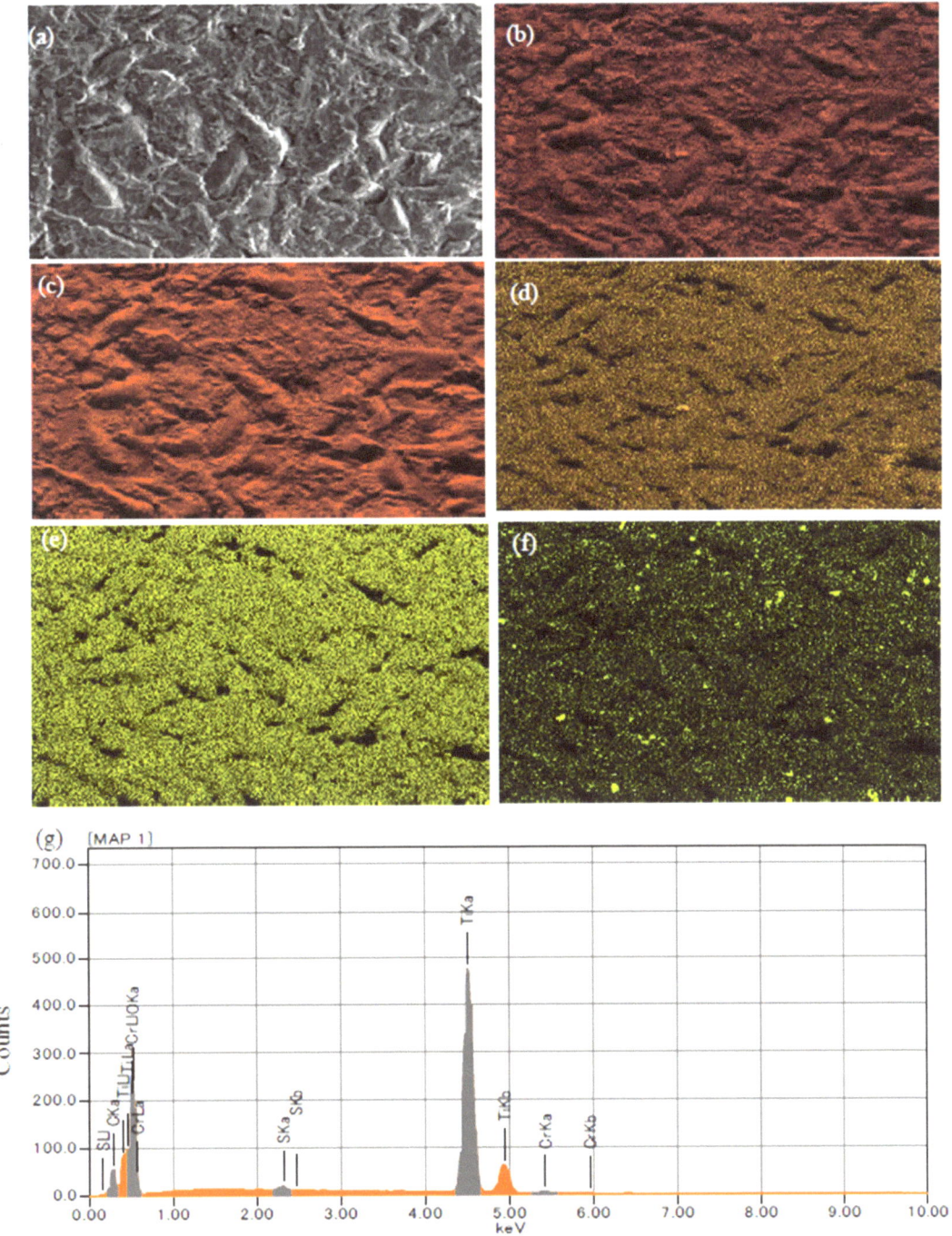

Figure 5. EDX elemental mappings of TiO$_2$/annealed pure Cr fracture coating: (**a**) SEM; (**b**) carbon; (**c**) oxygen; (**d**) sulfur; (**e**) titanium; (**f**) chromium; (**g**) map sum spectrum for room temperature pure Cr.

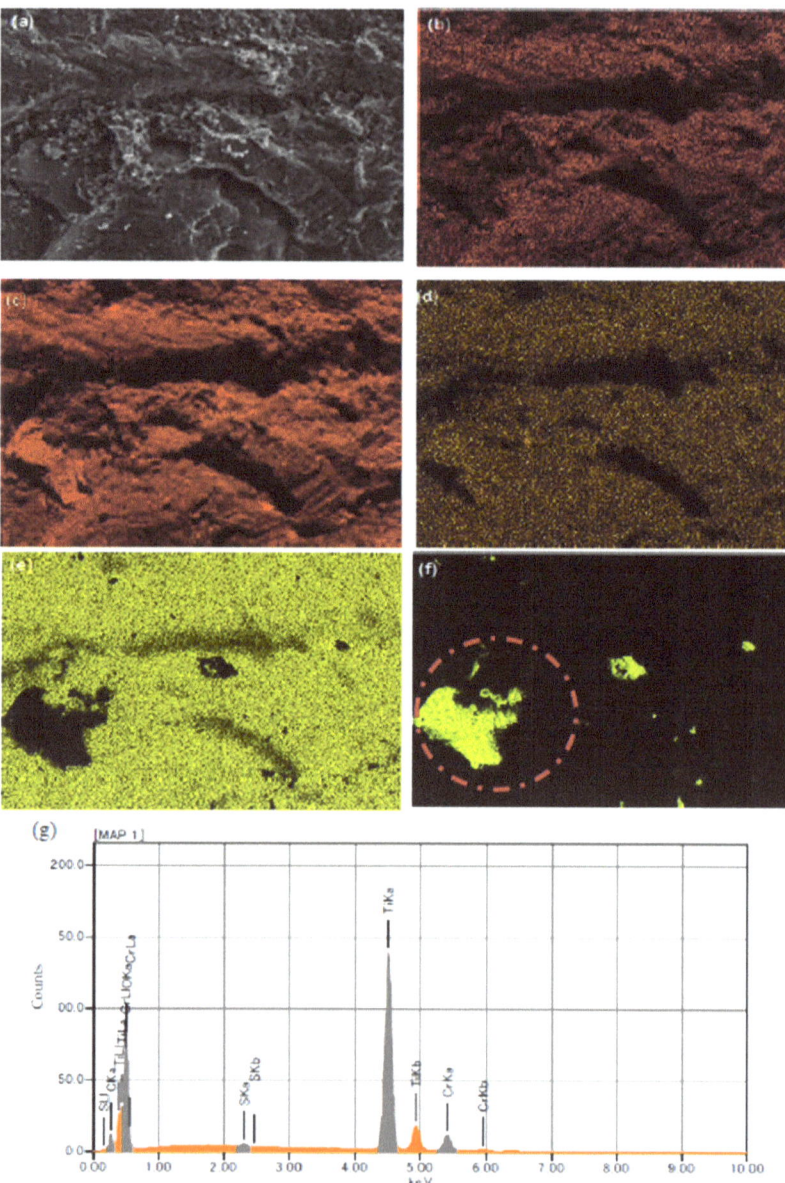

Figure 6. EDX elemental mappings of TiO$_2$/annealed pure Cr fracture coating: (**a**) SEM; (**b**) carbon; (**c**) oxygen; (**d**) sulfur; (**e**) titanium; (**f**) chromium; (**g**) map sum spectrum for 700 °C annealed pure Cr.

3.2. Depth Profile of Oxide Layer

Variations in the levels of oxygen and chromium within the substrates concerning their depths are illustrated in Figure 7a–c. The results demonstrated that there is a correlation between annealing temperature and the concentration of oxygen within the near-surface area. As the temperature of the substrate increased, the thickness of the pure chromium oxide layer also increased, leading to an increase in the adhesion strength of the TiO$_2$ coating upon pure chromium annealed substrate from room temperature to 700 °C annealed.

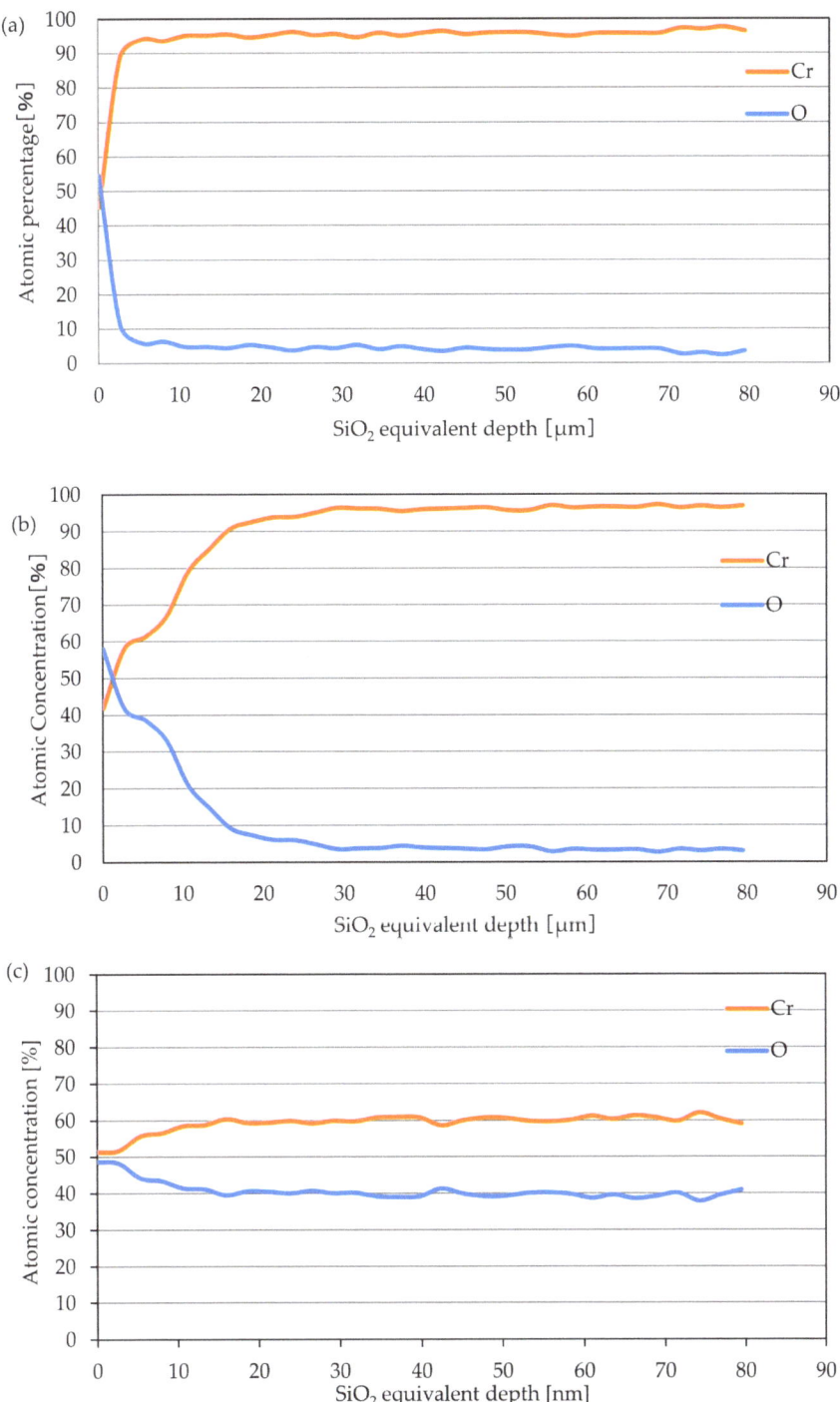

Figure 7. Depth profile analysis of pure Cr (**a**) Room temperature; (**b**) annealed 400 °C; and (**c**) annealed 700 °C.

Additionally, Figure 8a–c, which depicts the analysis of the oxide layer on SUS304, indicated that the thickness of the stainless-steel oxide layer also increased with the annealing temperature of the substrate. According to Ko et al. [27], when Cu/AlN and Al/ZrO$_2$ bonding couples are formed, the oxide layer becomes amorphous and atomic intermixing occurs at the interface due to chemical adhesion.

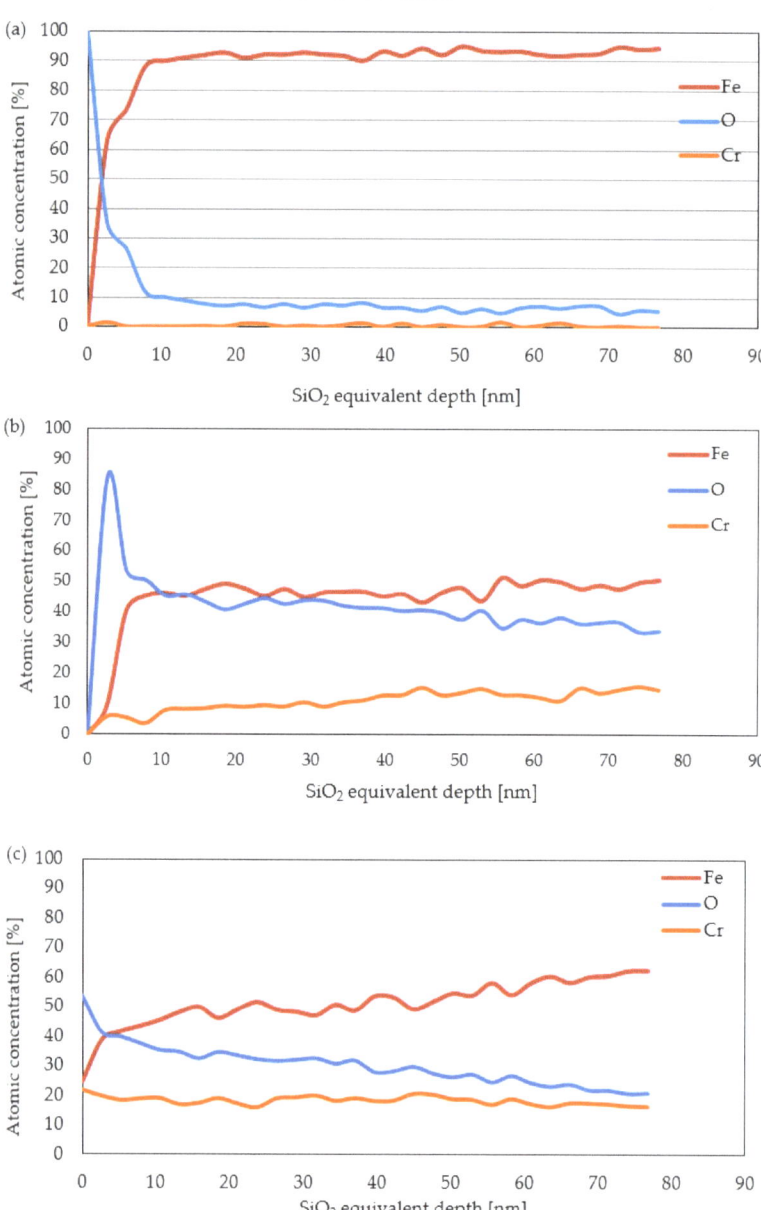

Figure 8. Depth profile analysis of SUS304 at (**a**) Room temperature; (**b**) annealed 400 °C; and (**c**) annealed 700 °C.

The results of this study indicated that the thicker oxide layers on the surface and higher temperatures led to increased adhesion strength of TiO$_2$ coating upon SUS304 and pure Cr. The bonding mechanism was influenced by the characteristics of Cr_2O_3 oxide between room temperature and 700 °C. Hence, further research would be conducted on the oxide's chemical composition.

3.3. Evaluation of the Chemical State of Iron, Chromium and Oxygen in the Oxide Layer

Figure 9a,b depict the oxygen content and chemical state of chromium, respectively. Figure 9a shows that the oxide layer on pure chromium at room temperature exhibits a dominant peak for chromium metal at 574.0 eV. In contrast, for pure chromium substrates annealed at 400 and 700 °C, Figure 9a illustrates the presence of a peak at 576.0 eV, corresponding to Cr_2O_3, across the outermost surface of the oxide layer. This indicates that Cr_2O_3 is the primary component of the oxide layer at these annealing temperatures.

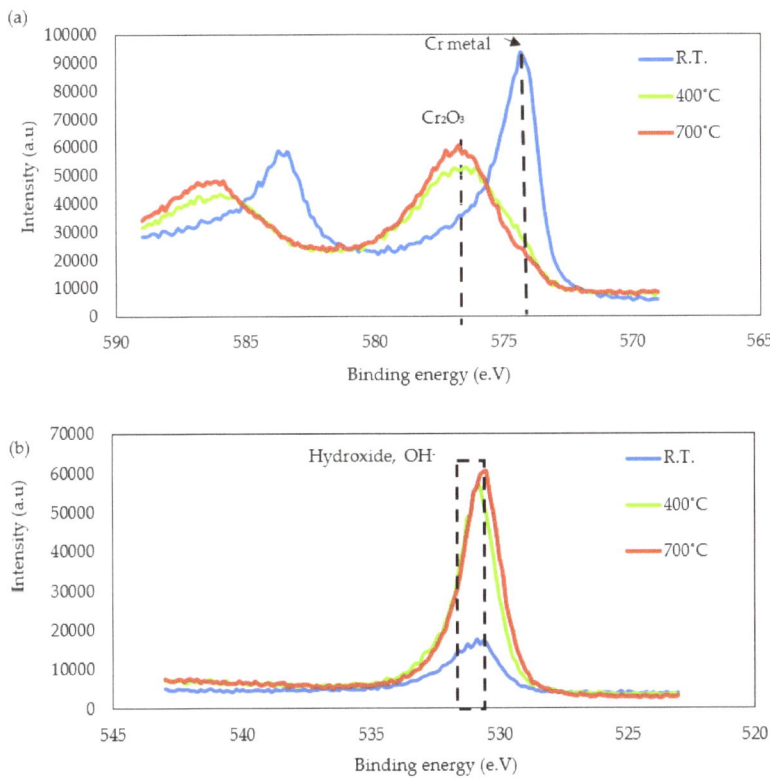

Figure 9. XPS spectra of (**a**) chromium; (**b**) oxygen for annealed pure chromium.

Figure 9b highlights the presence of hydroxide (OH^-) at the outermost surface of the oxide layer for pure chromium substrates annealed from room temperature to 700 °C. The red-dotted line represents a peak position of 531–532 eV, indicating the presence of hydroxide. The study findings suggest that the presence of hydroxide in the oxide layer, ranging from room temperature to 700 °C annealing temperatures, contributes to the observed trend of increasing adhesion strength of the coating with higher annealing temperatures.

Figure 10a–c provides a visual representation of the oxide layer chemical composition in SUS304. The chemical condition of ferum and chromium in SUS304 was identified via the 2P3/2 atomic orbital satellite peak. In Figure 10a, the location of the iron metal peak for stainless steel, which was roughly 706.7 eV [28], was observed on the outer surface of the room-temperature SUS304 substrate (a). When the SUS304 substrate was annealed at 400 °C and 700 °C, the outermost surface of the substrate showed a peak position of hematite Fe_2O_3 (Fe^{2+}) at around 709.3 eV [28]. This result suggested that there was a shift within the chemical state of the SUS304 substrate from a Fe metal state at room temperature to the hematite state after annealing at 400 °C and 700 °C.

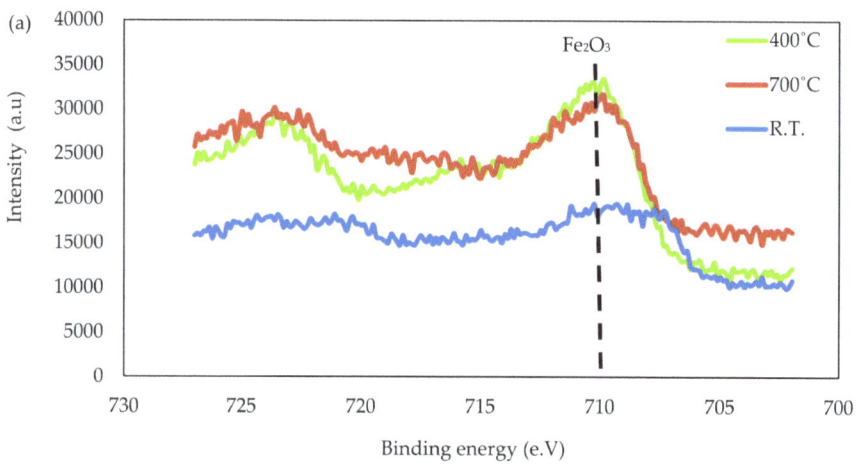

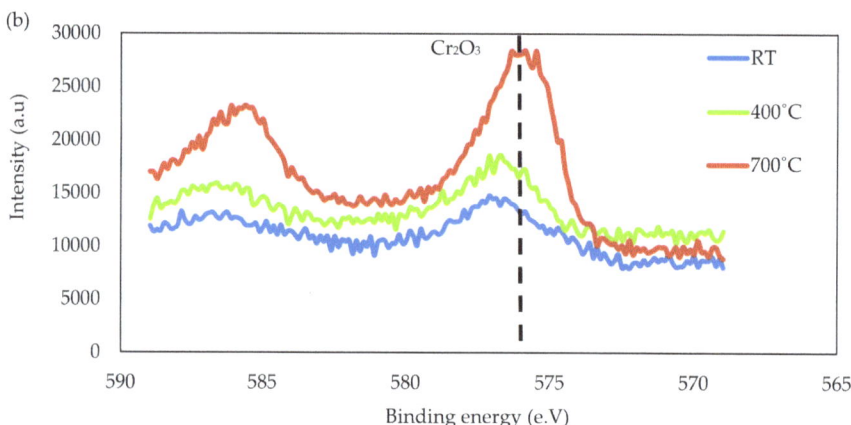

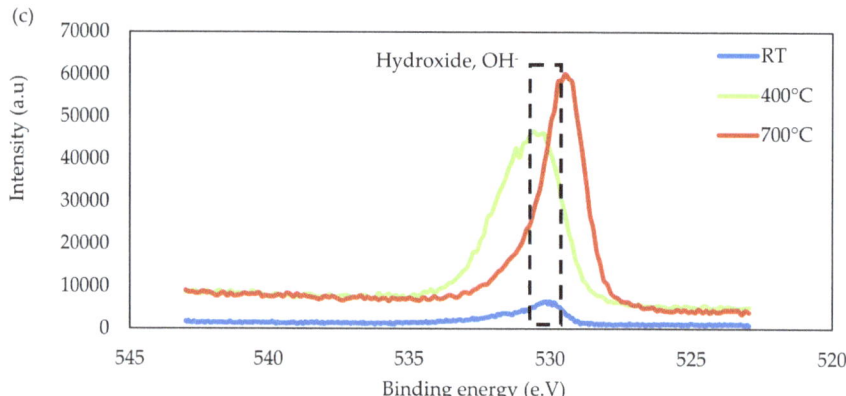

Figure 10. XPS spectra of SUS304: (**a**) Iron; (**b**) Chromium; and (**c**) oxygen.

The oxide layer of SUS304 at room temperature mainly consisted of chromium metal, as indicated by the notable peak at 574.0 eV [29]. For SUS304 substrates annealed at 400 and 700 °C, the peak position of Cr_2O_3 at 576.0 eV was observed across the outer surface of the oxide layer [28], demonstrating a shift from the metal state of chromium to the chromium oxide state. In Figure 10c, the red-dotted line demonstrated the presence of hydroxide, OH^-, between 531 and 532 eV [28], suggesting that the oxide layer for SUS304 at 400 and 700 °C was a combination of Cr_2O_3, Fe_2O_3 and hydroxide.

The process of cold spraying metal particles onto glass or ceramic substrates involves various factors, including the chemical characteristics of the substrates and the impact particles. In a study conducted by Song et al. [30], they examined the cold spraying of Al particles onto a glass substrate. This resulted in the formation of an 80 nm-thick interlayer between the Al particle and the glass substrate. The interlayer exhibited nanocrystalline grains and an amorphous phase, with a notable presence of sodium enrichment. The formation of this interlayer is believed to be a result of the high temperatures generated during the impact process, which induce a reaction between the Al particle and the glass substrate, leading to the formation of liquid stages at the interface.

The strong adhesion observed between the Al particles and the glass substrate is attributed to the high affinity of Al for oxygen within the substrate. The study findings regarding the adhesion strength of coatings on hard metal substrates align with the results reported by Song et al. They conducted a single particle study or wipe test to further investigate the influence of the chemical composition of the oxide layer on the substrate surface on the bonding mechanism.

3.4. Interface Oxide Layer TEM Analysis between TiO₂ Particle at Room Temperature and 700 °C Annealed Substrates

Figure 11 shows high-magnification images that illustrate the presence of an amorphous stage at the interlayer between annealed SUS304 and a single-particle TiO_2. The interlayer thickness was approximately 10 nm, confirming the existence of an interface oxide layer formed after cold-spraying TiO_2 onto SUS304.

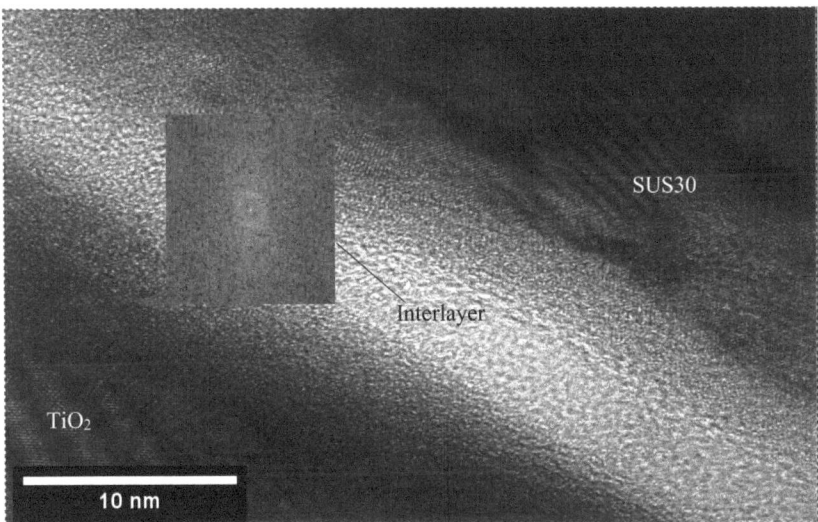

Figure 11. TiO_2/700 °C annealed SUS304 at the interlayer area and FTT image on the oxide layer.

In a study conducted by Kim et al., kinetic spraying was employed to deposit single titanium particles onto mirrored steel substrates. Their findings revealed the presence of a thin amorphous oxide layer at the interface between the particle and the substrate, even after experiencing severe plastic deformation due to particle impacts. This oxide layer left on the substrate surface after cold spraying acted as a bonding agent between the deposited particle and the substrate [31]. The bonding mechanism involved in this process was further investigated through TEM line analysis, focusing on both the room temperature and 700 °C annealed substrates.

The TEM line analysis shown in Figure 12a,b provided insights into the composition of a single-particle TiO_2 on annealed SUS304 at room temperature and 700 °C. The analysis revealed the presence of Ti, O and Cr atoms. The Ti and O atoms were attributed to the TiO_2 coating and hydroxide on the substrate surface, while the Cr atoms indicated the presence of the oxide layer on SUS304. At room temperature, the substrate exhibited Cr metal along with a combination of $Fe_2O_3^+$, Cr_2O_3 and OH^-. Annealing SUS304 at 700 °C resulted in the highest coating adhesion strength. The presence of Cr_2O_3 oxide may influence the bonding mechanism.

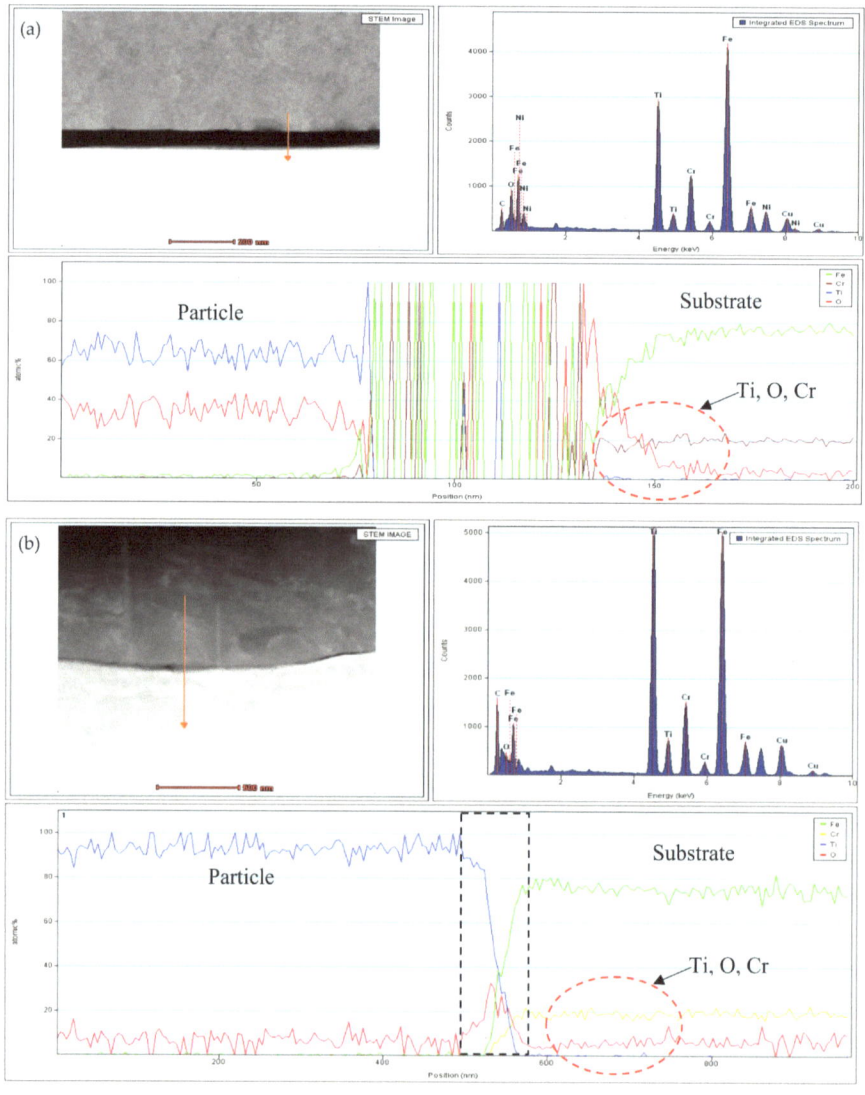

Figure 12. TEM line analysis of the TiO_2 at (**a**) room temperature and (**b**) 700 °C annealed SUS304.

In line with these findings, Song et al. [29] highlighted the importance of considering various factors, including the chemical characteristics of substrates and impact particles, when cold spraying metal particles onto glass or ceramics. Similarly, Drehmann et al. [22] investigated the effect of substrate pre-heating on the adhesion strength of Al particles on an AlN substrate with different levels of softness and hardness. They found that annealing the as-sprayed samples improved the adhesion strength. Increasing the temperature of the AlN substrate before spraying resulted in higher adhesion strength, which was attributed to the thermal energy input that triggered atomic mobility at the Al-AlN interface. Deformation-induced recrystallization near the interface also contributed to atomic mobility, reducing the grain orientation mismatch and improving adhesion strength.

3.5. Oxide Composition Evaluation at Interface of Coating Using X-ray Photoelectron Spectroscopy

Figure 13 provides a schematic diagram of the fracture coating surface on the 700 °C annealed substrate, as analyzed using the XPS method. The purpose of this analysis was to determine the chemical states of the elements present after the cold spray process. Both wide and narrow scan analyses were performed, and the results are presented in Figure 14a–c.

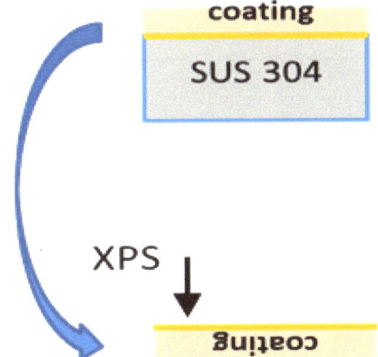

Figure 13. Fracture coating surface at 700 °C annealed.

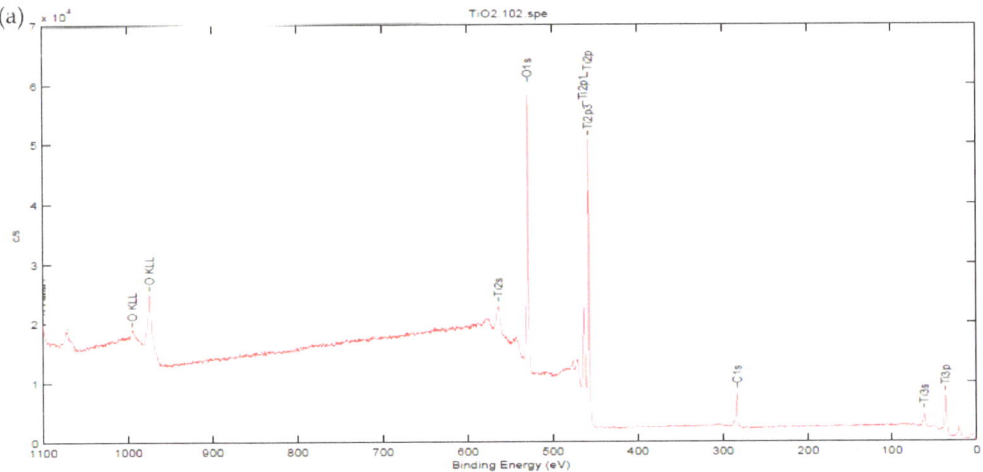

Figure 14. *Cont.*

(b)

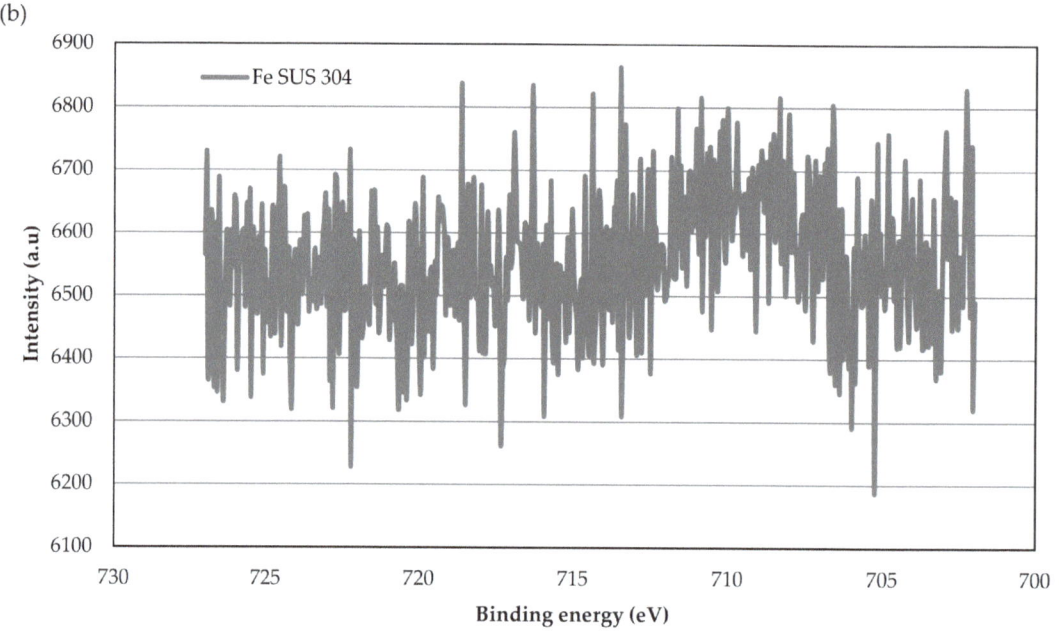

(c)

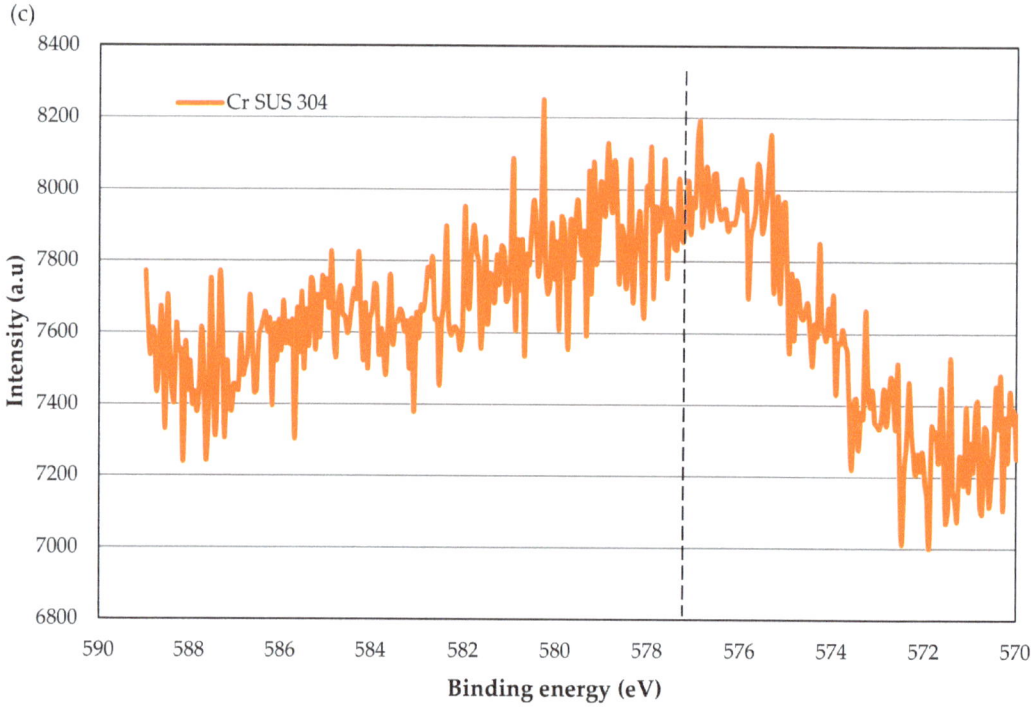

Figure 14. (**a**) Fracture coating interface SUS304 wide scan analysis and narrow scan analysis of coating fracture for (**b**) ferum and (**c**) chromium.

In Figure 14a, the wide-scan analysis of the coating interface reveals the presence of peaks corresponding to titanium, oxygen, carbon, and sodium elements. However, no traces of iron or chromium elements were detected during the wide-scan analysis. Moving to Figure 14b, the narrow analysis of the iron element shows the absence of an elemental peak at the coating interface. Conversely, Figure 14c demonstrates the detection of a chromium elemental peak at a position of 577 eV, indicating the presence of chromium hydroxyls (CrOOH). This suggests that the chemical state of chromium changed from Cr2O3 to CrOOH as a result of the reaction between the chromium oxide (Cr_2O_3) present on the oxide layer of the substrate surface and water molecules in the atmosphere during the cold spray process. It has been confirmed that the oxide layer's thickness increased as the substrate annealing temperature increased, resulting in a thicker layer of Cr_2O_3 on the substrate surface. This, in turn, contributes to the increase in coating adhesion strength. The highest coating adhesion strength was observed at 700 °C annealed substrates, with values of 2.63 MPa for SUS304 and 1.44 MPa for pure Cr. This finding indicates the importance of oxide thickness in influencing the increase in coating adhesion strength for SUS304.

Delamination and weak adhesion strength at interfaces between different materials pose significant challenges in various industries. Understanding the bonding mechanism is crucial to improving adhesion and overcoming these issues. The adhesion strength of cold spray coatings is influenced by factors such as adiabatic shear instability, static recrystallization, mechanical interlocking, and plastic deformation of colliding materials. The state of the substrate plays a significant role in these factors, affecting the bonding properties and characteristics of the cold-sprayed coatings.

In a study conducted by Salim et al. [32], it was found that varying spraying parameters had minimal impact on the adhesion strength of coatings, indicating that mechanical interlocking and substrate shear instability were not the primary bonding mechanisms. Instead, the adhesion strength was influenced by the hardness and oxidizability of the substrates. Modifying the surface chemistry of the substrates could enhance the adhesion strength of TiO_2 coatings. Chemical or physical bonding mechanisms were identified as the primary bonding mechanisms for ceramic coatings, supported by evidence of chemical bonding among TiO_2 particles in TEM images. Additionally, preheating increased the oxidizability of the substrate, which could decrease the adhesion strength of coatings, aligning with the chemical bonding mechanism.

According to Yamada et al., the agglomerated powder of TiO_2 consisted of main particles on the nanoscale with nanoporosity, resulting in a fractured surface with a dangling bond structure. Upon impact, the particles broke apart and then re-bonded, forming a more stable surface and enabling the bonding of newly impacting particles. The formation of the coating involved an interoxide reaction between $TiO_2\text{-}OH^-$ and the chromium oxide mixture ($Cr_2O_3 + Fe_2O_3 + OH^-$) on the top layer of annealed stainless steel. It was observed that the TiO_2 coating on the annealed SUS304 substrate increased in adhesion strength as the annealing temperature increased from room temperature to 700 °C. The adhesion of the impact can be influenced by the passivation layer, including its chemistry, thickness and structure. The impact bonding mechanism between cold-sprayed TiO_2 and SUS304 could be significantly influenced by the thickness of the passivation layer, which might seem unexpected at just 3 nm of growth [33–38].

This observation could be explained by considering the localized deformation of the interface that resulted in bonding, which is shown in the schematic diagram (Figure 15).

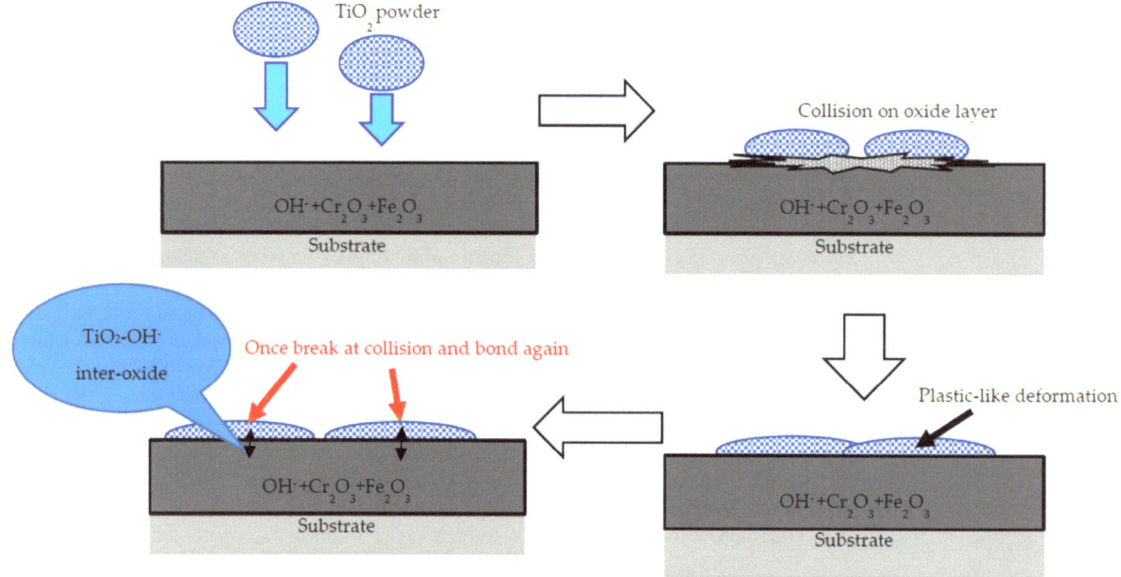

Figure 15. Schematic image of cold-sprayed TiO_2 deposition onto SUS304.

4. Conclusions

In this study, the influence of chromium oxide on the bonding mechanism between pure ceramic titanium dioxide and a SUS304 substrate was investigated. The characteristics of the pure chromium substrate were examined after annealing in an electric furnace at temperatures ranging from room temperature to 700 °C. The key findings can be summarized as follows:

- No detectable traces of iron were found at the coating interface, while a small amount of chromium hydroxyls was observed.
- The chemical state of chromium transformed from Cr_2O_3 to CrOOH due to a reaction between the pre-existing chromium oxide (Cr_2O_3) on the oxide layer of the substrate and water molecules present in the atmosphere during the cold spray process.

Author Contributions: Conceptualization, N.i.O. and M.Y.; methodology, N.i.O.; validation, N.i.O.; formal analysis, N.i.O.; investigation, N.i.O.; resources, N.i.O.; data curation, N.i.O.; writing—original draft preparation, N.i.O.; writing—review and editing, N.i.O., Y.Y., I.A.A.B., V.A.F. and T.A.R.; visualization, N.i.O.; supervision, M.Y.; project administration, N.i.O.; funding acquisition, S.A.b.S. All authors have read and agreed to the published version of the manuscript.

Funding: This research was funded by grant number PJP/2022/FTKMP/S01896.

Institutional Review Board Statement: Not applicable.

Informed Consent Statement: Not applicable.

Data Availability Statement: Not applicable.

Acknowledgments: The study is funded by the Ministry of Higher Education (MOHE) of Malaysia through the short grant (PJP), No: PJP/2022/FTKMP/S01896. The authors also would like to thank the Universiti Teknikal Malaysia Melaka (UTeM) for all the support.

Conflicts of Interest: The authors declare no conflict of interest.

References

1. Schmidt, K.; Assadi, H.; Gärtner, F.; Richter, H.; Stoltenho, T.; Kreye, H. Erratum to: From Particle acceleration to impact and bonding in cold spraying. *J. Therm. Spray Technol.* **2009**, *18*, 1038. [CrossRef]
2. Ziemian, C.W.; Wright, W.J.; Cipoletti, D.E. Influence of impact conditions on feedstock deposition behavior of cold-sprayed Fe-based metallic glass. *J. Therm. Spray Technol.* **2018**, *27*, 843–856. [CrossRef]
3. Vidaller, M.V.; List, A.; Gaertner, F.; Klassen, T.; Dosta, S.; Guilemany, J.M. Single impact bonding of cold sprayed Ti-6Al-4V powders on different substrates. *J. Therm. Spray Technol.* **2015**, *24*, 644–658. [CrossRef]
4. Welk, B.A.; Williams, R.E.A.; Viswanathan, G.B.; Gibson, M.A.; Liaw, P.K.; Fraser, H.L. Nature of the interfaces between the constituent phases in the high entropy alloy CoCrCuFeNiAl. *Ultramicroscopy* **2013**, *134*, 193–199. [CrossRef]
5. Assadi, H.; Kreye, H.; Gärtner, F.; Klassen, T. Cold spraying—A materials perspective. *Acta Mater.* **2016**, *116*, 382–407. [CrossRef]
6. Assadi, H.; Gärtner, F.; Stoltenho, T.; Kreye, H. Bonding mechanism in cold gas spraying. *Acta Mater.* **2003**, *51*, 4379–4394. [CrossRef]
7. Singh, S.; Singh, H.; Chaudhary, S.; Buddu, R.K. Effect of substrate surface roughness on properties of cold-sprayed copper coatings on SS316L steel. *Surf. Coat. Technol.* **2020**, *389*, 125619. [CrossRef]
8. Kumar, S.; Bae, G.; Lee, C. Influence of substrate roughness on bonding mechanism in cold spray. *Surf. Coat. Technol.* **2016**, *304*, 592–605. [CrossRef]
9. Alkhimov, A.P.; Papyrin, A.N.; Vyazemskogo, U.; Kosarev, V.F.; Nesterovich, N.I.; Shushpanov, M.M. Gas-Dynamic Spraying Method for Applying a Coating. U.S. Patent No 5,302,414, 12 April 1994.
10. Irissou, E.; Legoux, J.G.; Ryabinin, A.N.; Jodoin, B.; Moreau, C. Review on Cold Spray Process and Technology: Part I—Intellectual Property. *J. Therm. Spray Technol.* **2008**, *17*, 495–516. [CrossRef]
11. Goldbaum, D.; Poirier, D.; Irissou, E.; Legoux, J.-G.; Moreau, C. Review on cold spray process and technology US patents. In *Modern Cold Spray*; Springer: Cham, Switzerland, 2015; pp. 403–429.
12. Moridi, A.; Hassani-Gangaraj, S.M.; Guagliano, M.; Dao, M. Cold spray coating: Review of material systems and future perspectives. *Surf. Eng.* **2014**, *30*, 369–395. [CrossRef]
13. Hussain, T. Cold spraying of titanium: A review of bonding mechanisms, microstructure and properties. *Key Eng. Mater.* **2012**, *522*, 53–90. [CrossRef]
14. Borchers, C.; Stoltenhoff, T.; Gäartner, F.; Kreye, H.; Assadi, H. Deformation microstructure of cold gas sprayed coatings. *Mater. Res. Soc. Symp. Proc.* **2001**, *674*, 710. [CrossRef]
15. Nikbakht, R.; Seyedein, S.H.; Kheirandish, S.; Assadi, H.; Jodoin, B. Asymmetrical bonding in cold spraying of dissimilar materials. *Appl. Surf. Sci.* **2018**, *444*, 621–632. [CrossRef]
16. Hassani-Gangaraj, M.; Veysset, D.; Champagne, V.K.; Nelson, K.A.; Schuh, C.A. Adiabatic shear instability is not necessary for adhesion in cold spray. *Acta Mater.* **2018**, *158*, 430–439. [CrossRef]
17. Grujicic, M.; Zhao, C.I.; De Rosset, W.S.; Helfritch, D. Analysis of the impact velocity of powder particles in the cold-gas dynamic spray process. *Mater. Sci. Eng. A* **2004**, *368*, 222–230. [CrossRef]
18. Assadi, H.; Gärtner, F.; Klassen, T.; Kreye, H. Comment on "Adiabatic shear instability is not necessary for adhesion in cold spray". *Scr. Mater.* **2019**, *162*, 512–514. [CrossRef]
19. Hassani-Gangaraj, M.; Veysset, D.; Champagne, V.K.; Nelson, K.A.; Schuh, C.A. Response to comment on "Adiabatic shear instability is not necessary for adhesion in cold spray". *Scr. Mater.* **2019**, *162*, 515–519. [CrossRef]
20. Singh, S.; Kumar, M.; Sodhi, G.P.S.; Buddu, R.K.; Singh, H. Development of thick copper claddings on SS316L steel for in-vessel components of fusion reactors and copper-cast iron canisters. *Fusion Eng. Des.* **2018**, *128*, 126–137. [CrossRef]
21. Drehmann, R.; Grund, T.; Lampke, T.; Wielage, B.; Manygoats, K.; Schucknecht, T.; Rafaja, D. Splat formation and adhesion mechanisms of cold gas-sprayed Al coatings on Al₂O₃ substrates. *J. Therm. Spray Technol.* **2014**, *23*, 68–75. [CrossRef]
22. Wüstefeld, C.; Rafaja, D.; Motylenko, M.; Ullrich, M.; Drehmann, R.; Grund, T.; Lampke, T.; Wielage, B. Local heteroepitaxy as an adhesion mechanism in aluminium coatings cold gas sprayed on AlN substrates. *Acta Mater.* **2017**, *128*, 418–427. [CrossRef]
23. Dietrich, D.; Wielage, B.; Lampke, T.; Grund, T.; Kümme, S. Evolution of microstructure of cold-spray aluminum coatings on Al₂O₃ substrates. *Adv. Eng. Mater.* **2011**, *14*, 275–278. [CrossRef]
24. Yamada, m.; Isago, H.; Shima, K.; Nakano, H.; Fukumoto, M. Deposition of TiO₂ Ceramic Particles on Cold Spray Process. In Proceedings of the International Thermal Spray Conference & Exposition Raffles City Convention Centre, Singapore, 3–5 May 2010; Marple, G.M.B.R., Agarwal, A., Hyland, M.M., Lau, Y.-C., Li, C.-J., Lima, R.S., Eds.; ASM Thermal Spray Society: Singapore, 2010; pp. 172–176.
25. Moulder, J.F.; Stickle, W.F.; Sobol, P.E.; Bomben, K.D.; Chastain, J. *Handbook of X-ray Photoelectron Spectroscopy*; Perkin-Elmer Corporation Physical Electronics Division: Eden Prairie, MI, USA, 1993; p. 45.
26. Ko, K.H.; Choi, J.O.; Lee, H. The interfacial restructuring to amorphous: A new adhesion mechanism of cold-sprayed coatings. *Mater. Lett.* **2016**, *175*, 13–15. [CrossRef]
27. Mark, C.B. Advanced analysis of copper X-ray photoelectron spectra. *Surf. Interface Anal.* **2017**, *49*, 1325–1334.
28. Goodwin, H.B.; Gilbert, E.A.; Schwartz, C.M.; Greenidge, C.T. A preliminary study of the ductility of chromium. *J. Electrochem. Soc.* **1953**, *100*, 152–160. [CrossRef]
29. Minghui, S.; Hiroshi, A.; Seiji, K.; Kazuhiko, S. Reaction layer at the interface between aluminium particles and a glass substrate formed by cold spray. *J. Phys. D Appl. Phys.* **2013**, *46*, 195301.

30. Christoulis, D.K.; Guetta, S.; Guipont, V.; Jeandin, M. The influence of the substrate on the deposition of cold-sprayed titanium: An experimental and numerical study. *J. Therm. Spray Technol.* **2011**, *20*, 523–533. [CrossRef]
31. Arabgol, Z.; Vidaller, M.V.; Assadi, H.; Gärtner, F.; Klassen, T. Influence of thermal properties and temperature of substrate on the quality of cold-sprayed deposits. *Acta Mater.* **2017**, *127*, 287–301. [CrossRef]
32. Salim, N.T.; Yamada, M.; Nakano, H.; Shima, K.; Isago, H.; Fukumoto, M. The effect of post-treatments on the powder morphology of titanium dioxide (TiO$_2$) powders synthesized for cold spray. *Surf. Coat. Technol.* **2011**, *206*, 366–371. [CrossRef]
33. Ichikawa, Y.; Ogawa, K. Effect of substrate surface oxide film thickness on deposition behavior and deposition efficiency in the cold spray process. *J. Therm. Spray Technol.* **2015**, *24*, 1269–1276. [CrossRef]
34. Fukumoto, M.; Wada, H.; Tanabe, K.; Yamada, M.; Yamaguchi, E.; Niwa, A.; Sugimoto, M.; Izawa, M. Effect of substrate temperature on deposition behavior of copper particles on substrate surfaces in the cold spray process. *J. Therm. Spray Technol.* **2007**, *16*, 643–650. [CrossRef]
35. Noor irinah, O.; Yamada, M.; Yasui, T.; Fukumoto, M. On the role of substrate temperature into bonding mechanism of cold sprayed titanium dioxide. *IOP Conf. Ser.* **2020**, *920*, 012009. [CrossRef]
36. Viscusi, A.; Astarita, A.; Della Gatta, R.; Rubino, F. A Perspective review on the bonding mechanisms in cold gas dynamic spray. *Surf. Eng.* **2019**, *35*, 743–771. [CrossRef]
37. Dong, S.; Liao, H. Substrate pre-treatment by dry-ice blasting and cold spraying of titanium. *Surf. Eng.* **2018**, *34*, 1173–1180. [CrossRef]
38. Noor irinah, O.; Suhana, M.; Yusliza, Y.; Toibah, A.R.; Zaleha, M.; Syahriza, I.; Ilyani Akmar, A.B.; Santirraprahkash, S. A Preliminary study into the effect of oxide chemistry on the bonding mechanism of cold-sprayed titanium dioxide coatings on SUS316 stainless steel substrate. *J. Electrochem. Sci. Eng.* **2022**, *12*, 579–591.

Communication

Effect of Ball-Milled Feedstock Powder on Microstructure and Mechanical Properties of Cu-Ni-Al-Al$_2$O$_3$ Composite Coatings by Cold Spraying

Hongjin Liu [1,2,3], Mingkun Fu [1,2,3], Shaozhi Pang [1,2,3], Huaiqing Zhu [1,2,3], Chen Zhang [1,2,3], Lijun Ming [1,2,3], Xinyu Liu [1,2,3], Minghui Ding [1,2,3,*] and Yudong Fu [1,2,3]

[1] College of Material Science and Chemical Engineering, Harbin Engineering University, Harbin 150001, China; b9838@126.com (H.L.); fmk2002@163.com (M.F.); psz18845643023@163.com (S.P.); 17661047827@163.com (H.Z.); zhang.chen@hrbeu.edu.cn (C.Z.); minglijunabc@163.com (L.M.); 13114593371@163.com (X.L.); fuyudong@hrbeu.edu.cn (Y.F.)

[2] Key Laboratory of Super Light Material and Surface Technology Ministry of Education, Harbin Engineering University, Harbin 150001, China

[3] Institute of Surface/Interface Science and Technology, Harbin Engineering University, Harbin 150001, China

* Correspondence: mhding@hrbeu.edu.cn; Tel./Fax: +86-451-8251-8219

Abstract: Cu, Ni and Al powders mixed in a certain stoichiometric proportion were ground via ball milling and deposited as coatings using low pressure cold spraying (LPCS) technology. The effect of particle morphology on the powder structure as well as the microstructure, composition and mechanical properties of the coatings was studied. The results revealed a core–shell structure of ball-milled powders. Compared with a mechanically mixed (MM) coating, coatings after ball milling at a rotation speed of 200 rpm exhibited the most uniform composition distribution and a lower degree of porosity (by 0.29%). Moreover, ball milling at 200 rpm was conducive to a significant increase in the deposition efficiency of the sprayed powder (by 10.89%), thereby improving the microhardness distribution uniformity. The ball milling treatment improved the adhesion of the coatings, and the adhesion of the composite coating increased to 40.29 MPa with the increase in ball milling speed. The dry sliding wear tests indicated that ball milling treatment of sprayed powder significantly improved the wear properties of the coatings. The coating after ball milling at a speed of 250 rpm showed the lowest friction coefficient and wear rate, with values of 0.41 and 2.47×10^{-12} m^3/m, respectively. The wear mechanism of coatings changed from abrasive wear to adhesive wear with the increase in ball milling speed.

Keywords: ball milling; low-pressure cold spraying; Cu-based composite coatings; microstructure; deposition efficiency; mechanical properties

Citation: Liu, H.; Fu, M.; Pang, S.; Zhu, H.; Zhang, C.; Ming, L.; Liu, X.; Ding, M.; Fu, Y. Effect of Ball-Milled Feedstock Powder on Microstructure and Mechanical Properties of Cu-Ni-Al-Al$_2$O$_3$ Composite Coatings by Cold Spraying. *Coatings* **2023**, *13*, 948. https://doi.org/10.3390/coatings13050948

Academic Editor: Yasutaka Ando

Received: 25 April 2023
Revised: 9 May 2023
Accepted: 16 May 2023
Published: 18 May 2023

1. Introduction

Low-pressure cold spraying (LPCS) is a coating preparation technology based on supersonic fluid dynamics and high-speed impact dynamics [1–4]. Thanks to easy implementation and high efficiency, the method has broad application prospects in the fields of device repair, remanufacturing and additive manufacturing [5–7]. During spraying, the powder particles are accelerated to a supersonic speed under the action of a certain gas pressure, exerting influence on the surface of the substrate to produce severe plastic deformation and to form a coating [8–10]. Compared with traditional thermal spraying methods, the LPCS has the advantages of lower carrier gas temperature and less pronounced thermal impact on the substrate, thereby preventing the oxidation processes and ensuring low porosity and strong bonding between the particles in the coating [11–13]. Therefore, LPCS is suitable for depositing phase change sensitive materials and oxidizable materials [14,15].

The structure and properties of LPCS composite coatings can be adjusted and controlled via the following stages: the preparation of composite powders before spraying, the mixture of powders during spraying and the post-treatment of coatings (e.g., rolling, heat treatment, etc.). In particular, the morphology and structure of spraying powders are the important factors affecting the structure and performance of LPCS-produced coatings [16–18]. A thoroughly prepared powder can improve the deposition efficiency, structure and performance of the coating, which is essential for further repairing and remanufacturing using LPCS [19–21]. The common preparation routes of composite powders include mechanical mixing, spray drying, ball milling and sintering [22–27]. For example, Xiao [28] obtained a core–shell-structured WC-Co powder via ball milling and deposited it as a coating that possessed a uniform and dense structure with a porosity of 0.7% only. Li [29] produced a tin-reinforced Al5356 coating through ball milling. In all cases, ball-milled (BM) powders endowed the coatings with a denser and more uniform structure as well as a better wear resistance.

Cu/Ni/Al composite coating is a relatively complex coating system, there are many combinations of different materials (such as Cu/Al, Ni/Al, Cu/Ni), which are helpful for studying the structure and performance of coatings in different systems. The dispersion degree and morphology of spray powder have a significant impact on the structure and performance of the coatings. Therefore, this work aims to study $Cu-Ni-Al-Al_2O_3$ coatings fabricated via LPCS so as to establish the effect of BM powder morphology on their structure, morphology and mechanical properties.

2. Experimental Procedures

Commercial Cu (20–25 μm), Ni (20–25 μm) and Al (25–30 μm) powders were used as raw materials. Powders (56 wt.% Cu), (24 wt.% Ni) and (20 wt.% Al) were mechanically mixed and ball-milled for 4 h under an Ar atmosphere with a planetary ball mill (UBE-F2L, China) using ZrO grinding balls (10 mm, 5 mm and 3 mm). The rotation speeds were set to 150 rpm, 200 rpm and 250 rpm, and the ball-to-powder mass ratio was 10:1. To avoid excessive temperature rise during ball milling, the procedure was suspended for 10 min every half hour.

A LPCS system (TCY-LP-III, Beijing Tianchengyu New Material Technology Co., Ltd., Beijing, China) was employed for coating preparation. Before spraying, powders were mechanically mixed with Al_2O_3 powder (45–50 μm) at a mass ratio of 7:3 to improve the deposition efficiency and coating density [30,31]. Compressed air was used as accelerating gas at a pressure of 0.8–1.0 MPa and a temperature of about 550 °C. A standoff distance from the nozzle exit to the substrate surface was 15 mm. AZ91D magnesium alloy served as the substrates. Before powder spraying, the substrates were exposed to ultrasonic cleaning for 10 min with deionized water, acetone and ethanol. After that, blow drying and carborundum abrasion were performed to blast the cleaned substrates and roughen their surfaces so as to remove the oxide layer and allow the easier powder deposition [14]. In this paper, the Cu-based coatings were polished before testing.

A scanning electron microscope (SEM) (JSM-6480A, JEOL Ltd., Tokyo, Japan) was used to observe the surfaces and cross-sections of the composite coatings and the morphology of the frictions and wear areas. The energy dispersive spectrometer (EDS) coupled with the SEM instrument enabled the analysis of the element contents in the coatings. The images were acquired at the operating voltage of 20 kV and processed in image J software to assess the particle size of the powder and the porosity of coatings (porosity was measured using the Threshold function of image J software).

X-ray diffraction (XRD) (X'Pert Pro, PANalytical B.V., Almelo, the Netherlands) was applied to analyze the phase compositions of sprayed powders and Cu-based composite coatings. The measurements were carried out using a Cu target at the voltage of 40 kV and the current of 40 mA. The XRD profiles were calibrated by means of HighScore software and standard PDF cards.

A total of 20 g of spray powder was weighed and sprayed onto the AZ91D magnesium alloy substrate, and the powder deposition efficiency was calculated using the following formula:

$$DE = \frac{M_1 - M_2}{M_3} \times 100\%$$ (1)

where DE is the deposition efficiency of the powder; M_1 is the weight of the sample after spraying (g); M_2 is the weight of AZ91D magnesium alloy substrate (g); and M_3 is the weight (g) of the sprayed powder weighed.

The adhesion of coatings was measured on an electronic universal testing machine (WDW-100, Jinan Fangyuan Testing Instrument Co., Ltd., Jinan, China). Prior to the experiment, the mixed powder was sprayed onto the cylindrical AZ91D magnesium alloy base with a diameter of 20 mm. The obtained coating was then polished and bonded to another cylindrical magnesium alloy base. The pull-out test of the bonded samples was afterward carried out at a tensile speed of 0.2 mm/min, and the load corresponding to the coating pull-off was recorded. Each group of tests was repeated three times and the average value was taken. The bonding strength of the coatings was determined as follows:

$$\sigma = \frac{F}{A}$$ (2)

where σ is the bonding strength of the coating (MPa); F is the destructive force at which the coating is broken (N); and A is the area of the sample column (mm^2).

The microhardness of the Cu-based composite coatings was evaluated on a microhardness tester (HVS-1000B, Laizhou Huayin Testing Instrument Co., Ltd., Yantai, China) under a load of 100 g applied for 15 s. The average value was found after five hardness measurements.

A high-temperature friction and wear tester (HT-1000, Lanzhou Zhongke Kaihua Technology Development Co., Ltd., Lanzhou, China) was used to determine the wear resistance of coatings. The friction pair in the test was a GCr15 grinding ball with a diameter of 6 mm. During the experiment, the friction pair moved in a circular direction with a radius of 3 mm at a rotation speed of 280 r/min. The load was 5 N, and the test time was 10 min. After the completion of the friction and wear tests, the wear marks left on the sample were observed by SEM and their shape was assessed as well. At the same time, the volume wear rate of the samples was calculated from the formulae below [32]:

$$\Delta V = L0(r^2 \arcsin\frac{d}{2r} - \frac{d}{2}\sqrt{r^2 - (\frac{d}{2})^2})$$ (3)

$$W_N = \frac{\Delta V}{L}$$ (4)

where W_N is the volume wear rate of composite coatings (m^3/M); ΔV is the volume loss of composite coatings (m^3); L is the friction distance in the test (m); $L0$ is the circumference of the wear mark (m); r is the radius of the friction pair (m); and d is the wear mark width (m).

3. Results and Discussion

3.1. Characterization of Spray Powders

Figure 1 displays the microscopic topography of the sprayed powder after ball milling. According to Figure 1a, the powder at the rotation speed of 150 rpm still retained its original appearance and only a few particles were extruded and deformed therein. However, once the speed further increased to 200 and 250 rpm, the particles agglomerated and became irregular (Figure 1b,c). Figure 1d,e depicts the cross-sectional microstructures of sprayed powders after milling. It was found that powders possessed a core–shell structure, in which the Cu and Ni cores were covered by the Al shell. This could be attributed to the fact that during the ball milling, the softer Al powder continuously accumulated on the surface of the Cu and Ni powders to form a core–shell-structured powder. With the increase in rotation speed, the particle size in the powders decreased first and then increased. At the speed of 200 rpm, the particle size of the powder was about 13 μm. As soon as the speed increased to 250 rpm, the particle size approached 30 μm. At the same time, the particle shape became more irregular, which indicated that the particles underwent strong deformation and aggregation during milling.

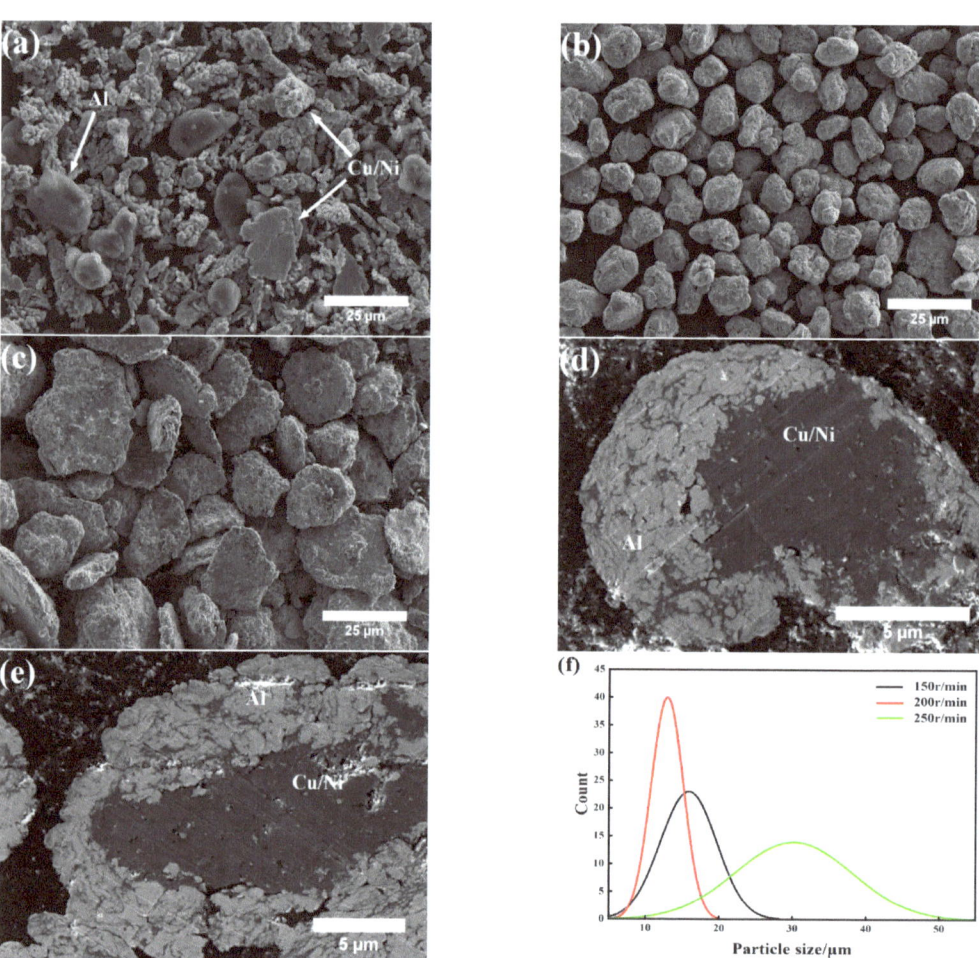

Figure 1. SEM images of powders at the rotation speed of (**a**) 150 rpm, (**b**) 200 rpm and (**c**) 250 rpm. (**d**) Cross-sectional microstructure of powder milled at 200 rpm. (**e**) Cross-sectional microstructure of powder milled at 250 rpm. (**f**) Particle size distributions in spray powders.

Figure 2 displays the XRD results of powders after ball milling at different rotation speeds. In all cases, the XRD profiles were quite similar to each other, meaning that the ball milling basically did not alter the phase structure of powders. A comparison of these XRD spectrograms with the XRD database (JCPDS cards nos. 00-004-0836, 00-004-0850, and 96-900-8461) revealed a stable presence of Cu, Ni and Al phases. Therefore, the ball milling could have only impacted the microstructure of powders conforming to the SEM images in Figure 1.

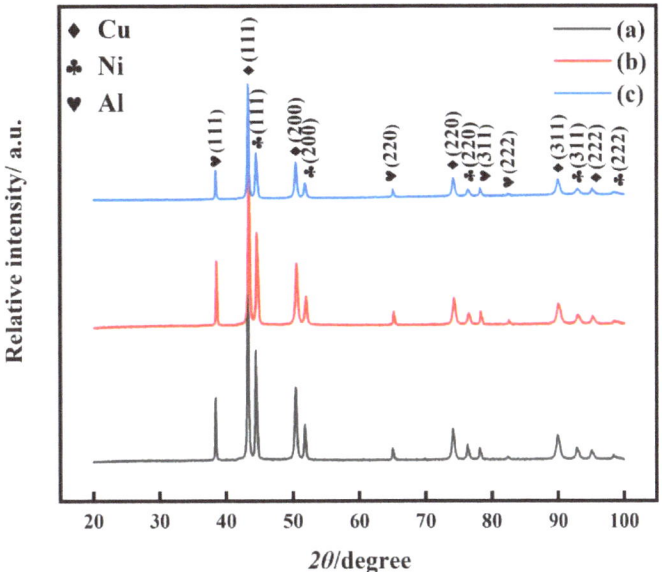

Figure 2. XRD pattern of spray powders at different rotation speeds: (**a**) 150 rpm, (**b**) 200 rpm and (**c**) 250 rpm.

3.2. Microstructures of Coatings

Figure 3 depicts the SEM images of coatings. In the mechanically mixed (MM) coating, the phases were homogeneously distributed (Figure 3a). In turn, the constituent phases in the ball-milled (BM) coatings were distributed in a more uniform manner, changing from isolated to staggered configurations (Figure 3c,d) because of the core–shell structure of powders. Figure 4 depicts the cross-sectional SEM images of coatings. In each case, the bonding interface between the coating and the substrate in the form of an irregular curve could be clearly observed. It was attributed to the severe plastic deformation of the powder particles after they collided with the substrate at the high speed and then combined together with the substrate.

Combining the SEM images of the surface and the cross-section images of coatings (Figures 3 and 4), it was implied that powders were firmly bonded to the substrates. No obvious pores and cracks were detected at the bonding interfaces and within the coatings. The overall porosity of the coatings was less than 1%. According to Table 1, the porosity at the rotation speed of 200 rpm was 0.29% (Table 1). However, once the speed rose to 250 rpm, the porosity increased to 0.76% (Table 1), which could be attributed to the fact that the particle size increased and the morphology became flat (Figure 1c,f), making the powder unsuitable for spraying and thereby reducing the density of the coating. At the same time, scarce Al_2O_3 particles were embedded in the coating, which played the role of compaction and secondary shot peening during the LPCS, thus reducing the porosity and increasing the compactness of the coating. However, while they possessed high hardness, the Al_2O_3 particles lacked any deformation ability and could not match the sprayed powder, causing the pore concentrations around them.

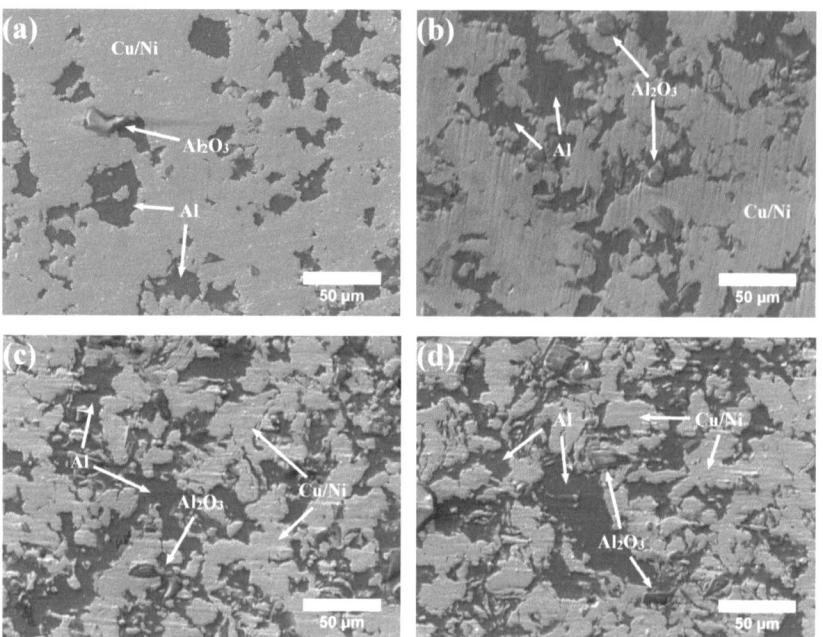

Figure 3. SEM images of coatings: (**a**) MM powder, (**b**) BM powder milled at 150 rpm, (**c**) BM powder milled at 200 rpm and (**d**) BM powder milled at 250 rpm.

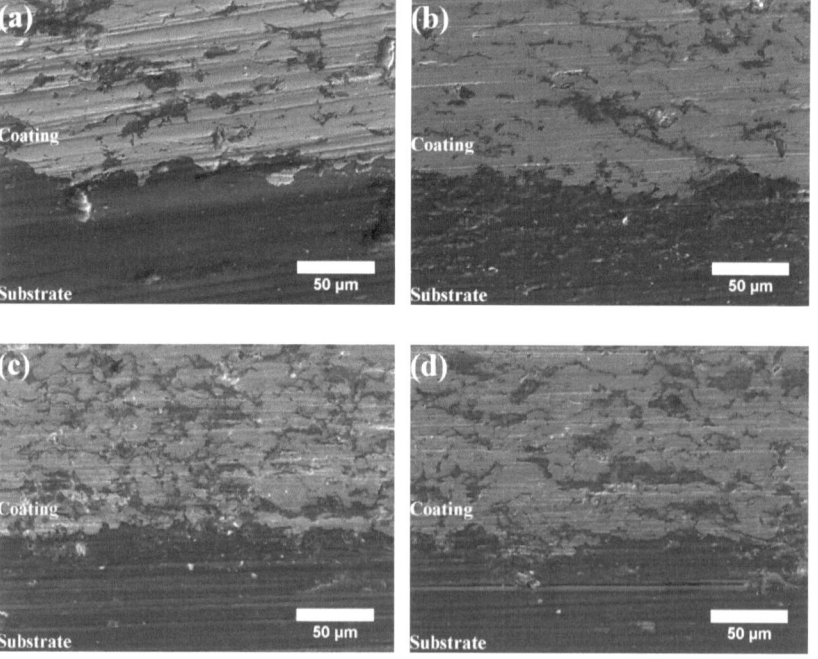

Figure 4. Cross-sectional SEM images of the coatings: (**a**) MM powder, (**b**) BM powder milled at 150 rpm, (**c**) BM powder milled at 200 rpm and (**d**) BM powder milled at 250 rpm.

Table 1. Material content and porosity of coatings.

Deposit	Cu (wt.%)	Ni (wt.%)	Al (wt.%)	Al$_2$O$_3$ (wt.%)	Porosity (%)
MM	84.2	9.34	4.82	1.65	0.58
BM (150 rpm)	70.54	18.46	8.72	2.28	0.41
BM (200 rpm)	63.17	19.70	14.39	2.74	0.29
BM (250 rpm)	59.45	12.01	26.30	2.24	0.76

Table 1 summarizes the EDS results on the coatings. The mass fraction of Al$_2$O$_3$ was calculated by using the mass fraction of O element. The mass fraction of Al$_2$O$_3$ should be slightly lower than the calculated value because a small amount of Al was oxidized during spraying. It was obvious that the BM powder increased the contents of Ni and Al in the coatings. Compared with Cu, the higher hardness of Ni made it difficult to deposit, while the smaller density of Al led to its lower kinetic energy during spraying, which was also not conducive to spraying. After the ball milling, on the one hand, the amounts of Ni and Al in the powder with a core–shell-structure dramatically increased during the co-deposition process; on the other hand, the Al shell strongly bonded to the substrate, which could make the coatings more compact, according to the porosity analysis.

Figure 5 depicts the XRD patterns of the coatings prepared from MM and BM powders at different rotation speeds. According to the data, the phase structures of the coatings were consistent with those of the sprayed powders (Figure 2), revealing neither oxidation nor phase transformation during the LPCS, as expected. Meanwhile, no diffraction peaks of Al$_2$O$_3$ appeared, indicating that a small amount of Al$_2$O$_3$ particles remaining in the coatings did not affect their structure.

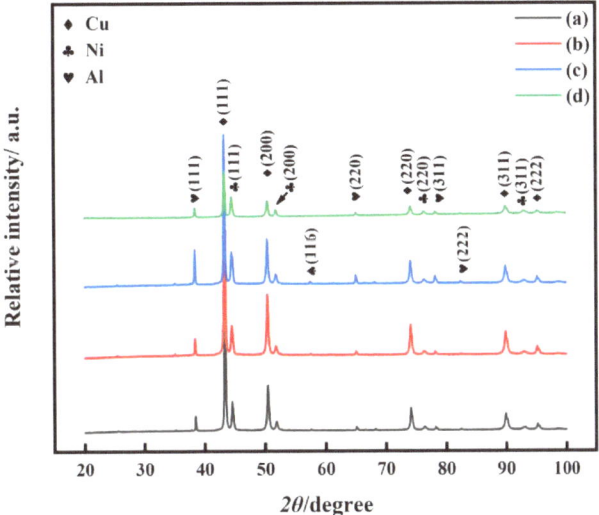

Figure 5. XRD patterns of coatings: (**a**) MM powder, (**b**) BM powder milled at 150 rpm, (**c**) BM powder milled at 200 rpm and (**d**) BM powder milled at 250 rpm.

3.3. Mechanical Performance of Coatings

As shown in Table 2, compared with the MM powder, ball milling significantly improved the deposition efficiency of the powder. In particular, the deposition rate of the ball-milled powder at the rotation speed of 200 rpm was 10.89% higher than that of the MM powder. The better deposition performance of the powder with a core–shell structure was attributed to the fact that the Cu and Ni cores possessed the sufficient kinetic energies. At the same time, the Al shell could have experienced severe plastic deformation. However, the deposition efficiency of the sprayed powder decreased to 9.36% only at a rotation speed of 250 rpm, indicating that the powder was not suitable for spraying at this time.

Table 2. Deposition efficiency, hardness, adhesion, friction coefficient and volume wear rate of coatings.

Deposit	Deposition Efficiency (%)	Hardness ($HV_{0.1}$)	Adhesion (MPa)	Friction Coefficient	Wear Rate ($\times 10^{-12}$ m^3/m)
MM Powder	30.71 ± 2.13	155.76 ± 6.71	31.17 ± 2.93	0.56 ± 0.051	8.43
150 rpm	36.23 ± 2.49	149.88 ± 3.21	31.45 ± 3.03	0.51 ± 0.045	10.19
200 rpm	41.60 ± 3.02	153.03 ± 1.34	40.29 ± 3.95	0.49 ± 0.035	4.92
250 rpm	9.36 ± 1.31	136.55 ± 10.00	37.44 ± 3.18	0.41 ± 0.024	2.47

The hardness of coatings is given in Table 2. It was established that the impact of ball milling on the rigidity of coatings was not obvious, meaning that the work hardening of the powders due to plastic deformation in the deposition process exceeded the effect of ball milling on the powders themselves. Meanwhile, in addition to the work hardening, the powder dispersion uniformity also exerted strong influence on the hardness of the coating. Figure 6 displays the hardness through the specimens as a function of distance from the substrate. It was evident that the hardness of the MM coating and BM (250 rpm) coating fluctuated to a large extent, while slightly changing in BM coatings treated at 150 and 200 rpm. This indicated the improvement in internal uniformity of the two latter coatings. A drastic variation in the hardness of the coating processed at 250 rpm could be ascribed to the excessive aggregation and uneven distribution of Al elements in the outer layer of the powder during ball milling (Figure 3d). At the same time, the high porosity of the coating (Table 1) was another important factor leading to the hardness fluctuations throughout the coating.

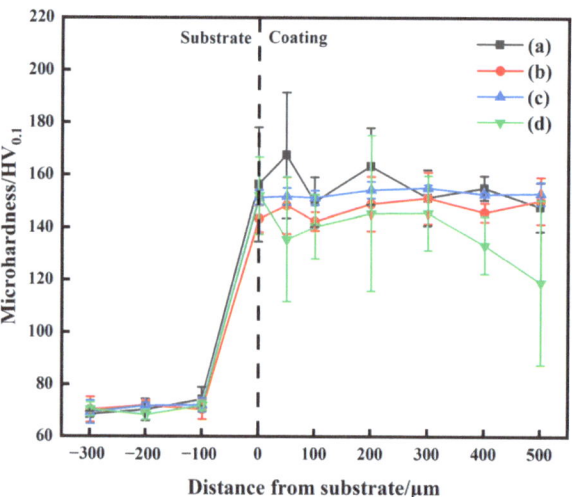

Figure 6. Microhardness of coatings: (**a**) MM powder, (**b**) BM powder milled at 150 rpm, (**c**) BM powder milled at 200 rpm and (**d**) BM powder milled at 250 rpm.

The adhesion properties of coatings are shown in Figure 7. It was found that the adhesion of the BM coating treated at 150 rpm was the same as that of the MM coating. With the increase in the rotation speed, the adhesion of coatings increased to a large extent, which was related to the microstructural peculiarities of the relevant powders (Figure 1a–c). In particular, the adhesion of the BM coating was 40.29 MPa at the speed of 200 rpm (Table 2), which was 29.26% higher than that of the MM coating. This was because the core–

shell-structured powders improved the adhesion of coatings due to the stronger mechanical engagement ability between the Al shell and the substrate [33]. In the tensile testing, all the coatings were broken in the middle, indicating that the specimens underwent cohesive failure. In a word, the bonding strength between the coatings and the substrates was higher than the cohesion strength of the coatings themselves.

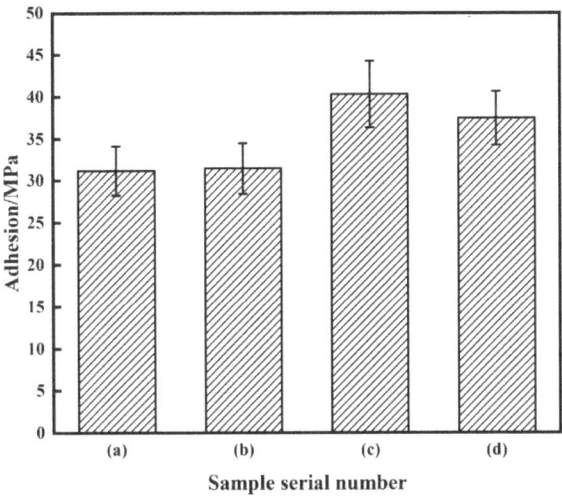

Figure 7. Adhesion of coatings: (**a**) MM powder, (**b**) BM powder milled at 150 rpm, (**c**) BM powder milled at 200 rpm and (**d**) BM powder milled at 250 rpm.

Figure 8 depicts the evolution of friction coefficients of composite coatings with the friction time. After a short running-in period (about 1 min), the friction coefficients of coatings reached a relatively stable state. The average friction coefficients estimated from the friction times between 2 and 10 min are listed in Table 2. In particular, the friction coefficients of BM composite coatings were lower than that of the MM coating, which was associated with a more uniform composition distribution in the former coatings, and the higher the rotation speed, the lower the friction coefficient. In addition, the wear rate (WR) of BM coatings decreased with the increase in milling speed. For example, the WR of the coating at 250 rpm was 2.47×10^{-12} m^3/m, being only about one-third of that of the MM coating, despite the composite coating yielding lower hardness values [34,35]. This means that the ball milling of spraying powders under the optimal conditions were able to significantly improve the wear resistance of coatings and reduce their friction coefficient.

To elucidate the wear mechanisms of the coatings, the corresponding SEM images were further analyzed (Figure 9). The wear surfaces of the BM coating (150 rpm) and MM coating were characterized by deep grooves and scratches (Figure 9a,b). On the other hand, some hard particles peeled off due to wear providing abrasives for further abrasive wear. These particles remained between the friction pair and the coating to cut the coating, thus forming plenty of grooves and furrows parallel to the friction direction on the wear surface. Therefore, both coatings experienced abrasive wear [36]. Once the ball-milling speed reached 200 rpm and 250 rpm, continuous peeling cracks and delamination could be observed, which were attributed to the overall fracture and peeling of the coating material as well as plastic deformation during the friction process, leading to adhesive wear. Therefore, the wear mechanism of coatings changed from abrasive wear to adhesive wear with the increase in rotation speed.

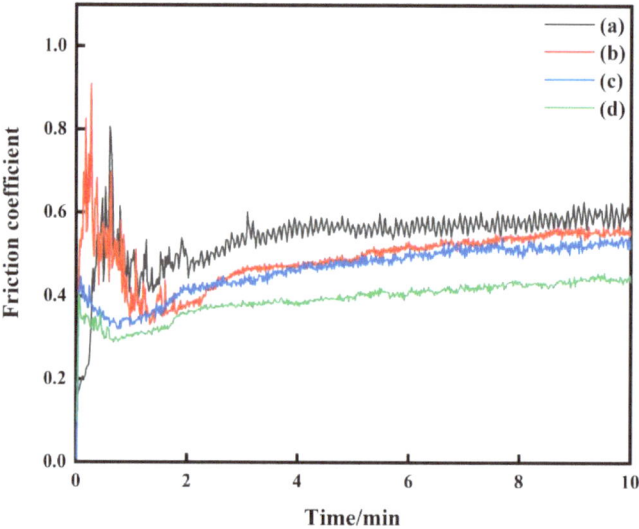

Figure 8. Evolution of friction coefficients of composite coatings with the friction time: (**a**) MM powder; (**b**) BM powder milled at 150 rpm; (**c**) BM powder milled at 200 rpm; (**d**) BM powder milled at 250 rpm.

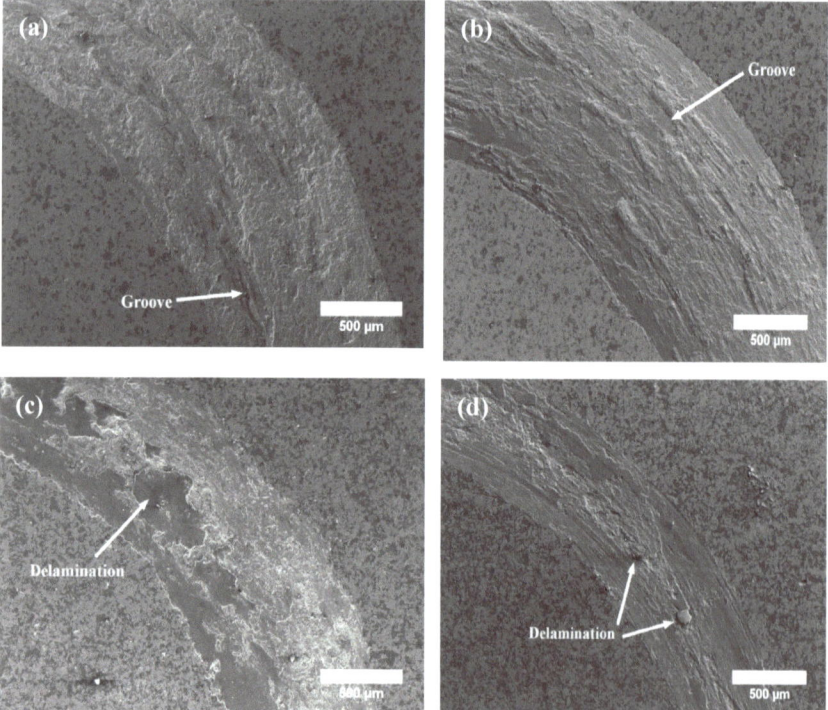

Figure 9. SEM images of wear tracks: (**a**) MM powder, (**b**) BM powder milled at 150 rpm, (**c**) BM powder milled at 200 rpm and (**d**) BM powder milled at 250 rpm.

4. Conclusions

Copper-based mechanically mixed (MM) and ball-milled (BM) composite Cu-Ni-Al-Al_2O_3 coatings were deposited on AZ91D Mg alloy substrates using LPCS. According to the SEM images of BM powders, a core–shell structure with uniformly mixed components formed at the rotation speed above 200 rpm. The relevant coatings exhibited a uniform and dense structure with a low degree of porosity (0.29%) as well as strong mechanical bonding to the substrate. Aside from this, the deposition efficiency of BM powder at 200 rpm could reach 40.61% and exceeded 10.89% than that of MM powder, which was attributed to the ability of the Al shell of the sprayed powder to undergo more severe plastic deformation and deposit into coatings, and the change in microhardness with the coating thickness was the smallest, indicating that the uniform coating composition and the coating hardness was stable at around 153.03 $HV_{0.1}$. The presence of the Al shell in BM powders exerted a positive effect on the adhesive properties of the composite coatings, resulting in the higher adhesion of BM coatings than of MM coatings. Meanwhile, the friction coefficients and WR values of the BM coatings were inferior to those of the MM coatings at rotation speeds of higher than 200 rpm. Based on the structural analysis and mechanical parameters of coatings, the ball milling of spraying powders at 200 rpm ensured the uniform deposition of coatings and enhanced their mechanical characteristics.

Author Contributions: Conceptualization, H.L. and S.P.; methodology, M.F.; software, H.Z.; validation, H.L. and M.D.; formal analysis, H.Z.; investigation, H.L.; resources, S.P.; data curation, H.L., C.Z., L.M., X.L. and M.D.; writing—original draft preparation, H.L. and M.D.; writing—review and editing, H.L.; visualization, H.L.; supervision, M.D. and Y.F.; project administration, H.L.; funding acquisition, M.D. All authors have read and agreed to the published version of the manuscript.

Funding: This work was funded by the Natural Science Foundation of Heilongjiang Province (LH2021E029).

Institutional Review Board Statement: Not applicable.

Informed Consent Statement: Not applicable.

Data Availability Statement: The data presented in this study are available on request from the corresponding author.

Conflicts of Interest: The authors declare no conflict of interest.

References

1. Li, W.; Assadi, H.; Gaertner, F.; Yin, S.A. Review of Advanced Composite and Nanostructured Coatings by Solid-State Cold Spraying Process. *Crit. Rev. Solid State Mater. Sci.* **2018**, *44*, 109–156. [CrossRef]
2. Li, C.J.; Li, W.Y. Deposition characteristics of titanium coating in cold spraying. *Surf. Coat. Technol.* **2003**, *167*, 278–283. [CrossRef]
3. Yu, P.; Xie, Y.; Yin, S.; Lupoi, R. Fabrication of Ta-Ag composite deposits via cold spray: Investigation of bonding mechanism and deposition behavior. *J. Adv. Join. Process.* **2022**, *6*, 100127. [CrossRef]
4. Singh, R.; Kondás, J.; Bauer, C.; Cizek, J.; Medricky, J.; Csaki, S.; Čupera, J.; Procházka, R.; Melzer, D.; Konopík, P. Bulk-like ductility of cold spray additively manufactured copper in the as-sprayed state. *Addit. Manuf. Lett.* **2022**, *3*, 100052. [CrossRef]
5. Luo, X.T.; Li, C.J. Tailoring the composite interface at lower temperature by the nanoscale interfacial active layer formed in cold sprayed cBN/NiCrAl nanocomposite. *Mater. Des.* **2018**, *140*, 387–399. [CrossRef]
6. Prasad, K.; Khalik, M.A.; Hutasoit, N.; Rahman Rashid, R.A.; Rashid, R.; Duguid, A.; Palanisamy, S. Printability of low-cost pre-heat-treated ball milled Al7075 powders using compressed air assisted cold spray additive manufacturing. *Addit. Manuf. Lett.* **2022**, *3*, 100046. [CrossRef]
7. Luo, X.T.; Wei, Y.K.; Wang, Y.; Li, C.J. Microstructure and mechanical property of Ti and Ti6Al4V prepared by an in-situ shot peening assisted cold spraying. *Mater. Des.* **2015**, *85*, 527–533. [CrossRef]
8. Sun, W.; Chu, X.; Lan, H.; Huang, R.; Huang, J.; Xie, Y.; Huang, J.; Huang, G. Current Implementation Status of Cold Spray Technology: A Short Review. *J. Therm. Spray Technol.* **2022**, *31*, 848–865. [CrossRef]
9. Zahiri, S.H.; Fraser, D.; Jahedi, M. Recrystallization of Cold Spray-Fabricated CP Titanium Structures. *J. Therm. Spray Technol.* **2008**, *18*, 16–22. [CrossRef]
10. Gleason, M.A.; Sousa, B.C.; Tsaknopoulos, K.; Grubbs, J.A.; Hay, J.; Nardi, A.; Brown, C.A.; Cote, D.L. Application of Mass Finishing for Surface Modification of Copper Cold Sprayed Material Consolidations. *Materials* **2022**, *15*, 2054. [CrossRef]

11. List, A.; Gärtner, F.; Mori, T.; Schulze, M.; Assadi, H.; Kuroda, S.; Klassen, T. Cold Spraying of Amorphous Cu50Zr50 Alloys. *J. Therm. Spray Technol.* **2014**, *24*, 108–118. [CrossRef]
12. Luo, X.T.; Yang, E.J.; Shang, F.L.; Yang, G.J.; Li, C.X.; Li, C.J. Microstructure, Mechanical Properties, and Two-Body Abrasive Wear Behavior of Cold-Sprayed 20 vol.% Cubic BN-NiCrAl Nanocomposite Coating. *J. Therm. Spray Technol.* **2014**, *23*, 1181–1190. [CrossRef]
13. Pitchuka, S.B.; Boesl, B.; Zhang, C.; Lahiri, D.; Nieto, A.; Sundararajan, G.; Agarwal, A. Dry sliding wear behavior of cold sprayed aluminum amorphous/nanocrystalline alloy coatings. *Surf. Coat. Technol.* **2014**, *238*, 118–125. [CrossRef]
14. Tao, Y.; Xiong, T.; Sun, C.; Kong, L.; Cui, X.; Li, T.; Song, G.L. Microstructure and corrosion performance of a cold sprayed aluminium coating on AZ91D magnesium alloy. *Corros. Sci.* **2010**, *52*, 3191–3197. [CrossRef]
15. Gardon, M.; Latorre, A.; Torrell, M.; Dosta, S.; Fernández, J.; Guilemany, J.M. Cold gas spray titanium coatings onto a biocompatible polymer. *Mater. Lett.* **2013**, *106*, 97–99. [CrossRef]
16. Yin, S.; Zhang, Z.; Ekoi, E.J.; Wang, J.J.; Dowling, D.P.; Nicolosi, V.; Lupoi, R. Novel cold spray for fabricating graphene-reinforced metal matrix composites. *Mater. Lett.* **2017**, *196*, 172–175. [CrossRef]
17. Silva, F.S.D.; Bedoya, J.; Dosta, S.; Cinca, N.; Cano, I.G.; Guilemany, J.M.; Benedetti, A.V. Corrosion characteristics of cold gas spray coatings of reinforced aluminum deposited onto carbon steel. *Corros. Sci.* **2017**, *114*, 57–71. [CrossRef]
18. Poirier, D.; Legoux, J.G.; Drew, R.A.L.; Gauvin, R. Consolidation of Al$_2$O$_3$/Al Nanocomposite Powder by Cold Spray. *J. Therm. Spray Technol.* **2010**, *20*, 275–284. [CrossRef]
19. Trautmann, M.; Ahmad, H.; Wagner, G. Influencing the Size and Shape of High-Energy Ball Milled Particle Reinforced Aluminum Alloy Powder. *Materials* **2022**, *15*, 3022. [CrossRef]
20. Vidyuk, T.M.; Dudina, D.V.; Korchagin, M.A.; Gavrilov, A.I.; Bokhonov, B.B.; Ukhina, A.V.; Esikov, M.A.; Shikalov, V.S.; Kosarev, V.F. Spark plasma sintering treatment of cold sprayed materials for synthesis and structural modification: A case study using TiC-Cu composites. *Mater. Lett. X* **2022**, *14*, 100140. [CrossRef]
21. Chen, J.; An, Y.; Liu, G.; Chen, G.; Zhao, X.; Jia, L. Tribological Performance and Thermal Stability of a Novel Cold Sprayed Nanostructured Ni-based Lubrication Coating. *J. Therm. Spray Technol.* **2022**, *31*, 1702–1711. [CrossRef]
22. Zhang, D.L. Processing of advanced materials using high-energy mechanical milling. *Prog. Mater. Sci.* **2004**, *49*, 537–560. [CrossRef]
23. Chen, X.; Li, C.; Bai, X.; Liu, H.; Xu, S.; Hu, Y. Microstructure, Microhardness, Fracture Toughness, and Abrasive Wear of In-Situ Synthesized TiC/Ti-Al Composite Coatings by Cold Spraying Combined with Heat Treatment. *Coatings* **2021**, *11*, 1034. [CrossRef]
24. El-Eskandarany, M.S.; Ali, N.; Banyan, M.; Al-Ajmi, F. Cold Gas-Dynamic Spray for Catalyzation of Plastically Deformed Mg-Strips with Ni Powder. *Nanomaterials* **2021**, *11*, 1169. [CrossRef]
25. Zhang, Q.; Li, C.J.; Wang, X.R.; Ren, Z.L.; Li, C.X.; Yang, G.J. Formation of NiAl Intermetallic Compound by Cold Spraying of Ball-Milled Ni/Al Alloy Powder Through Postannealing Treatment. *J. Therm. Spray Technol.* **2008**, *17*, 715–720. [CrossRef]
26. Li, W.Y.; Li, C.J. Characterization of cold-sprayed nanostructured Fe-based alloy. *Appl. Surf. Sci.* **2010**, *256*, 2193–2198. [CrossRef]
27. Ghelichi, R.; Bagherifard, S.; Mac Donald, D.; Brochu, M.; Jahed, H.; Jodoin, B.; Guagliano, M. Fatigue strength of Al alloy cold sprayed with nanocrystalline powders. *Int. J. Fatigue* **2014**, *65*, 51–57. [CrossRef]
28. Luo, X.T.; Li, C.X.; Shang, F.L.; Yang, G.J.; Wang, Y.Y.; Li, C.J. WC-Co Composite Coating Deposited by Cold Spraying of a Core-Shell-Structured WC-Co Powder. *J. Therm. Spray Technol.* **2014**, *24*, 100–107. [CrossRef]
29. Li, W.Y.; Zhang, G.; Zhang, C.; Elkedim, O.; Liao, H.; Coddet, C. Effect of Ball Milling of Feedstock Powder on Microstructure and Properties of TiN Particle-Reinforced Al Alloy-Based Composites Fabricated by Cold Spraying. *J. Therm. Spray Technol.* **2008**, *17*, 316–322. [CrossRef]
30. Lee, Y.T.R.; Ashrafizadeh, H.; Fisher, G.; McDonald, A. Effect of type of reinforcing particles on the deposition efficiency and wear resistance of low-pressure cold-sprayed metal matrix composite coatings. *Surf. Coat. Technol.* **2017**, *324*, 190–200. [CrossRef]
31. Spencer, K.; Fabijanic, D.M.; Zhang, M.X. The use of Al–Al$_2$O$_3$ cold spray coatings to improve the surface properties of magnesium alloys. *Surf. Coat. Technol.* **2009**, *204*, 336–344. [CrossRef]
32. Yang, Y.; Wu, H. Improving the wear resistance of AZ91D magnesium alloys by laser cladding with Al–Si powders. *Mater. Lett.* **2009**, *63*, 19–21. [CrossRef]
33. Li, W.Y.; Yang, K.; Zhang, D.D.; Zhou, X.L.; Guo, X.P. Interface behavior of particles upon impacting during cold spraying of Cu/Ni/Al mixture. *Mater. Des.* **2016**, *95*, 237–246. [CrossRef]
34. Zhang, L.Y.; Yang, S.M.; Lv, X.; Jie, X.H. Wear and Corrosion Resistance of Cold-Sprayed Cu-Based Composite Coatings on Magnesium Substrate. *J. Therm. Spray Technol.* **2019**, *28*, 1212–1224. [CrossRef]
35. Chen, W.Y.; Yu, Y.; Cheng, J.; Wang, S.; Zhu, S.Y.; Liu, W.M.; Yang, J. Microstructure, Mechanical Properties and Dry Sliding Wear Behavior of Cu-Al$_2$O$_3$-Graphite Solid-Lubricating Coatings Deposited by Low-Pressure Cold Spraying. *J. Therm. Spray Technol.* **2018**, *27*, 1652–1663. [CrossRef]
36. Wang, Y.; Zhu, Y.; Li, R.; Wang, H.; Tian, L.; Li, H. Microstructure and Wear Behavior of Cold-Sprayed Cu-BNNSs Composite Coating. *J. Therm. Spray Technol.* **2021**, *30*, 1482–1492. [CrossRef]

Article

Influence of the Gas Preheating Temperature on the Microstructure and Electrical Resistivity of Copper Thin Films Prepared via Vacuum Cold Spraying

Kai Ma, Qing-Feng Zhang, Hui-Yu Zhang, Chang-Jiu Li and Cheng-Xin Li *

State Key Laboratory for Mechanical Behavior of Materials, School of Materials Science and Engineering, Xi'an Jiaotong University, Xi'an 710049, China
* Correspondence: licx@mail.xjtu.edu.cn

Abstract: Vacuum cold spraying (VCS) has emerged as an environmentally sustainable method for fabricating ceramic and metal films. A high particle impact velocity is a critical factor in the deposition of metal particles during the VCS process, which can be significantly enhanced through gas preheating. This study employs Computational Fluid Dynamics (CFD) simulations to investigate the substantial impact of gas preheating temperature on particle impact velocity and temperature. Elevating the gas temperature leads to higher particle impact velocity, resulting in severe deformation and the formation of dense copper films. The experimental results indicate improvements in both film compactness and electrical properties with gas preheating. Remarkably, the electrical resistivity of the copper film deposited at a gas preheating temperature of 350 °C was measured at 4.4×10^{-8} Ω·m. This study also examines the evolution of cone-shaped pits on the surface of copper films prepared on rough substrates. VCS demonstrates a self-adaptive repair mechanism when depositing metal films onto rough ceramic substrates, making it a promising method for ceramic surface metallization.

Keywords: vacuum cold spraying (VCS); aerosol deposition (AD) method; copper thin films; gas preheating; electrical resistivity

Citation: Ma, K.; Zhang, Q.-F.; Zhang, H.-Y.; Li, C.-J.; Li, C.-X. Influence of the Gas Preheating Temperature on the Microstructure and Electrical Resistivity of Copper Thin Films Prepared via Vacuum Cold Spraying. *Coatings* **2023**, *13*, 1870. https://doi.org/10.3390/coatings13111870

Academic Editor: Alexandru Enesca

Received: 27 September 2023
Revised: 26 October 2023
Accepted: 30 October 2023
Published: 31 October 2023

1. Introduction

In recent years, there has been a rapid development in microelectromechanical systems (MEMS) products, which has garnered significant attention in relation to the design and fabrication of miniaturized micro–nano devices. Metal materials, specifically copper and silver, have gained widespread utilization in the manufacturing of electrical and heat-dissipating components due to their remarkable electrical and thermal conductivity which is due to their exceptional electrical and thermal conductivity [1,2]. Consequently, there is an urgent need to investigate metallization processes to meet the evolving demands of micro–nano device manufacturing technology. Up to this point, methods such as electroplating [3], chemical deposition [4], chemical vapor deposition [5], and magnetron sputtering [6] have commonly been employed for metallization processes. Nevertheless, these methods present certain environmental concerns and exhibit limitations in terms of thick film deposition efficiency. To address these challenges, it is imperative to develop a more straightforward, efficient, and environmentally friendly metallization process for manufacturing purposes.

Vacuum cold spraying (VCS), also known as the aerosol deposition (AD) method [7,8], has attracted attention as a dry film manufacturing process technology. It enables the rapid preparation of ceramic films, ranging in thickness from several to tens of microns, on various substrates such as ceramics, metals, glass, and even polymers [9–14]. Currently, there are limited reports on the deposition of metal coatings using AD, but these reports often exhibit higher electrical resistivity [15–17]. The formation of the ceramic films is attributed to the reduction of the crystallite size caused by a fracture and plastic deformation in the VCS process [7,18]. Given the ductile nature of metal materials in comparison to brittle ceramics, it is anticipated that the deposition mechanism of metal particles significantly differs from that of ceramic particles. In cold spraying, the bonding of metal particles is due to plastic deformation and adiabatic shear instability occurring at high impact velocities [19]. It has been confirmed that the impact velocity and temperature of the particles are the most critical factors for the deposition of metal particles [20,21]. However, the high impact velocity is attributed to the high gas pressure of tens of atmospheres and the high gas temperature of hundreds of Celsius. Due to the low gas pressure of one atmosphere or even lower and room temperature gas used in the VCS process, these pose a formidable challenge to achieving metal films as dense as those produced via cold spraying [17,22]. The increase in gas temperature will not only effectively increase the particle impact velocity, but also enable in situ particle heating [23,24]. Thus, elevating the gas temperature emerges as an effective strategy to enhance the preparation of copper thin films via VCS.

In this study, the influence of gas temperature on gas flow and particle acceleration was investigated through Computational Fluid Dynamics (CFD) simulations. The copper thin films were fabricated on alumina substrates using the VCS process, comparing cases with and without gas preheating. The electrical resistivity and microstructure of the copper films were characterized. This study not only contributes to a deeper understanding of the VCS technique but also provides a foundation for the design and development of high-performance copper thin films in advanced electronic and microelectronic devices.

2. Materials and Methods

Commercial copper powder (Shanghai St-nano Science and Technology Co., Ltd., Shanghai, China) was used as feedstock material for the experiments. The copper powder particles, with mean particle sizes of 1.8 µm, exhibit a spherical morphology and are devoid of porosity, as shown in Figure 1. Alumina slides ($12 \times 12 \times 2$ mm) with a surface roughness of Ra 1.0 µm were used as substrates. Before spraying, the substrates was ultrasonically cleaned in ethanol to obtain a clean surface.

Figure 1. The morphology of the copper powder.

The spray process was conducted utilizing a home-developed vacuum cold spraying system by Xi'an Jiaotong University, as previously detailed [25–27]. Notably, a redesigned de Laval nozzle, featuring a throat diameter of 0.7 mm and an exit diameter of 2.5 mm, was utilized for this study. Helium (99.99%) was employed as the carrier gas, flowing at a rate of 5 L/min. To modulate the gas temperature, a home-made heating device was positioned between the gas pipe and the nozzle in the vacuum deposition chamber. This device allowed for control of the gas temperature, ranging from 20 °C to 350 °C. Diverse thicknesses of copper films were achieved through variations in the number of scanning passes. A comprehensive list of the key process parameters employed in the VCS process is provided in Table 1.

Table 1. Deposition parameters of the vacuum cold spraying.

Parameter	Unit	Value
Gas flow rate	L/min	5
Chamber pressure	Pa	<300
Distance from nozzle exit to the substrate	mm	5
Nozzle traversal speed	mm/s	2
Gas temperature	°C	20–350

After the vacuum cold spraying experiments, the top view and cross-sectional microstructures of the copper films deposited at different temperatures were examined using field-emission scanning electron microscopy (FE-SEM, MIRA3 LMH, TESCAN, Brno-Kohoutovice, Czech Republic). The resistivity of the VCS-deposited copper films was measured using a 4-point probe (RTS-9, Four Probes Technology Co., Ltd., Shenzhen, China).

A commercially available CFD code, Fluent (17.0 Fluent Inc., New York, NY, USA), was used to predict steady gas flow and analyze particle acceleration behavior during VCS. Thanks to the axisymmetric nozzle used in this study, a two-dimensional axisymmetric model, as shown in Figure 2a, was established to save computational time [28,29]. The dimension of the nozzle in this simulation was consistent with the experimental nozzle used. The substrate was placed 5 mm away from the nozzle exit. The computational domain was meshed by structured grids with 23,000 nodes to achieve a grid-independent solution, as shown in Figure 2b,c. The gas inlet was chosen as the pressure inlet with a pressure value of 0.05 MPa and four temperature values of 293 K, 423 K, 523 K, and 623 K, while the outlet pressure and outlet temperature were constant and equal to 300 Pa and 300 K. The reference atmosphere was at a pressure of 0 Pa, and the heat transfer process between the nozzle wall and the gas flow was not considered.

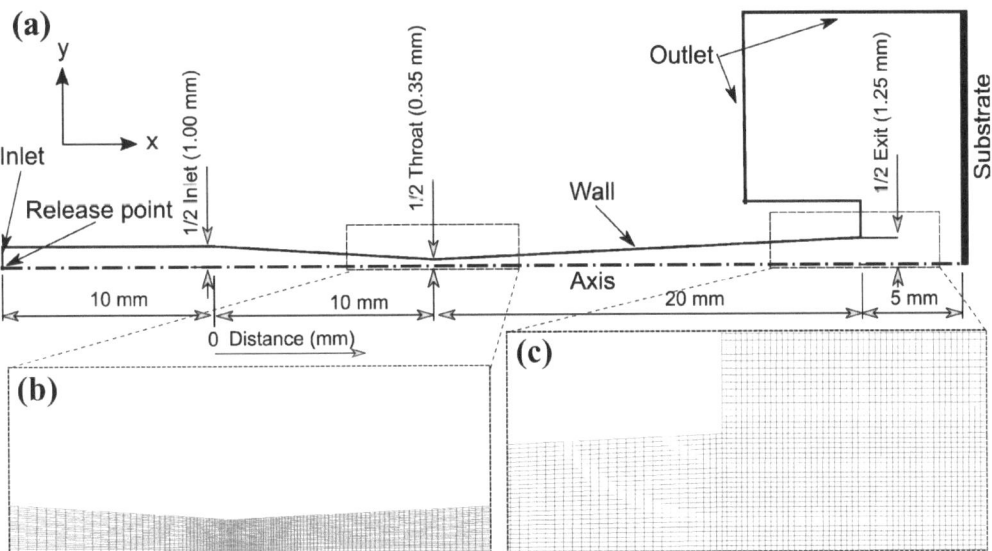

Figure 2. (**a**) Schematic diagram of the computational domain and the boundary conditions, with an enlarged view of mesh at (**b**) nozzle throat, and (**c**) nozzle exit.

The gas was taken as an ideal and compressible fluid, and a coupled implicit method based on the density was used to solve the flow field. For simulation accuracy, the shear-stress transport k-ω model was employed to simulate the turbulence flow [10,30]. The discrete phase model available in Fluent with Lagrangian one-way coupling was used to compute the acceleration of copper particles.

Furthermore, the heat transfer between the gas flow and copper particles was taken into account. The spherical particles were released at a point in the VCS gas stream 10 mm upstream of the nozzle inlet with an initial temperature of 300 K (see Figure 2a). The Stochastic-Tracking type model available in Fluent was used to consider turbulence effects, in which the Discrete Random Walk model can be employed to predict the particle distribution and velocity.

3. Results and Discussion

3.1. Effect of Gas Preheating Temperature on Particle Acceleration

For an ideal gas, a one-dimensional isentropic model provides a straightforward and easily calculable means to estimate gas flow properties [31]. In this isentropic mode, the Mach number at the nozzle axis is solely dependent on nozzle size (expansion ratio) and gas species. When considering a specific nozzle and a defined working gas, elevating the inlet gas temperature emerges as the most effective method to significantly enhance the velocity characteristics of the flow field [32]. This approach has found widespread application in augmenting the quality of coatings in cold spraying, especially for materials with powders possessing a higher critical velocity [33,34]. Consequently, the inlet gas temperature is of considerable significance in the preparation of metal films via VCS.

Figure 3a–d illustrate the contour plots of gas flow velocity in four scenarios characterized by distinct inlet gas pressures (293 K, 423 K, 523 K, and 623 K). In addition, Figure 3e,f present plots depicting the gas flow velocity and temperature profiles along the nozzle's central axis as functions of the distance from the nozzle inlet. Remarkably, it becomes evident that, with increasing inlet gas temperature, the gas flow velocity distributions in these four cases share a similar pattern. As anticipated, the gas flow velocity is initially low at the inlet and increases towards the nozzle throat. The most significant acceleration in gas flow velocity occurs primarily at the nozzle throat, where the transition from subsonic to supersonic velocity takes place. In all four cases, the flow pressure at the nozzle exit remains higher than the chamber pressure (300 Pa) (refer to Table 2), ensuring the continued increase in supersonic flow velocity. As depicted in Figure 3f, the rise in gas flow velocity correlates with a proportional decrease in gas temperature. This phenomenon has also been observed in cold spraying applications [35–37]. Upon reaching the substrate, the supersonic flow experiences a sharp transformation, with the gas flow velocity plummeting from supersonic to zero.

Conversely, as the gas flow velocity decreases, the gas temperature increases rapidly in accordance with the declining flow velocity. In both cold spraying [38,39] and VCS [10,30], a bow shock with a low-velocity, high-density region commonly forms near the substrate's surface. As the inlet gas temperature escalates from 293 K to 623 K, the gas temperature at the bow shock increases from 309 K to 671 K. Concurrently, the thickness of the bow shock exhibits a slight increment, rising from 0.90 mm to 1.15 mm. These fluctuations in flow temperature and bow shock thickness significantly impact particle interactions, particularly for small particles, a subject we will delve into later.

The gas flow velocity and gas temperature demonstrate a conspicuous upward trend as the inlet gas temperature rises. This trend aligns with the outcomes predicted by isentropic theory (see Table 2). As the inlet gas temperature elevates from 293 K to 623 K, the simulated gas flow velocity at the nozzle exit's center surges from 1572 m/s to 2136 m/s. Due to the influence of viscous effects in the flow, the simulated gas flow velocities in all four cases are somewhat lower than those predicted by the isentropic theory. Consequently, as illustrated in Figure 4a, the Mach numbers along the nozzle's central axis in these four cases fall short of the Mach numbers calculated using the isentropic theory.

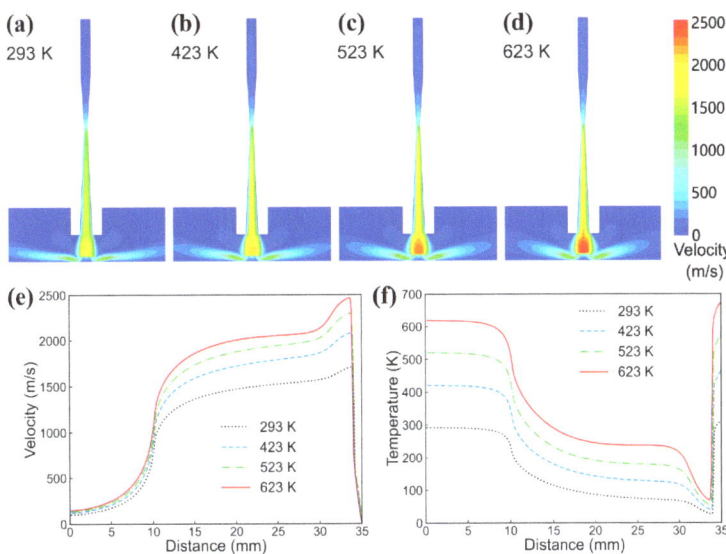

Figure 3. CFD gas flow velocity results with different inlet gas temperatures of (**a**) 293 K, (**b**) 423 K, (**c**) 523 K, and (**d**) 623 K. And changes of (**e**) velocity and (**f**) temperature of flow along the nozzle centerline.

Table 2. A comparison of CFD simulation results and isentropic nozzle theory for four inlet gas temperatures.

Inlet Gas Temperature (K)	Velocity at Nozzle Exit Center (m/s)		Maximum Flow Velocity (m/s)	Pressure at Nozzle Exit Center (Pa)	
	Isentropic	Simulation		Isentropic	Simulation
293	1668	1572	1709	412	875
423	2004	1868	2082	412	1082
523	2228	2031	2302	412	1208
623	2432	2136	2468	412	1295

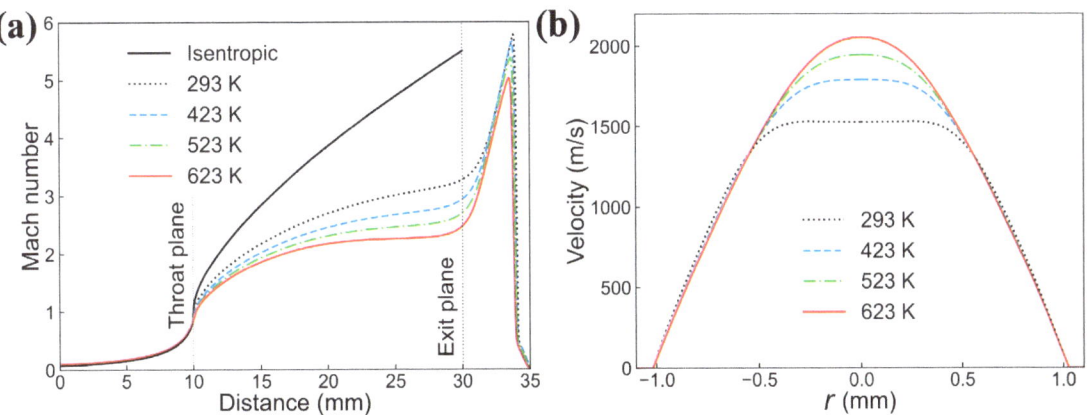

Figure 4. (**a**) Mach number along the nozzle center line and (**b**) gas flow velocity along the radial line 5 mm upstream from the nozzle exit plane.

Furthermore, as the inlet gas temperature increases, the Mach number inside the nozzle experiences a contrasting decrease. Concurrently, the boundary layer near the nozzle's inner wall thickens, as depicted in Figure 4b. Given the typical usage of slit nozzles or small-sized de Laval nozzles in VCS, the boundary layer is a critical factor that cannot be disregarded. Therefore, it is advisable to employ a near-wall model, as opposed to a wall function, to address the near-wall region in CFD simulations. Considering the chamber pressure (300 Pa) near the gas pressure at the nozzle exit center, as calculated using the isentropic theory (441 Pa), the maximum simulated gas velocity in all four cases closely aligns with that predicted by the isentropic theory.

Figure 5 illustrates the variation in the mean impact velocity and mean impact temperature of copper particles as a function of particle size. These particles experience two distinct stages: an acceleration stage inside the nozzle and a subsequent deceleration stage within the bow shock. Despite the rapid acceleration of small particles by the carrier gas, their velocity experiences a significant reduction upon passing through the bow shock due to their limited inertia. Consequently, there exists an optimal particle size that yields the highest particle impact velocity. This phenomenon is also observed in cold spraying simulations [40]. Notably, as the inlet gas temperature increases, the optimal particle size for achieving maximum particle velocity undergoes a slight increase. This can be attributed to the heightened gas stagnation pressure and gas density within the bow shock, leading to a more pronounced deceleration effect on smaller particles.

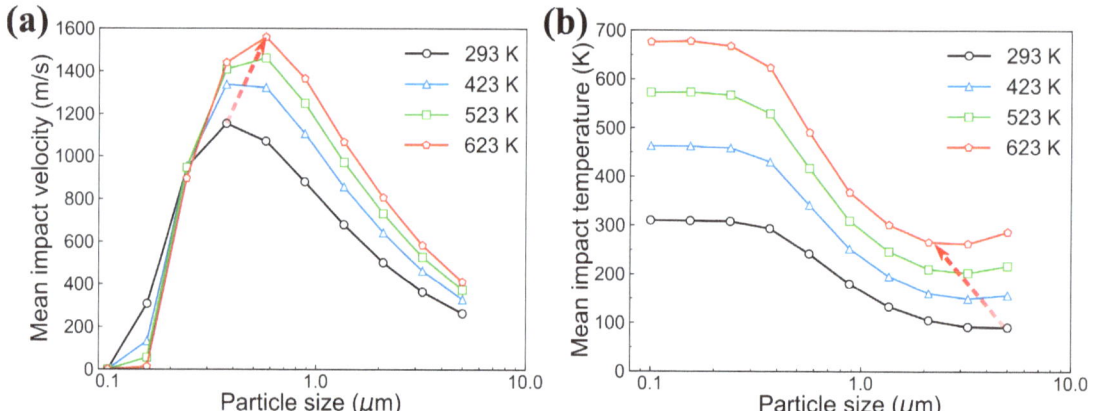

Figure 5. (**a**) Particle mean impact velocity and (**b**) mean impact temperature for different particle diameters at gas temperatures from 293 K to 623 K. (The arrows represent the variation in particle size corresponding to the lowest impact temperature.)

Furthermore, when the inlet gas temperature is raised, particles smaller than 0.2 μm exhibit lower particle impact velocities. This behavior is because smaller particles, possessing lower mass, can more readily respond to changes in gas temperature. Consequently, these small particles (<0.2 μm) attain an impact temperature very similar to the gas temperature of the bow shock (as depicted in Figure 5b). In contrast, the temperature of slightly larger particles (>2 μm) decreases with the gas temperature inside the nozzle. However, their brief time of residence within the bow shock is insufficient to acquire a significant amount of heat. Consequently, there exists a particle size corresponding to the lowest particle impact temperature. In summary, elevating the gas temperature before particle ejection can significantly enhance both the particle impact velocity and impact temperature, thereby promoting increased deformation during particle collisions.

3.2. Microstructure and Electrical Properties of Copper Films Deposited at Different Gas Temperatures

To investigate the influence of gas temperature on the preparation of VCS-deposited copper films, experiments were conducted at gas temperatures of 150 °C, 250 °C, and 350 °C with gas preheating, as well as at 20 °C without gas preheating. Figure 6 presents the surface and cross-sectional morphology of the copper films prepared on the alumina substrate at these four different temperatures. Remarkably, even without gas preheating, the copper film can be successfully deposited onto the alumina substrate. However, numerous spherical copper particles with insufficient deformation and noticeable gaps between them are evident on the film surface in Figure 6a,b. Without gas preheating, the copper particles lack the necessary impact velocity, resulting in insufficient deformation of the copper particles, making it difficult to achieve effective bonding. In the case of deposition at a gas preheating temperature of 150 °C, as depicted in Figure 6c,d, the number of undeformed particles on the film surface decreases, although gaps between the particles are still observable. With deposition at a gas preheating temperature of 250 °C (Figure 6e,f), the number of undeformed particles and gaps between particles on the film surface further diminishes, signifying a notable improvement in copper particle adhesion. When the gas preheating temperature reaches 350 °C (Figure 6g,h), the particles' outlines and the gaps between them are hardly discernible on the film surface. In other words, the copper particles undergo substantial flattening through plastic deformation and tightly adhere to form a dense copper film.

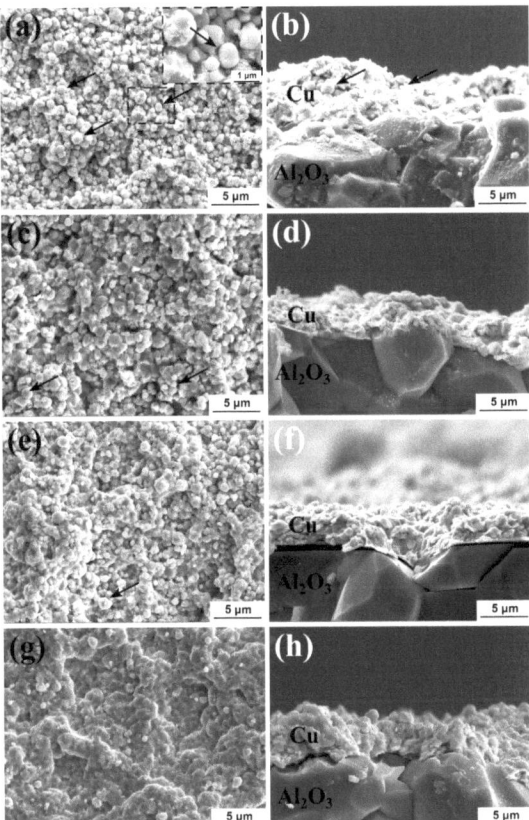

Figure 6. Microstructure morphologies of copper films deposited on Al_2O_3 substrates at gas preheating temperatures of (**a,b**) 20 °C, (**c,d**) 150 °C, (**e,f**) 250 °C, and (**g,h**) 350 °C. Black arrows mark undeformed copper particles.

From the simulation results in Section 3.1, the microstructural differences observed can be attributed to two key factors. Firstly, higher gas temperatures increase particle velocity during VCS, imparting greater kinetic energy to the particles and leading to more pronounced plastic deformation. This phenomenon parallels observations made during cold spray [20,21]. Secondly, elevated gas temperatures facilitate easier particle deformation [41,42]. In summary, increasing the gas temperature proves more conducive to the preparation of a dense VCS-deposited copper film.

Figure 7 presents the electrical resistivity of the copper films as a function of gas preheating temperature. The electrical resistivity decreases as the gas temperature increases. Specifically, at a gas preheating temperature of 20 °C, the resistivity of copper film exhibits its highest value, approximately 1.3×10^{-6} $\Omega \cdot m$. This elevation in resistivity is attributed to the low particle binding and the presence of gaps between the particles. Notably, as the gas preheating temperature increases, the resistivity significantly decreases, eventually reaching approximately 4.4×10^{-8} $\Omega \cdot m$ at a gas preheating temperature of 350 °C. This value is only about 2.6 times that of bulk copper (1.7×10^{-8} $\Omega \cdot m$) [43], surpassing the previously reported AD copper film [17]. These results suggest that VCS-deposited copper films hold promise for future applications in ceramic surface metallization.

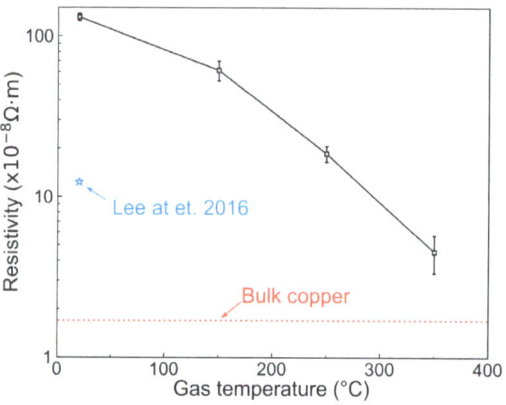

Figure 7. Electrical resistivity of copper films deposited at different gas temperatures. (The dotted line is the resistivity of bulk copper, and asterisk data comes from the literature [17]).

The electrical resistivity of VCS-deposited copper films is closely linked to both particle bonding and film compactness. Higher gas temperatures (higher than 250 °C) result in increased particle impact velocity and temperature, which, in turn, enhance the microcontacts among particles, ultimately leading to the formation of a denser copper film. Additionally, the presence of Cu_2O and CuO oxides in copper films plays a crucial role in elevating the electrical resistivity of the copper film above that of bulk copper [17]. Consequently, future research should focus on controlling the oxygen content of sub-micron copper powder and preventing oxidation during particle deposition to further optimize the electrical properties of these films.

3.3. Copper Films Grown on Alumina Substrates

To investigate the growth of the copper films during the VCS process, the film thickness was measured after each scan. Figure 8 displays the film thickness as a function of the number of scans. When performing a single scan, as shown in Figure 9a, the copper particles completely cover the surface of the alumina substrate, forming a continuous copper thin film with an initial thickness of approximately 4.7 μm. Subsequently, the film thickness exhibits a linear increase as the number of scans increases. The rate of thickness growth reaches approximately 4.3 μm per scan, resulting in a high deposition rate for a 10×10 mm^2 area, equivalent to 1.03 μm per minute. By the time 15 scans are completed, the thickness of the copper film reaches approximately 63 μm. In essence, vacuum cold spraying can be employed to produce copper thin films in the range of several microns in thickness, and thicker copper films to the order of tens of microns or even more, by increasing the number of scans.

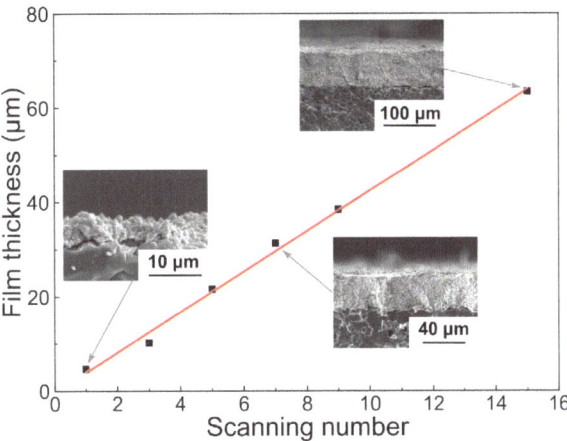

Figure 8. Thickness of copper films deposited at different scanning numbers.

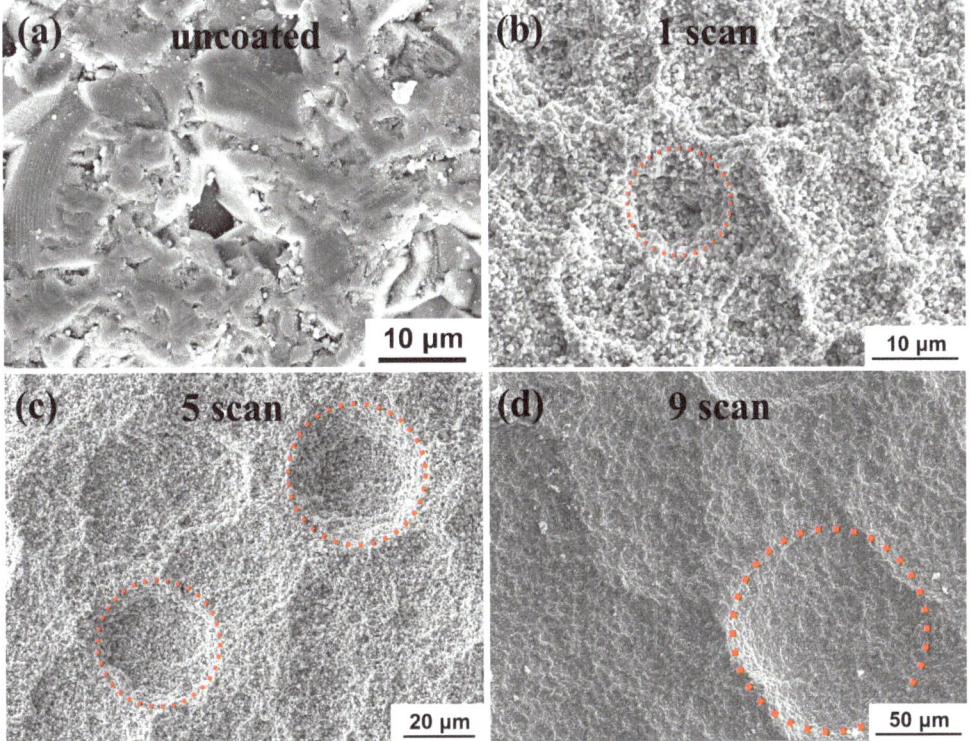

Figure 9. Surface topography of copper films on an alumina substrate with different scanning numbers: (**a**) uncoated, (**b**) 1 scan, (**c**) 5 scans and (**d**) 9 scans. The red dot circle indicates one-shaped pits.

Figure 9 presents surface microstructure images of the copper films on the alumina substrate as the number of scans increases. While a single scan results in complete coverage of the alumina surface by copper particles, the surface of the copper films exhibits significant roughness, accompanied by the presence of cone-shaped pits (highlighted by red dotted circles in Figure 9a). This phenomenon arises due to the thin nature of the coating, which inherits the irregularities of the substrate surface. As the number of scans increases, the thickness of the copper film steadily increases, and the edges of the cone-shaped pits expand outward, leading to an increase in their diameter. As the film thickness increases from 5 μm to 38 μm, the diameter of the cone-shaped pits expands from approximately 11 μm to around 91 μm, as shown in Figure 10. With further increases in film thickness, the cone-shaped pits gradually vanish, and their edges become increasingly difficult to discern.

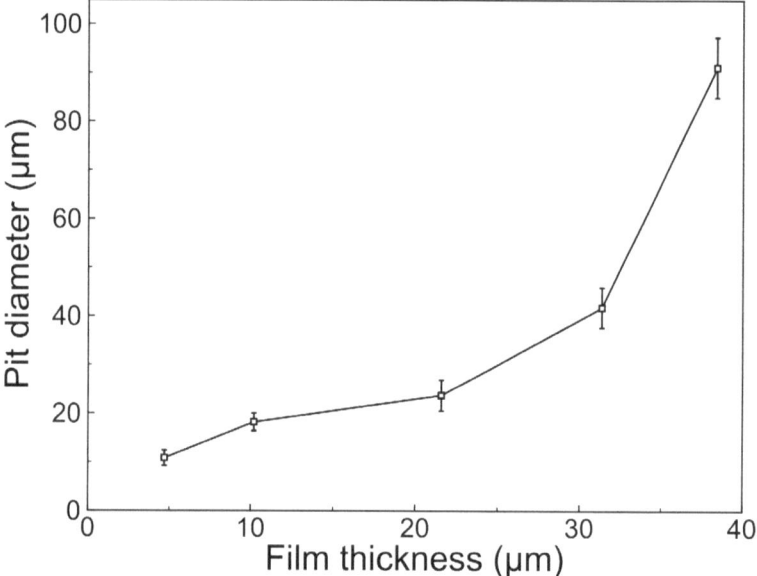

Figure 10. Diameter of the cone-shaped pits on copper films with different film thicknesses.

The cross-sectional view of one of these cone-shaped pits, as shown in Figure 11a, confirms that it originates from a pit on the substrate's surface. Furthermore, the pit diameter undergoes more pronounced changes with the thickness of the copper films, as depicted in Figure 10. As the pit diameter increases, the pit walls become flatter, ultimately resulting in the gradual disappearance of these larger pits. Consequently, observing large pits on the surface of copper films with a thickness exceeding 40 μm becomes challenging. Figure 11b displays the SEM image of the peeled surface of the copper film from the alumina substrate. This image reveals that small pits on the substrate surface can be adequately filled by an individual or a limited number of copper particles (highlighted in blue). However, for slightly larger pits, multiple particles are required for filling (highlighted in red). Furthermore, larger pits, such as those within the red circle in Figure 9a and the pits in Figure 11a, result in the persistence and evolution of cone-shape pits. Mechanical interlocking is the primary factor contributing to the adhesion of metal coatings to ceramic substrates. The presence of certain irregularities on the substrate surface further enhances the substrate's ability to achieve a robust bond with the copper film [17,44,45].

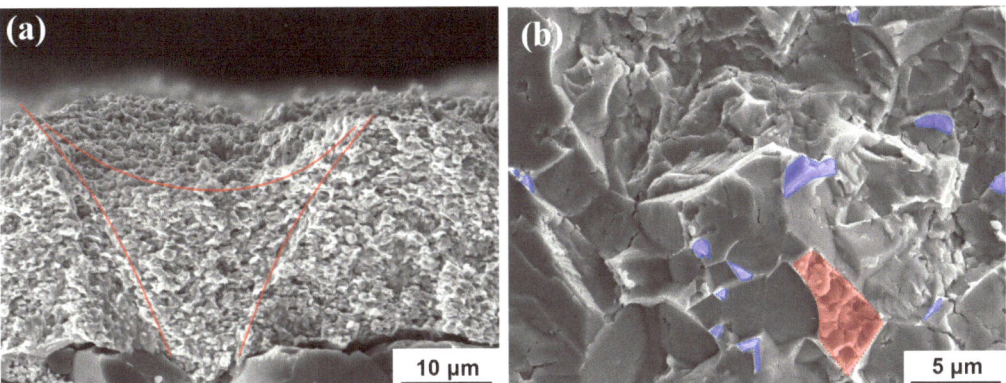

Figure 11. (**a**) Cross-sectional SEM image of the cone-shape pit on the copper film, and (**b**) SEM image of the peeled surface of the copper film from the substrate. The blue highlighting represents small pits filled with a single or a limited number of copper particles, while the red highlighting indicates large pits filled with multiple copper particles.

Figure 12 provides a schematic representation of copper film growth on a rough alumina substrate. The substrate exhibits numerous small pits measuring several hundred nanometers in size, along with larger pits of a few microns, as depicted in Figure 12a. Unlike the deposition mechanism observed in brittle particles such as ceramics, which involves the reduction of crystallite size during the VCS process [7,46], ductile metals primarily undergo plastic deformation during deposition [17,19,22]. When copper particles collide with the high-hardness alumina substrate, they experience high stress levels, resulting in substantial plastic deformation. Consequently, the valleys in the rough alumina substrate become covered with deformed copper particles, forming a mechanical interlock between the copper films and the alumina substrate (Figure 12b).

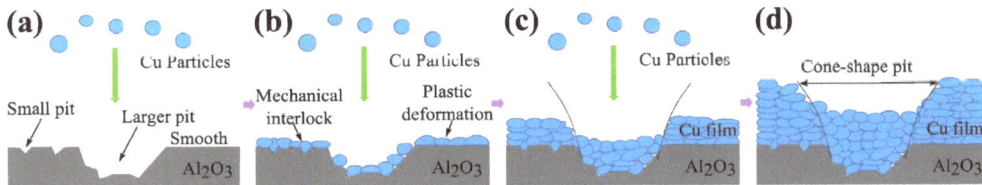

Figure 12. The schematic diagram of the copper film grown on a rough alumina substrate. (**a**) Uncoated Al_2O_3 substrate, (**b**) initial deposition process, (**c**) subsequent deposition process and (**d**) deposition completion resulting in the formation of a copper film.

However, there are some larger pits whose sizes exceed that of individual copper particles. While copper particles are also deposited inside these pits, they tend to remain within the pits and become incorporated into the film (Figure 12c). It has been reported that the impact angle plays a significant role in the deposition efficiency of copper in the cold spray process, with lower deposition efficiency observed at smaller impact angles [47]. As the thickness of the copper films increases, the cone-shaped pits expand outward due to the lower deposition efficiency of copper particles on the sidewalls of these pits (Figure 12d). Once the thickness of the copper films reaches a certain threshold, the side walls of the cone-shaped pits tend to become flatter, eventually leading to the gradual disappearance of these pits. Consequently, there exists a self-adaptive repair mechanism when depositing metal films on a rough substrate using VCS.

4. Conclusions

In this study, the results from CFD simulations reveal that even in a chamber with pressure below 300 Pa, the bow shock near the substrate still exerts an influence on the impact velocity of ultrafine particles. Moreover, there exists an optimal particle size that yields the highest particle impact velocity. Concurrently, the significant impact of gas temperature on particle impact temperature and impact velocity is confirmed. Compared to room temperature conditions, appropriately increasing the gas temperature enhances particle velocity and softens the particles, leading to substantial particle deformation and the formation of a dense copper film. Additionally, copper films were successfully fabricated on alumina substrates using the vacuum cold spray (VCS) process. The electrical resistivity of the copper films decreases as the gas preheating temperature increases. Specifically, the electrical resistivity decreases from 1.3×10^{-6} $\Omega \cdot$m to 4.4×10^{-8} $\Omega \cdot$m as the gas temperature increases from 20 °C to 350 °C. In summary, vacuum cold spraying demonstrates potential for applications in ceramic surface metallization.

Furthermore, the thickness of the copper films exhibits a linear increase with an increase in the number of scans. This implies that VCS can be employed to produce thin metal films in the range of several microns and thicker metal films ranging from tens of microns to even thicker layers. When copper films are prepared on rough alumina substrates, cone-shaped pits are observed on the film surface. As the copper films grow, the pit diameter increases and eventually disappears. It has been confirmed that the evolution of these pits is primarily influenced by the impact angle. On the high-hardness, rough alumina substrate, copper particles undergo plastic deformation and fill the valleys on the rough surface, creating a mechanical interlocking effect. For larger pits whose sizes exceed the particle size, their profiles are transferred to the film surface and tend to flatten as the film thickness increases. Consequently, there exists a self-adaptive repair mechanism when depositing metal films on rough ceramic substrates using the VCS process.

Author Contributions: Conceptualization, K.M. and C.-X.L.; Methodology, K.M., Q.-F.Z., H.-Y.Z. and C.-X.L.; Validation, K.M. and Q.-F.Z.; Formal analysis, K.M.; Investigation, K.M. and Q.-F.Z.; Resources, C.-J.L.; Data curation, Q.-F.Z.; Writing—original draft, K.M.; Writing—review & editing, H.-Y.Z. and C.-J.L.; Visualization, K.M.; Supervision, C.-X.L.; Funding acquisition, C.-X.L. All authors have read and agreed to the published version of the manuscript.

Funding: This research was funded by the State Key Laboratory for the Mechanical Behavior of Materials, Xi'an Jiaotong University. All the samples were fabricated and characterized in the thermal spray laboratory (TSL-XJTU).

Institutional Review Board Statement: Not applicable.

Informed Consent Statement: Not applicable.

Data Availability Statement: Not applicable.

Conflicts of Interest: The authors declare no conflict of interest.

References

1. Roa, S.; Sirena, M. Size Effects on the Optimization of the Mechanical Resistance and the Electrical Conductivity of Cu Thin Films. *Mater. Today Commun.* **2021**, *28*, 102572. [CrossRef]
2. Kao, Z.-K.; Hung, Y.-H.; Liao, Y.-C. Formation of Conductive Silver Films via Inkjet Reaction System. *J. Mater. Chem.* **2011**, *21*, 18799–18803. [CrossRef]
3. Zhang, Y.; Hang, T.; Dong, M.; Wu, Y.; Ling, H.; Hu, A.; Li, M. Effects of 2-mercaptopyridine and Janus Green B as levelers on electrical resistance of electrodeposited copper thin film for interconnects. *Thin Solid Films* **2019**, *677*, 39–44. [CrossRef]
4. Yang, L.; Zhang, D.; Zhang, X.; Tian, A. Electroless copper deposition and interface characteristics of ionic electroactive polymer. *J. Mater. Res. Technol.* **2021**, *11*, 849–856. [CrossRef]
5. Bouhafs, C.; Pezzini, S.; Geisenhof, F.R.; Mishra, N.; Mišeikis, V.; Niu, Y.; Struzzi, C.; Weitz, R.T.; Zakharov, A.A.; Forti, S.; et al. Synthesis of large-area rhombohedral few-layer graphene by chemical vapor deposition on copper. *Carbon* **2021**, *177*, 282–290. [CrossRef]

6. Velicu, I.-L.; Ianoş, G.-T.; Porosnicu, C.; Mihăilă, I.; Burducea, I.; Velea, A.; Cristea, D.; Munteanu, D.; Tiron, V. Energy-enhanced deposition of copper thin films by bipolar high power impulse magnetron sputtering. *Surf. Coat. Technol.* **2019**, *359*, 97–107. [CrossRef]

7. Akedo, J. Room temperature impact consolidation and application to ceramic coatings: Aerosol deposition method. *J. Ceram. Soc. Jpn.* **2020**, *128*, 101–116. [CrossRef]

8. Hanft, D.; Exner, J.; Schubert, M.; Stöcker, T.; Fuierer, P.; Moos, R. An overview of the aerosol deposition method: Process fundamentals and new trends in materials applications. *J. Ceram. Sci. Technol.* **2015**, *6*, 147–181. [CrossRef]

9. Akedo, J. Room temperature impact consolidation (RTIC) of fine ceramic powder by aerosol deposition method and applications to microdevices. *J. Therm. Spray Technol.* **2008**, *17*, 181–198. [CrossRef]

10. Chun, D.M.; Choi, J.O.; Lee, C.S.; Ahn, S.H. Effect of stand-off distance for cold gas spraying of fine ceramic particles under low vacuum and room temperature using nano-particle deposition system (NPDS). *Surf. Coat. Technol.* **2012**, *206*, 2125–2132. [CrossRef]

11. Lin, S.C.; Wu, W.J. Fabrication of PZT MEMS energy harvester based on silicon and stainless-steel substrates utilizing an aerosol deposition method. *J. Micromech. Microeng.* **2013**, *23*, 125028. [CrossRef]

12. Park, H.; Kim, J.; Lee, C. Dynamic fragmentation process and fragment microstructure evolution of alumina particles in a vacuum kinetic spraying system. *Scr. Mater.* **2015**, *108*, 72–75. [CrossRef]

13. Hwang, G.-T.; Annapureddy, V.; Han, J.H.; Joe, D.J.; Baek, C.; Park, D.Y.; Kim, D.H.; Park, J.H.; Jeong, C.K.; Park, K.-I.; et al. Self-powered wireless sensor node enabled by an aerosol-deposited PZT flexible energy harvester. *Adv. Energy Mater.* **2016**, *6*, 1600237. [CrossRef]

14. Wang, L.-S.; Li, C.-X.; Ma, K.; Zhang, S.-L.; Yang, G.-J.; Li, C.-J. La0.8Sr0.2Ga0.8Mg0.2O3 electrolytes prepared by vacuum cold spray under heated gas for improved performance of SOFCs. *Ceram. Int.* **2018**, *44*, 13773–13781. [CrossRef]

15. Kim, I.-S.; Jang, C.-I.; Cho, M.-Y.; Lee, Y.-S.; Koo, S.-M.; Lee, D.-W.; Oh, J.-M. Deformation Mechanism of Powder-Spray-Coated Cu Particle at Room Temperature. *Jpn. J. Appl. Phys.* **2020**, *59*, 095504. [CrossRef]

16. Lee, D.-W.; Cho, M.-Y.; Kim, I.-S.; Kim, Y.-N.; Lee, D.; Koo, S.-M.; Park, C.; Oh, J.-M. Experimental and Numerical Study for Cu Metal Coatings at Room Temperature via Powder Spray Process. *Surf. Coat. Technol.* **2018**, *353*, 66–74. [CrossRef]

17. Lee, D.-W.; Kwon, O.-Y.; Cho, W.-J.; Song, J.-K.; Kim, Y.-N. Characteristics and Mechanism of Cu Films Fabricated at Room Temperature by Aerosol Deposition. *Nanoscale Res. Lett.* **2016**, *11*, 162. [CrossRef] [PubMed]

18. Liu, Y.; Wang, Y.; Suo, X.; Gong, Y.; Li, C.; Li, H. Impact-induced bonding and boundary amorphization of TiN ceramic particles during room temperature vacuum cold spray deposition. *Ceram. Int.* **2016**, *42*, 1640–1647. [CrossRef]

19. Grujicic, M.; Zhao, C.L.; DeRosset, W.S.; Helfritch, D. Adiabatic shear instability based mechanism for particles/substrate bonding in the cold-gas dynamic-spray process. *Mater. Des.* **2004**, *25*, 681–688. [CrossRef]

20. Schmidt, T.; Gärtner, F.; Assadi, H.; Kreye, H. Development of a generalized parameter window for cold spray deposition. *Acta Mater.* **2006**, *54*, 729–742. [CrossRef]

21. Ning, X.-J.; Jang, J.-H.; Kim, H.-J. The effects of powder properties on in-flight particle velocity and deposition process during low pressure cold spray process. *Appl. Surf. Sci.* **2007**, *253*, 7449–7455. [CrossRef]

22. King, P.C.; Bae, G.; Zahiri, S.H.; Jahedi, M.; Lee, C. An experimental and finite element study of cold spray copper impact onto two aluminum substrates. *J. Therm. Spray Technol.* **2010**, *19*, 620–634. [CrossRef]

23. Fukanuma, H.; Ohno, N.; Sun, B.; Huang, R. In-flight particle velocity measurements with DPV-2000 in cold spray. *Surf. Coat. Technol.* **2006**, *201*, 1935–1941. [CrossRef]

24. Binder, K.; Gottschalk, J.; Kollenda, M.; Gärtner, F.; Klassen, T. Influence of Impact Angle and Gas Temperature on Mechanical Properties of Titanium Cold Spray Deposits. *J. Therm. Spray Technol.* **2011**, *20*, 234–242. [CrossRef]

25. Li, J.; Zhang, Y.; Ma, K.; Pan, X.-D.; Li, C.-X.; Yang, G.-J.; Li, C.-J. Microstructure and transparent super-hydrophobic performance of vacuum cold-sprayed Al2O3 and SiO2 aerogel composite coating. *J. Therm. Spray Technol.* **2018**, *27*, 471–482. [CrossRef]

26. Liu, Y.; Wang, Y.Y.; Yang, G.J.; Feng, J.J.; Kusumoto, K. Effect of nano-sized TiN additions on the electrical properties of vacuum cold sprayed SiC coatings. *J. Therm. Spray Technol.* **2010**, *19*, 1238–1243. [CrossRef]

27. Ma, K.; Li, C.-J.; Li, C.-X. Narrow and Thin Copper Linear Pattern Deposited by Vacuum Cold Spraying and Deposition Behavior Simulation. *J. Therm. Spray Technol.* **2021**, *30*, 571–583. [CrossRef]

28. Li, W.-Y.; Liao, H.; Douchy, G.; Coddet, C. Optimal design of a cold spray nozzle by numerical analysis of particle velocity and experimental validation with 316L stainless steel powder. *Mater. Des.* **2007**, *28*, 2129–2137. [CrossRef]

29. Suo, X.; Yin, S.; Planche, M.-P.; Liu, T.; Liao, H. Strong effect of carrier gas species on particle velocity during cold spray processes. *Surf. Coat. Technol.* **2015**, *268*, 90–93. [CrossRef]

30. Li, C.; Singh, N.; Andrews, A.; Olson, B.A.; Schwartzentruber, T.E.; Hogan, C.J. Mass, momentum, and energy transfer in supersonic aerosol deposition processes. *Int. J. Heat Mass Transf.* **2019**, *129*, 1161–1171. [CrossRef]

31. Grujicic, M.; Zhao, C.L.; Tong, C.; DeRosset, W.S.; Helfritch, D. Analysis of the impact velocity of powder particles in the cold-gas dynamic-spray process. *Mater. Sci. Eng. A* **2004**, *368*, 222–230. [CrossRef]

32. Yin, S.; Meyer, M.; Li, W.; Liao, H.; Lupoi, R. Gas flow, particle acceleration, and heat transfer in cold spray: A review. *J. Therm. Spray Technol.* **2016**, *25*, 874–896. [CrossRef]

33. Goldbaum, D.; Shockley, J.M.; Chromik, R.R.; Rezaeian, A.; Yue, S.; Legoux, J.-G.; Irissou, E. The effect of deposition conditions on adhesion strength of Ti and Ti6Al4V cold spray splats. *J. Therm. Spray Technol.* **2012**, *21*, 288–303. [CrossRef]

34. Huang, R.; Fukanuma, H. Study of the influence of particle velocity on adhesive strength of cold spray deposits. *J. Therm. Spray Technol.* **2012**, *21*, 541–549. [CrossRef]
35. Tang, W.; Liu, J.; Chen, Q.; Zhang, X.; Chen, Z. The effects of two gas flow streams with initial temperature and pressure differences in cold spraying nozzle. *Surf. Coat. Technol.* **2014**, *240*, 86–95. [CrossRef]
36. Wang, X.; Zhang, B.; Lv, J.; Yin, S. Investigation on the Clogging Behavior and Additional Wall Cooling for the Axial-Injection Cold Spray Nozzle. *J. Therm. Spray Technol.* **2015**, *24*, 696–701. [CrossRef]
37. Meng, X.; Zhang, J.; Zhao, J.; Liang, Y.; Zhang, Y. Influence of Gas Temperature on Microstructure and Properties of Cold Spray 304SS Coating. *J. Mater. Sci. Technol.* **2011**, *27*, 809–815. [CrossRef]
38. Pattison, J.; Celotto, S.; Khan, A.; O'neill, W. Standoff distance and bow shock phenomena in the cold spray process. *Surf. Coat. Technol.* **2008**, *202*, 1443–1454. [CrossRef]
39. Yin, S.; Wang, X.F.; Li, W.Y. Computational analysis of the effect of nozzle cross-section shape on gas flow and particle acceleration in cold spraying. *Surf. Coat. Technol.* **2011**, *205*, 2970–2977. [CrossRef]
40. Assadi, H.; Schmidt, T.; Richter, H.; Kliemann, J.-O.; Binder, K.; Gärtner, F.; Klassen, T.; Kreye, H. On parameter selection in cold spraying. *J. Therm. Spray Technol.* **2011**, *20*, 1161–1176. [CrossRef]
41. Yin, S.; Wang, X.; Suo, X.; Liao, H.; Guo, Z.; Li, W.; Coddet, C. Deposition behavior of thermally softened copper particles in cold spraying. *Acta Mater.* **2013**, *61*, 5105–5118. [CrossRef]
42. Schmidt, T.; Gaertner, F.; Kreye, H. New developments in cold spray based on higher gas and particle temperatures. *J. Therm. Spray Technol.* **2006**, *15*, 488–494. [CrossRef]
43. Matula, R.A. Electrical resistivity of copper, gold, palladium, and silver. *J. Phys. Chem. Ref. Data* **1979**, *8*, 1147–1298. [CrossRef]
44. Qin, J.; Huang, Q.; Wang, X.; Suo, X.; Wang, J.; Li, H. Interfacial metal/ceramic bonding mechanism for metallization of ceramics via cold spraying. *J. Mater. Process. Technol.* **2021**, *288*, 116845. [CrossRef]
45. Singh, S.; Raman, R.K.S.; Berndt, C.C.; Singh, H. Influence of Cold Spray Parameters on Bonding Mechanisms: A Review. *Metals* **2021**, *11*, 2016. [CrossRef]
46. Chun, D.M.; Ahn, S.H. Deposition mechanism of dry sprayed ceramic particles at room temperature using a nano-particle deposition system. *Acta Mater.* **2011**, *59*, 2693–2703. [CrossRef]
47. Li, C.; Li, W.; Wang, Y.; Fukanuma, H. Effect of spray angle on deposition characteristics in cold spraying. In *Thermal Spray*; Digital Library: New York, NY, USA, 2003; pp. 91–96.

Article

Cracking Behavior of Atmospheric Plasma-Sprayed 8YSZ Thermal Barrier Coatings during Thermal Shock Test

Jibo Huang [1], Wen Sun [2], Renzhong Huang [2] and Wenhua Ma [3],*

1   School of Materials Science and Engineering, South China University of Technology, Guangzhou 510640, China
2   Guangzhou Institute of Hubei Chaozhuo Aviation Technology Co., Ltd., Guangzhou 510530, China
3   Center for Industrial Analysis and Testing, Guangdong Academy of Sciences, Guangzhou 510650, China
*   Correspondence: mahua8011@hotmail.com; Tel.: +86-20-3723-8633

Abstract: The failure of plasma-sprayed thermal barrier coatings (TBCs) during service is usually related to the cracking behavior. In this study, plasma-sprayed TBCs were prepared with two kinds of agglomerated sintered yttria-stabilized zirconia (YSZ) powders with different particle sizes. The evolution of mechanical properties and crack propagation behavior of the coatings during the whole life stage were studied by a thermal shock test. The effect of powder particle size on the cracking behavior of the TBCs during thermal shock was analyzed from the aspect of pore structure, mechanical properties, and stress state of the coatings. The crack propagation and coalescence in the direction parallel to the substrate in the coating is the main factor leading to the spalling failure of the coating during thermal shock. Although the coating prepared by fine YSZ has higher fracture toughness, the lower strain tolerance will increase the cracking driving force on the crack tip of the coating during thermal shock, and the cracks in the coating propagate merge at a faster rate during thermal shock. The larger porosity and pore size of the coating prepared by coarse YSZ help the coating suffer less thermal stress during thermal shock. Although the existence of pores reduces the fracture toughness of the coating to a certain extent, the increase of strain tolerance reduces the crack growth rate in the coating, so the coating has a longer life.

Keywords: plasma-spraying; thermal barrier coating; powder size; thermal shock; cracking

Citation: Huang, J.; Sun, W.; Huang, R.; Ma, W. Cracking Behavior of Atmospheric Plasma-Sprayed 8YSZ Thermal Barrier Coatings during Thermal Shock Test. *Coatings* **2023**, *13*, 243. https://doi.org/10.3390/coatings13020243

Academic Editor: Charafeddine Jama

Received: 30 December 2022
Revised: 16 January 2023
Accepted: 18 January 2023
Published: 20 January 2023

1. Introduction

Thermal spraying coatings are often used to strengthen the surface of key components in aerospace, petrochemical and metallurgical fields to enhance their high temperature performance, wear and corrosion resistance [1–4]. Plasma-sprayed yttria-stabilized zirconia (YSZ) thermal barrier coating (TBCs) is a kind of functional coating which is widely used in thermal protection and thermal insulation of hot-end parts of gas turbine and aero-engine. The application of TBCs can improve the working temperature of the engine, thus improving the efficiency and performance [5,6]. Although the TBC technology has been successfully applied in gas turbine and aero-engine for many years, the coating still faces the risk of failure in a harsh service environment [7,8]. Once the coating peels off prematurely, the alloy protected by it will be directly exposed to excessive temperature, which poses a hidden danger and threat to the safety and reliability of the whole engine [9]. The failure of the TBCs during service is usually related to the cracking behavior of the coatings [10–12]. Therefore, the study of crack propagation behavior of the coatings is helpful to understand the failure mechanism of coatings and guide the development of new TBCs with long life.

The service process of TBCs is often accompanied by thermal shocks. Under the action of thermal stress, the coating is prone to crack near the TC/TGO/BC interface, resulting in the spalling of the ceramic layer and the failure of the coating [11,13,14]. The cracking resistance of the coating is very important to maintain the structural integrity of the coating during service, so in the development and preparation process of the TBCs, it is necessary to improve the fracture toughness of the ceramic layer so as to strengthen the resistance to interlayer cracking of the coating [15–17]. The feedstock powder morphology is an important parameter affecting the properties of TBCs. The powder size affects the deposition, solidification, and crystallization of molten droplets by influencing the particle state in the plasma, thus affecting the pore structure and mechanical properties of the coating [18,19]. Dwivedi et al. [19] studied the effect of the particle size of YSZ powder on the fracture toughness of plasma-sprayed coatings. Their results show that the fracture toughness of the coating decreases with the increase in powder size. The coating prepared with fine particle size powder has dense microstructure, low porosity and high fracture toughness. In addition to the cracking resistance, the pore structure of the coating is also an important factor affecting the cracking behavior of TBCs during thermal shock service [20–22]. The microstructure of plasma-sprayed coating contains a variety of pores, such as unbonded defects, macropores and microcracks [23,24]. The pores in the coating can reduce the thermal conductivity, increase the strain tolerance of the coating, and reduce the thermal stress during service, thus affecting the cracking behavior of the coating [20,25,26].

The cracking failure of plasma-sprayed TBCs under thermal shock largely depends on the joint action of cracking resistance and cracking driving force of the coatings [27]. The selection of particle size of sprayed powder is an important step in the TBC preparation process. Previous studies have shown that the coatings prepared by fine YSZ powder have denser structure and higher fracture toughness because of more sufficient melting of the fine powder during plasma-spraying [18]. However, on the other hand, increasing the strain tolerance of the coating with a porous structure is also very important to improve the thermal shock cracking resistance of the TBCs [28,29]. Up to now, it is still not very clear how the particle size of the powder responds to the cracking and failure behavior of the coating during thermal shock. In this study, plasma-sprayed TBCs were prepared with two kinds of agglomerated sintered YSZ powders with particle sizes of 5–45 μm and 15–70 μm. The evolution of mechanical properties and crack propagation behavior of the coatings in the whole life stage were studied by thermal shock experiments. The effect of powder particle size on the cracking failure behavior of the coating during thermal shock was analyzed from the point of view of pore structure, mechanical properties and stress state.

2. Materials and Methods

2.1. Coating Preparation

The TBCs samples were made of disc-shaped nickel-base alloy (IN-738) with thickness of 3 mm and diameter of 25.4 mm as substrate. Before spraying the coating, the surface of the substrate was treated by sandblasting and ultrasonic cleaning with acetone. The metal bonding layer coating and the ceramic top coat were deposited successively on the prepared substrate surface by atmospheric plasma-spraying (APS) process. The bond coat was prepared by NiCrAlY powder with a particle size of 5–70 μm. The ceramic coatings were sprayed with two kinds of 8YSZ (8wt.% Y_2O_3 stabilized ZrO_2) powders with different particle sizes, which are called fine and coarse powders, respectively. Two kinds of 8YSZ powders are obtained by screening the same batch of powders. The morphology and particle size distribution of the two kinds of YSZ powders are shown in Figure 1. It can be seen that the two kinds of powders show good sphericity. From the enlarged surface and section, it can be seen that the powders are formed by agglomeration and sintering of submicron particles. The particle size distribution range of fine powder measured by laser particle size analyzer is 5–45 μm, and D50 is about 25 μm. The measured distribution range of coarse particle size powder is 15–70 μm, and D50 is about 45 μm.

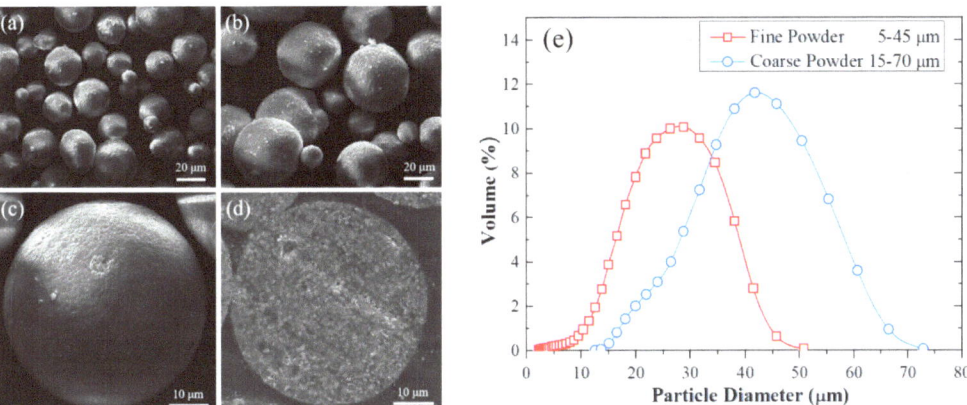

Figure 1. Morphology and particle size distribution of the two kinds of YSZ powders: (**a**) fine powder; (**b**) coarse powder; (**c**) magnification of powder surface morphology; (**d**) magnification of powder cross-section morphology and (**e**) particle size distribution.

The YSZ ceramic coat and NiCrAlY bond coat of all samples were prepared by a commercial atmospheric plasma-spraying equipment (APS-2000, Beijing Aeronautical Manufacturing Technology Research Institute, Beijing, China). Figure 2 shows the Schematic diagram of the YSZ coating deposited by plasma-spraying process. In the process of spraying, argon was used as the primary gas to form the plasma flame arc, and hydrogen was used as the auxiliary gas. The pressure of the main gas and auxiliary gas was controlled at 0.4 and 0.25 MPa respectively. The primary gas flow was controlled in 47 L/min, and the spraying voltage was controlled by adjusting the flow rate of auxiliary hydrogen, so as to adjust the arc power. According to the difference of melting point of YSZ ceramic and NiCrAlY metal powder, the power of plasma-spray gun was controlled to 36 and 30 kW, and the spraying distance was controlled to 70 and 100 mm respectively when spraying ceramic coat and bond coat. The moving speed of the spray gun was maintained at 150 mm/s during the coating spraying process. The two kinds of YSZ powders were prepared under the same powder feeding conditions, argon was used as carrier gas was controlled at 9 L/min, and the rotational speed of powder feeder was controlled at 1.5 r/min. The thickness of the deposited coating was controlled by spraying times. All TBC samples were prepared into a ceramic coat with a thickness of 300 μm and a bond coat with a thickness of 150 μm.

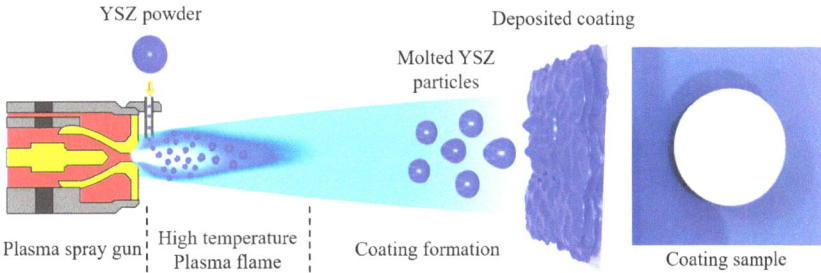

Figure 2. Schematic diagram of the YSZ coating deposited by plasma-spraying process and images of the prepared coating samples.

2.2. Evaluation of Coating Properties and Characterizations

The thermal shock experiment adopted the assessment temperature of 1050 °C. Each thermal shock included holding in a muffle furnace for 30 min and rapid cooling in water at room temperature. After the cooled sample was dried, the next thermal cycle was continued, and the thermal shock process was repeated until the coating fails. To study the crack evolution behavior of the TBCs, the thermal shocked TBC sample to different life stages was used to investigate the crack propagation behavior of the coating in the whole life stage.

The cross-sectional structure of coatings was observed using a scanning electron microscope (SEM, ZEISS EVO MA15, Carl Zeiss SMT Ltd., Cambridge, UK) in the backscattered and secondary electron modes with an acceleration voltage of 20 kV. The pore information of the coating was obtained by measuring the statistical cross-section image of the coating by the image analysis software Image-J 1.50i (National Institutes of Health, USA, https://imagej.nih.gov/ij (accessed on 17 January 2023)). In the process of statistical cracks, the image of the coating structure was continuously taken along the cross-section of the coating under an optical microscope (OM, ZEISS Observer A1 m, Carl Zeiss SMT Ltd., Cambridge, UK) under a 100× magnification, and then the image is spliced into an image containing the continuous cross-section of the coating. The crack information in the coating was obtained by count the length and number of cracks on the coating cross-section. In order to avoid the influence of the crack caused by the edge effect on the statistical results of the internal crack information of the coating, only the cracks in the 12.7 mm region (about half the diameter of the coating sample) at the center of the cross-section of the coating were counted. The crack density of the coating was obtained by dividing the number of cracks by the statistical length 12.7 mm. The cracking behavior of coatings in the thermal cycling was obtained by characterizing cracks in the samples after different cycles of thermal cycles.

The phase analyses of the YSZ feedstocks and coatings before and after thermal cyclic testing were examined by X-ray diffraction (XRD, D/max2550VB/PC, RIGAKU, Tokyo, Japan) using filtered Cu Kα radiation at an accelerated voltage of 40 kV and a current of 100 mA. Diffraction angles were set in the range of 10° to 80° with a step width of 0.02°. The hardness and fracture toughness of the coatings prepared by two kinds of particle size YSZ at different life stages were measured by a Vickers indenter. The indentation load of 300 g was used to measure the hardness, and the holding time was set to 15 s. The fracture toughness of the coating was measured by the indentation method to determine the crack propagation length of the coating under the action of indentation and calculate the crack extension force (G_C). To make the coating crack under the action of indentation, the indentation load used to measure the fracture toughness was 1000 g. According to the indentation morphology and the crack generated, the critical crack growth energy release rate G_C was calculated according to the following formula [21,30]:

$$G_C = 6.115 \times 10^{-4} \cdot a^2 \cdot P/c^3 \tag{1}$$

where a is the half-diagonal length of the indentation, c is the crack length from the center of the indentation, and P is the intention load.

3. Results and Discussion

3.1. Microstructure and Fracture Toughness of Coating

Figure 3 shows the microstructure and pore size distribution information of the as-sprayed coatings prepared by two kinds of YSZ powders. The coating prepared by coarse particle size powder has higher porosity than that prepared by fine particle size YSZ. In particular, there are a large number of large pores in the coating prepared by coarse particle size powder. The particle size of the powder affects the deposition, solidification, and crystallization of molten droplets during plasma-spraying process, thus affecting the microstructure of the coatings [31]. Fine YSZ powder typically results in good melting during deposition, which reduces the volume of inter-splats gaps and voids during coating formation. As a result, the coating prepared by fine YSZ powder has lower porosity and smaller pores than that prepared by coarse YSZ powder.

Figure 3. Typical microstructure and pore size distribution of the as-sprayed coatings prepared by the fine and coarse YSZ powder.

Figure 4 shows the change of porosity of the coating prepared by two kinds of particle size YSZ powder during thermal shock. With the increase of thermal shock cycles, the porosity of the two kinds of coatings decreased rapidly at first, and then tended to stabilize at a certain level. It is worth noting that the porosity of the coating prepared by coarse YSZ powder is always higher than that of the coating prepared by fine YSZ. Previous studies have reported that plasma-sprayed coatings will be significantly sintered within a few hours of thermal shock, resulting in a decrease in the porosity of the coatings [32,33]. The sintering rate of the coating is related to the size of the pores in the coating, in which the large pores are difficult to be healed by sintering [34–36]. From the microstructure of the coating, it can be seen that the pores in the coating prepared by fine YSZ powder are finer than those of coarse YSZ powder, so the coating is easier to be sintered. The coating prepared by coarse powder contains more pores of larger size, so the sintering degree of the coating is not as serious as that of fine powder, and the coating still has relatively high porosity after thermal shock.

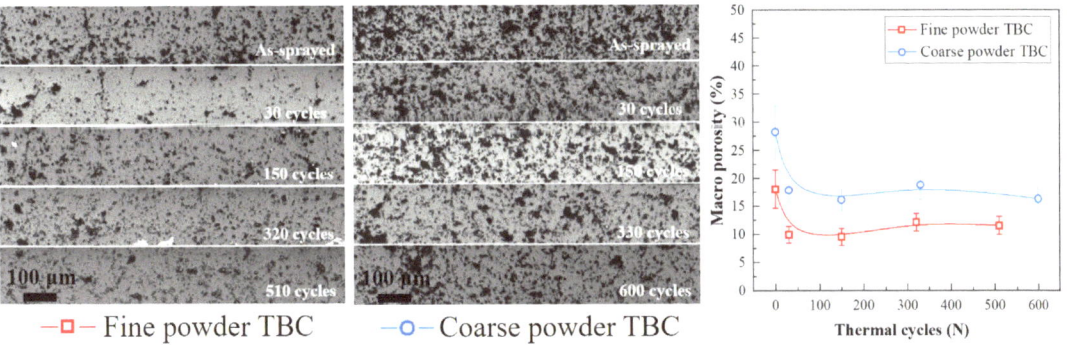

Figure 4. Evolution of porosity of two kinds of coatings during thermal cycling test.

Figure 5 shows the XRD patterns of two kinds of YSZ powders and corresponding coatings. From the diffraction angle of 27° to 33°, it can be seen that there is a small amount of monoclinic phase (m phase) in the original YSZ powders of both sizes. However, after the two kinds of powders are melted and solidified to form the coating during plasma-spraying, there is no monoclinic phase exist in the coating. The structure of zirconia in both coatings is tetragonal. Therefore, the particle size of YSZ powder does not affect the phase structure of the coating, and the influence of phase structure can be excluded in the subsequent analysis of the cracking behavior of the two kinds of coating.

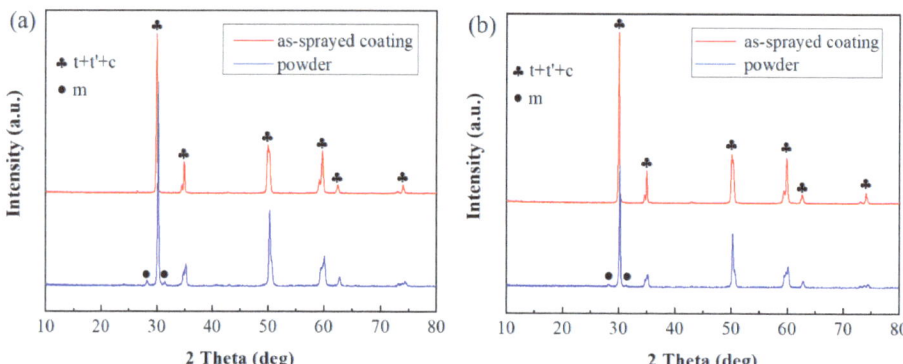

Figure 5. XRD patterns of two kinds of YSZ powder and corresponding coating: (**a**) fine powder; (**b**) coarse powder.

Fracture toughness is very important to resist cracking damage of coatings during thermal shock. Coatings with high fracture toughness are usually desired in coating design and preparation. Figure 6 shows the change of fracture toughness of the coatings during thermal shock. The fracture toughness of the coating prepared by two kinds of YSZ powder both increases with the increase of thermal shock time, and the fracture toughness of the coating prepared by fine particle size YSZ is always higher than that of the coating prepared by coarse YSZ powder during thermal shock. The porosity is an important factor affecting the fracture toughness of the coating [19]. The difference of fracture toughness between the two kinds of coatings is caused by the pore structure of the coatings. Finer YSZ powder resulted in better melting, compaction, and thus a denser structure. Coatings produced using coarse YSZ powder resulted in unmelted particle population, and a high level of porosity, including large interlamellar separations. The existence of pores reduces the cohesive bonding strength of the splats in the coating. The porosity of the coating prepared by fine YSZ powder is lower than that prepared by coarse YSZ during the whole thermal shock process, so the coating with a denser structure has higher fracture toughness.

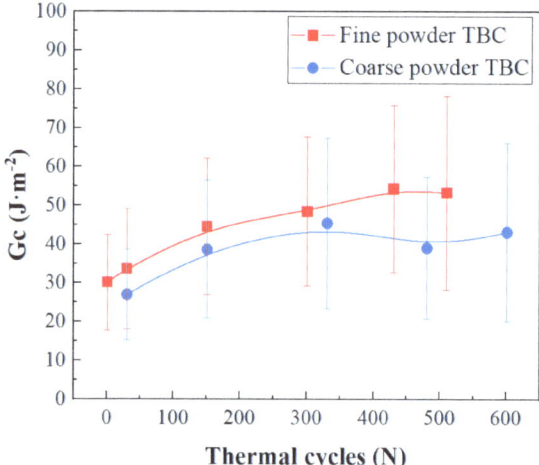

Figure 6. Evolution of fracture toughness of two kinds of coatings during thermal cycling test.

3.2. Thermal Shock Spalling Behavior of Coating

In the thermal shock test, with the increase of cycle times, the thermal stress produced during thermal shock leads to the formation and propagation of cracks in the coating, which is characterized by the spalling of the coating macroscopically. Figure 7 shows the changes of macroscopic morphology and spalling ratio of the coatings prepared by two kinds of particle size YSZ during thermal shock experiments. When the two kinds of coatings were subjected to about 400 thermal shocks, only a little spalling occurred at the edge of the samples. After that, the spalling area of the coating increases with the increase of thermal shock cycles. The coating prepared by fine YSZ exfoliated in a large area during 400–500 thermal shock, and the spalling rate increased rapidly, from about 10% to 70%, or even completely. In contrast, the spalling rate of the coating prepared by coarse YSZ still increased gently after 400 thermal shocks, and there was no sharp spalling of the coating. Peeling only occurred at the edge of the coating after 600 thermal shocks.

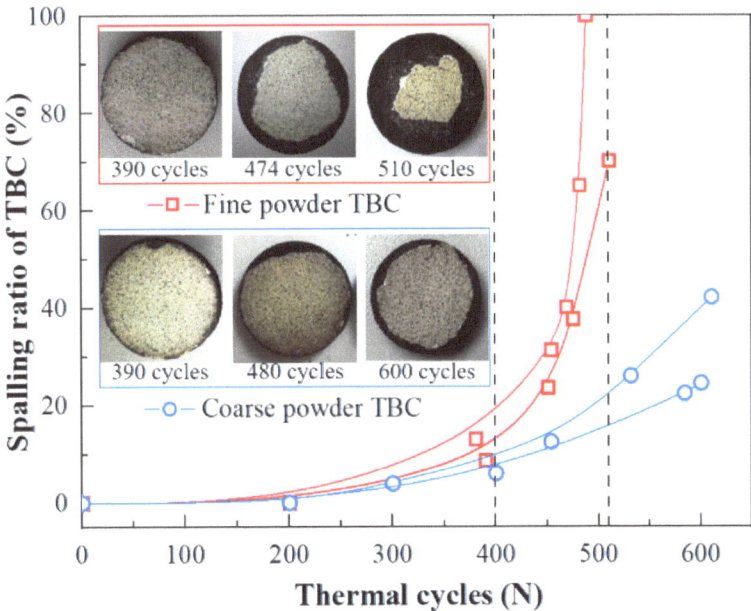

Figure 7. Evolution of macroscopic morphology and spalling ratio of the two kinds of TBCs during thermal cycling.

3.3. Cracking Behavior of Coating

Figure 8 shows the variation of average crack length and crack density with thermal shock cycles in the coatings prepared by two kinds of YSZ powders. In the early stage of thermal shock (less than 400 cycles), the crack length of the two coatings increases uniformly with the increase of thermal shock cycles. After 400 thermal shocks, the crack length in the coating prepared by fine YSZ powder increases rapidly, and the average crack length increases to more than 200 μm after 500 thermal shocks. From the cross-section of the coating, it can be found that there is a penetrating transverse crack in the ceramic layer above the bond coat in the coating. Corresponding to the rapid increase of crack length, the coating exfoliates rapidly in large area macroscopically. Similar results in previous research also shows that the crack in the plasma-sprayed TBCs will expand to about 200 μm in length at 80% of the whole life, and then the coating will peel off rapidly and lead to the failure [11]. The average crack length of the coating prepared by coarse YSZ powder was about 80 μm after 200 thermal shocks. During the subsequent thermal shock, the crack length in the coating did not increase obviously, and even after 600 thermal shocks, the average crack length in the coating was only about 100 μm. The penetrating long crack formed by crack merging was not observed in the cross-section of the coating. As a result, the coating did not peel off rapidly after 500 thermal shocks.

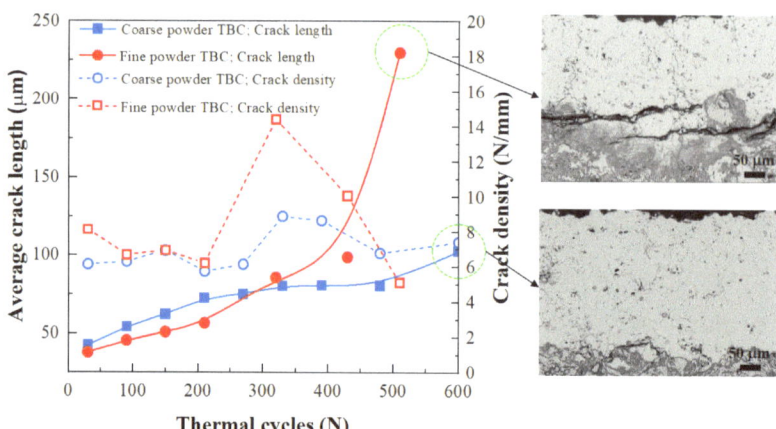

Figure 8. Evolution of crack length and density in the two kinds of coatings during thermal cycling test and typical crack morphology in coatings after thermal cycling.

The evolution of crack density is similar to that found by Jordan et al. [11]. The crack density in the coating prepared by two kinds of YSZ changed little in the first 200 thermal shocks. With the increase of cycles, the crack density in the coating shows a saddle-shaped trend, especially the coating prepared by fine YSZ is more obvious. After 200 thermal shocks, the crack density in the coating prepared by fine YSZ increases rapidly with the increase of cycle times, and reaches the maximum at 320 thermal shocks. Since then, as the thermal shock cycle continues to increase, the crack density in the coating decreases gradually. However, the crack density of the coating prepared by coarse YSZ did not change significantly during the whole thermal shock process.

The change trend of crack density is related to the crack evolution behavior in the coating, and the crack propagation in the coating can be inferred from the change of crack density and length. In practice, cracks exist and propagate in a three-dimensional form in the coating. The crack observation and statistics were obtained from the two-dimensional image of the cross-section, so it is necessary to accurately interpret the actual crack evolution state of the specimen corresponding to the crack information in the two-dimensional section.

Figure 9 shows a schematic diagram of the crack evolution behavior in the coating during thermal shock, so as to understand the relationship between the crack behavior in the coating and the statistical crack density. Figure 9a indicates the initiation of No. 1, 2, 3 and 4 cracks in the coating at the early stage of thermal shock. The crack is disc-shaped and propagates in the shape of emission in the coating. Crack No. 1 of the four cracks in this stage is located on the cut and observed section of the specimen and is observed in the crack statistics, while cracks No. 2, 3 and 4 cannot be observed because they are not on the cutting line. With the increase of thermal shock cycle, the crack in the coating propagates, and the crack state after propagation is shown in Figure 9b. After the No.1 crack propagates, the crack length observed from the cutting section increases, which reflects that the crack length increases in the crack statistics. After the No. 2 and No. 3 cracks near the cutting section propagate, the crack range extends to the observed cross-section, which reflects the statistical result that the crack density increases. It should be noted that some of these cracks may have just extended to the observed cross-section, so the observed crack length is shorter, while the actual crack length in the coating is larger. In addition, some cracks (such as No. 4) may not be observed in the cutting section after propagation, so such cracks will not be counted. As the thermal shock cycle continues to increase, the cracks in the coating will further expand and merge, such as cracks No. 2 and No. 3 in Figure 9c. After the crack merges, the number of cracks observed in the cross-section decreases and long cracks appear.

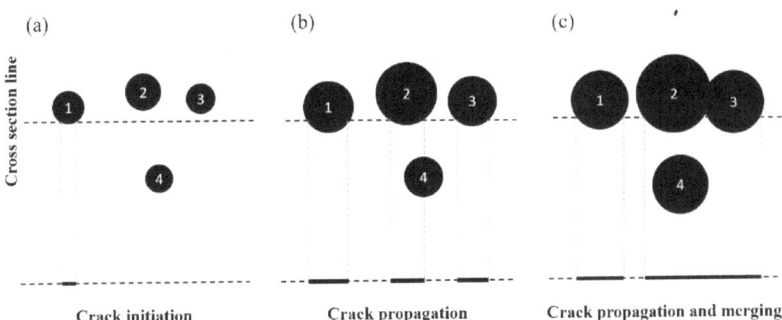

Figure 9. Schematic diagram of crack evolution behavior in the coating during thermal cycling: (**a**) crack initiation; (**b**) crack propagation and (**c**) crack coalescence.

Based on the above analysis, it can be known that the increase of crack density in the coating section during thermal shock reflects the rapid crack propagation in the coating or the initiation of new cracks near the cross-section. The decrease of crack density in the later stage of thermal shock reflects the coalescence of cracks in the coating. When a large number of cracks in the coating merge, the coating will peel off rapidly and fail.

In Figure 8, the crack density of the coating prepared by fine YSZ increases rapidly during 200–300 thermal shocks, indicating that the cracks in the coating propagate at a faster rate, and a large number of cracks are observed in the cross-section. In this process, the accelerated increase of crack length in the coating also reflects that the crack in the coating is in the stage of active propagation. The sharp decrease of crack density in the coating after 300 thermal shocks indicates that more and more cracks merge in the coating, and accordingly, the statistical crack length in the cross-section increases rapidly. From the statistical results of crack length distribution in the coating after thermal shock in Figure 10, it can be seen that not only the number of shorter cracks in the cross-section of the coating increased significantly after 320 thermal shocks, but also a large number of cracks with a length of more than 200 μm appeared in the coating. After 510 thermal shocks, the number of short cracks (<200 μm) in the coating decreased, but the number of long cracks increased significantly. At this time, there is a penetrating horizontal crack in the coating, so the coating peels off rapidly and fails. Therefore, 200 μm can be regarded as the critical crack length of thermal shock failure of the coating. When the length of a large number of cracks in the coating exceeds 200 μm, the cracks in the coating will merge, resulting in the rapid spalling of the coating in the form of instability.

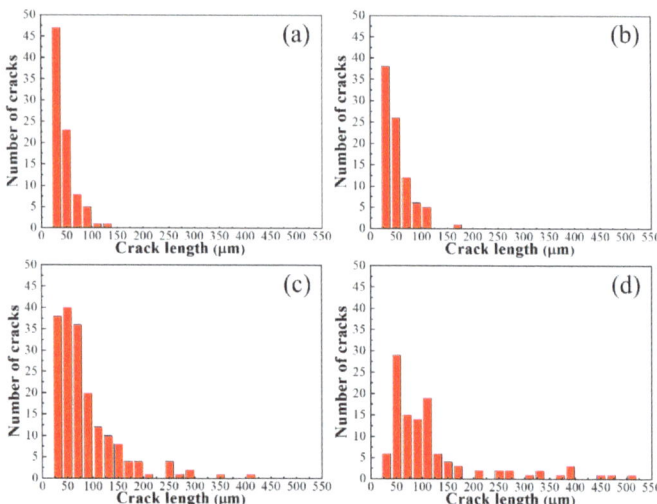

Figure 10. Distribution of crack length in the coating prepared by fine YSZ powder after different thermal cycles: (**a**) 90 cycles; (**b**) 150 cycles; (**c**) 320 cycles and (**d**) 510 cycles.

The crack density and average length of the coating prepared by coarse YSZ powder did not change significantly after 300 thermal shocks, indicating that the crack initiation and propagation behavior in the coating is slow. The statistics of crack length distribution in Figure 11 confirmed that the coating did not form many long cracks after 330 and 480 thermal shocks, and the crack distribution in the coating was similar to that in the initial stage of thermal shock. These results show that the cracks in the coating prepared by fine YSZ powder undergo a process of rapid propagation and coalescence in about 500 thermal shocks, which leads to the failure of the coating. However, the crack length and number of the coating prepared by coarse YSZ are stable during the whole thermal shock process, and there is no rapid crack propagation and coalescence after 480 thermal shocks, so the coating shows a longer thermal shock life.

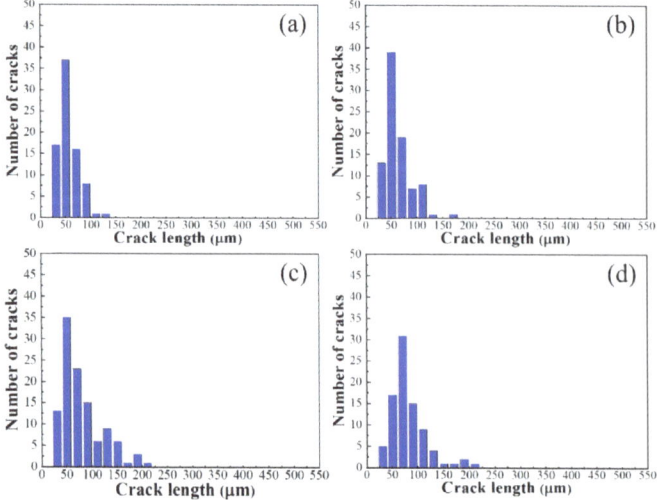

Figure 11. Distribution of crack length in the coating prepared by coarse YSZ powder after different thermal cycles: (**a**) 90 cycles; (**b**) 150 cycles; (**c**) 330 cycles and (**d**) 480 cycles.

3.4. Effect of Strain Tolerance on Thermal Stress of Coating

Fracture toughness is a key factor affecting the cracking behavior of coatings. Before thermal shock experiments, according to the results of fracture toughness of two kinds of coatings, it is expected that the crack growth rate of the coating prepared with fine YSZ will be slower than that of the coating prepared with coarse YSZ, resulting in a longer thermal shock life. However, the thermal shock test results are contrary to the expectation, which shows that in addition to the fracture toughness, there are other factors that affect the cracking behavior of the coating.

In this study, the difference between the coatings prepared by the two kinds of YSZ powder is mainly reflected in the pore structure. Although the dense structure of the coating prepared by fine particle size YSZ powder has better fracture toughness, the decrease of porosity may also reduce the strain tolerance of the coating. The strain tolerance of the coating is also an important factor affecting the cracking behavior of TBCs during thermal shock conditions [26,28].

The hardness of the coating can reflect its strain tolerance. The higher the hardness of the coating is, the worse the strain tolerance is. During the thermal shock process, the hardness evolution of the coatings prepared by two kinds of particle size YSZ is shown in Figure 12. The hardness of the two kinds of coatings increased after thermal shock, and the hardness of the coating prepared by fine YSZ powder was higher than that of the coating prepared by coarse one during the whole thermal shock process. The hardness of the coating is related to its porosity. The literature has shown that the hardness and elastic modulus of the coating decrease with the increase of porosity [29,37]. In this study, the porosity of the coating prepared by coarse YSZ is higher than that of the coating prepared by fine particle size YSZ during the whole thermal shock process, so the coating has higher strain tolerance.

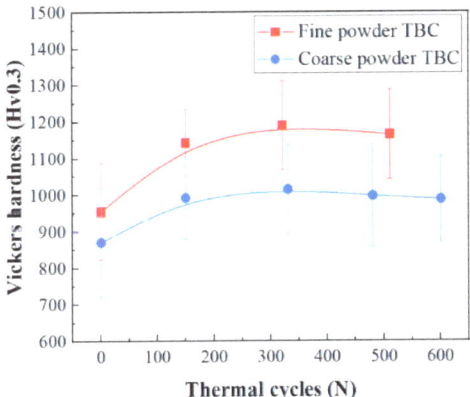

Figure 12. Hardness evolution of two kinds of YSZ coatings during thermal cycling.

In addition to the anti-cracking ability of the coating itself, the thermal stress formed during thermal shock is also the key to affect the cracking behavior of the TBCs. Finite element simulation method was used to analyze the effect of the strain tolerance of the ceramic layer on the thermal stress at the crack tip of the coating.

Figure 13 shows the establishment of the finite element geometric model and the boundary conditions. The model was established according to the actual coating samples, including a Ni-based alloy substrate with a thickness of 3 mm, a bond coat layer with an average thickness of 150 μm and a top layer of YSZ ceramics with a thickness of 300 μm. A wavy interface similar to the actual shape was introduced between the bonding layer and the ceramic layer. From the cross-sectional morphology of the coating after thermal shock, it can be seen that the cracking leading to the spalling of the coating occurs in the ceramic layer within 100 μm above the interface between the bonding layer and the ceramic layer. According to the morphology of the crack in the coating (Figure 13a), a semi-elliptical crack was set up in the YSZ layer near the bond coat on the right side of the finite element model, and the thermal stress at the crack tip during thermal shock was investigated. The symmetrical boundary condition was adopted on the left side of the finite element model, the bottom boundary restricts the movement of the Y axis, and the upper boundary is set as a free boundary. Considering that the material below the right boundary crack is continuous, the multi-point constraint boundary condition (MPC) was adopted to make the boundary have the same displacement in the X-axis direction. Since there is a penetrating vertical crack in the upper part of the right boundary crack, it was set as a free boundary. The model includes nickel-based alloy matrix, NiCrAlY bonding layer and YSZ ceramic layer. The material property parameters of each layer used in the finite element simulation were taken from the literature [38,39], as shown in Table 1. To analyze the influence of the strain tolerance of the ceramic layer on the cracking of the coating, the elastic modulus of the YSZ layer is taken as a variable in the finite element simulation. The elastic modulus of 20, 40 and 60 GPa were selected to calculate the thermal stress of crack tip in the coating during thermal shock.

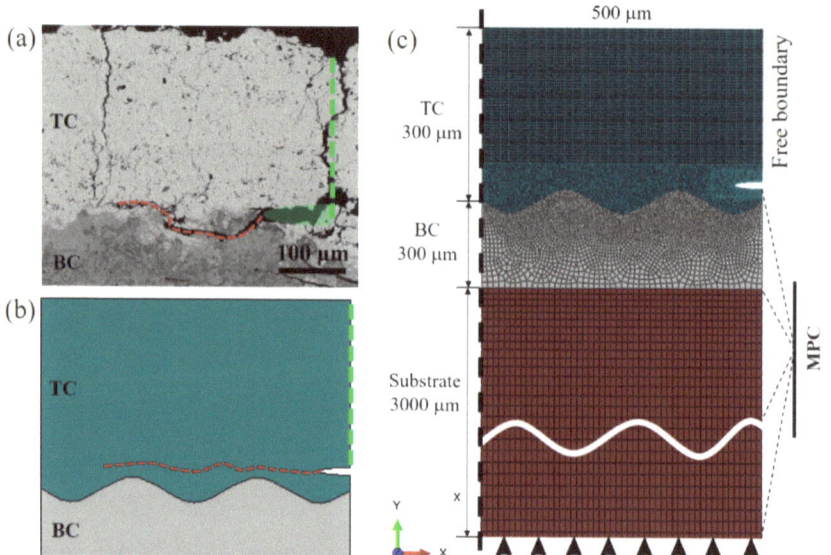

Figure 13. The geometric model, boundary conditions and meshes of the finite element model. (**a**) SEM images of cracking characteristics in the coating; (**b**) simplified model extracted from coating structure and (**c**) mesh structure and boundary conditions of the finite element model.

Table 1. Temperature dependent physical properties of substrate, bond coat and YSZ topcoat.

Materials	Temperature (°C)	Elastic Modulus (GPa)			Thermal Expansion Coefficient (10^{-6} °C^{-1})	Poisson's Ratio	Yield Strength (GPa)
Substrate	20	123			12	0.36	888
	420	108			12.5	0.37	887
	820	92			14.7	0.38	910
	1020	84			15.8	0.38	470
BC	20	152			12.3	0.32	426
	420	145			15.2	0.33	396
	820	109			16.3	0.35	284
	1020	72			17.6	0.35	150
YSZ	20	20	40	60	9.7	0.2	-
	420	20	40	60	9.7	0.2	-
	820	20	40	60	10	0.2	-
	1020	20	40	60	10.4	0.2	-

In the process of finite element simulation, a temperature load of simulating thermal shock was applied to the model to calculate the thermal stress caused by the thermal expansion mismatch of the coating. The temperature loading process includes heating the sample from room temperature (20 °C) to 1050 °C for 120 s, then holding 30 min at high temperature, and cooling to room temperature after 120 s. In the process of thermal shock, the cracking parallel to the interface direction of the bond coat/ceramic layer is the main reason for the spalling of the coating, so the stress S22 at the crack tip perpendicular to the cracking direction was paid more attention. Figure 14 shows the cloud diagram of the stress distribution of S22 in the coating at the end of the heating stage during the thermal shock process, which shows that there is an obvious stress concentration at the crack tip. The greater the stress there is, the greater the driving force of crack propagation is, the more serious the cracking in the coating will be.

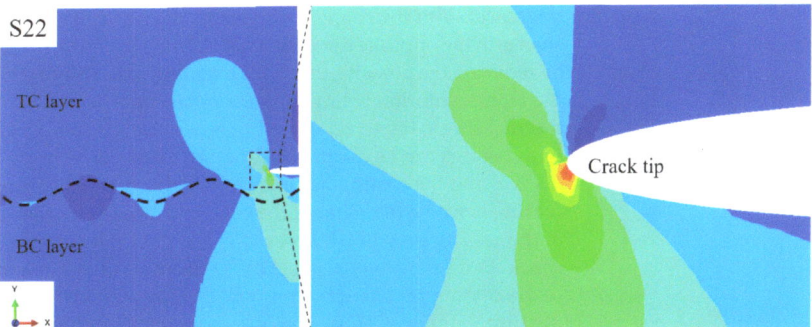

Figure 14. Stress (S22) distribution in coating at the end of heating stage during thermal cycling.

The evolution of temperature and crack tip stress S22 of TBCs during the whole thermal shock process is shown in Figure 15. In the thermal shock heating stage, the stress at the crack tip increases with the increase of the sample temperature, and the maximum stress appears at the end of the heating stage. The stress in this process is a tensile stress perpendicular to the direction of crack propagation. The magnitude of the stress is an important factor driving the crack propagation. The elastic modulus of the ceramic layer has a significant effect on the stress level at the crack tip. When the elastic modulus increases from 20 GPa to 40 GPa and 60 GPa, the maximum stress S22 at the crack tip increases from 800 MPa to 1200 MPa and 1480 MPa, respectively.

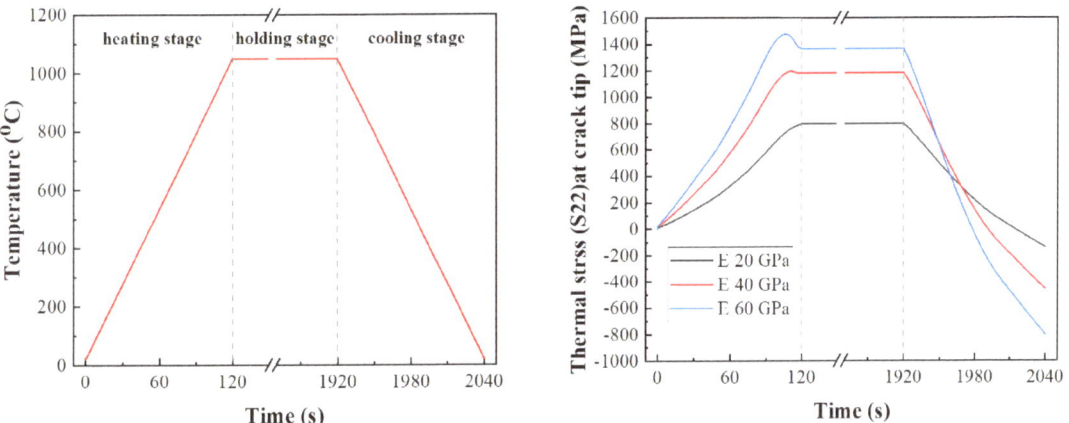

Figure 15. Evolution of stress S22 at the crack tip in the coating during thermal cycling.

In this simulation, the stress state of the coating during thermal shock was studied by elastic-plastic analysis, and the effect of elastic modulus of ceramic layer on the crack tip stress of the coating was investigated. The results show that reducing the elastic modulus of the ceramic coating plays a very significant role in alleviating the crack driving force in the coating. From the difference of porosity and hardness, it can be seen that the coating prepared by coarse particle size YSZ has lower elastic modulus than the coating prepared by fine particle size YSZ, so the coating bears less thermal stress at the crack tip during thermal shock.

3.5. Thermal Shock Resistant Analysis of Coating

The cracking behavior of TBCs during thermal shock depends on two factors; the cracking resistance of the coating itself and the cracking driving force formed during thermal shock. The porosity of the coating will affect its fracture toughness and strain tolerance. With the change of porosity, the fracture toughness and strain tolerance often change in the opposite direction, that is, when the fracture toughness of the coating increases by reducing the porosity, the strain tolerance of the coating will become worse. Considering the crack evolution behavior, porosity and mechanical properties of the coating during thermal shock, the effect of YSZ powder particle size on the thermal shock failure of the two coatings can be clarified. Although the porosity of the coating prepared by fine YSZ is small and the coating has relatively high fracture toughness, the strain tolerance of this coating is worse than that of the coating prepared by coarse YSZ. Therefore, the crack tip of the coating prepared by coarse YSZ is driven by greater cracking force during thermal shock. Since the adverse effect of the decrease of strain tolerance caused by low porosity of the coating prepared by fine YSZ is more significant than that on the improvement of fracture toughness of the coating, so the crack in the coating propagates at a faster rate during thermal shock, and the coating shows a shorter thermal shock life.

In addition, the pore size of the coating prepared by fine YSZ is smaller. These smaller pores are easier to be sintered at high temperature, resulting in a further reduction of the strain tolerance of the coating. Therefore, with the increase of thermal shock cycle, the cracking driving force of the crack tip in the coating becomes larger and larger. However, the pore size of the coating prepared by coarse YSZ is larger, and there are still many pores that cannot be sintered after thermal shock, which restrain the strain tolerance degradation of the coating to a certain extent. The study of Rad et al. [40] shows that the pores in the ceramic layer will significantly affect the stress distribution in the coating. The dispersed pores can alleviate stress concentration in the coating. In this experiment, the coating prepared by coarse YSZ powder was subjected to less thermal stress during thermal shock thanks to its higher porosity and larger pore size. Although the porosity reduces the fracture toughness of the coating to some extent, the increase of strain tolerance reduces the crack growth rate in the coating, so the coating prepared by coarse YSZ powder shows a longer thermal shock life.

4. Conclusions

The YSZ coatings with different microstructures were prepared by using two kinds of agglomerated sintered powders with particle sizes of 5–45 µm and 15–70 µm. The cracking failure behavior of the coatings during thermal shock was studied. The main conclusions are as follows:

(1)   The effect of particle size of YSZ powder on the microstructure of the coating is mainly reflected in the porosity and pore size. The porosity of the coating prepared by fine powder is low, the pore size is small, and the small pores in the coating are easy to be sintered during thermal shock, which further reduces the porosity of the coating. The porosity and pore size of the coating prepared by coarse powder are relatively larger, the larger pores are difficult to be sintered after thermal shock, and the porosity of the coating is relatively high.

(2) The particle size of YSZ powder does not affect the phase structure of the atmospheric plasma-sprayed 8YSZ coating. The YSZ coatings prepared by coarse and fine particles are tetragonal zirconia.

(3) The particle size of the powder leads to the difference of the pore structure of the coating, which in turn affects the fracture toughness and strain tolerance of the coating. The change of porosity influences the fracture toughness and strain tolerance of the coating in opposite direction, that is, when the fracture toughness of the coating is increased by reducing the porosity, the strain tolerance of the coating will become worse. The coating prepared by fine powder has high fracture toughness because of its compact structure, but the strain tolerance of the coating is poor. Due to the existence of many pores, the fracture toughness of the coating prepared by coarse powder is relatively lower, but the strain tolerance of the coating is better.

(4) The fracture toughness and strain tolerance of the ceramic layer play an important role in the thermal shock life of the coating. Although the coating prepared by fine powder has relatively high fracture toughness, its lower strain tolerance will increase the cracking driving force on the crack tip of the coating during thermal shock. In the process of thermal shock, the cracks of the coating propagate and merge at a faster speed, and the thermal shock life is short. Although the larger porosity and pore size in the coating prepared by coarse particle size powder reduce the fracture toughness of the coating to a certain extent, the increase of sintering resistance and strain tolerance reduces the crack growth rate in the coating, and the coating has a longer life.

(5) The results of this study provide guidance for the development and design of high performance TBCs. When adjusting the pore structure of the coating, attention should be paid to the improvement of anti-cracking ability and strain tolerance at the same time.

Author Contributions: Formal analysis, J.H.; investigation, J.H.; resources, J.H.; data curation, W.S., R.H. and W.M.; writing—original draft preparation, J.H.; writing—review and editing, W.S., R.H. and W.M. All authors have read and agreed to the published version of the manuscript.

Funding: This work was supported by "the Fundamental Research Funds for the Central Universities, South China University of Technology".

Institutional Review Board Statement: Not applicable.

Informed Consent Statement: Not applicable.

Data Availability Statement: All data that support the findings of this study are included within the article.

Conflicts of Interest: The authors declare no conflict of interest.

References

1. Babu, P.S.; Madhavi, Y.; Krishna, L.R.; Sivakumar, G.; Rao, D.S.; Padmanabham, G. Thermal Spray Coatings for Erosion–Corrosion Resistant Applications. *Trans. Indian Inst. Met.* **2020**, *73*, 2141–2159. [CrossRef]
2. Cinca, N.; Lima, C.R.C.; Guilemany, J.M. An overview of intermetallics research and application: Status of thermal spray coatings. *J. Mater. Res. Technol.* **2013**, *2*, 75–86. [CrossRef]
3. Hardwicke, C.U.; Lau, Y.-C. Advances in thermal spray coatings for gas turbines and energy generation: A review. *J. Therm. Spray Technol.* **2013**, *22*, 564–576. [CrossRef]
4. Paleu, C.C.; Munteanu, C.; Istrate, B.; Bhaumik, S.; Vizureanu, P.; Bălțatu, M.S.; Paleu, V. Microstructural Analysis and Tribological Behavior of AMDRY 1371 (Mo–NiCrFeBSiC) Atmospheric Plasma Spray Deposited Thin Coatings. *Coatings* **2020**, *10*, 1186. [CrossRef]
5. Thakare, J.G.; Pandey, C.; Mahapatra, M.; Mulik, R.S. Thermal barrier coatings—A state of the art review. *Met. Mater. Int.* **2021**, *27*, 1947–1968. [CrossRef]
6. Wee, S.; Do, J.; Kim, K.; Lee, C.; Seok, C.; Choi, B.-G.; Choi, Y.; Kim, W. Review on mechanical thermal properties of superalloys and thermal barrier coating used in gas turbines. *Appl. Sci.* **2020**, *10*, 5476. [CrossRef]
7. Darolia, R. Thermal barrier coatings technology: Critical review, progress update, remaining challenges and prospects. *Int. Mater. Rev.* **2013**, *58*, 315–348. [CrossRef]

8. Mehta, A.; Vasudev, H.; Singh, S.; Prakash, C.; Saxena, K.K.; Linul, E.; Buddhi, D.; Xu, J. Processing and Advancements in the development of thermal barrier coatings: A Review. *Coatings* **2022**, *12*, 1318. [CrossRef]
9. Kumar, V.; Balasubramanian, K. Progress update on failure mechanisms of advanced thermal barrier coatings: A review. *Prog. Org. Coat.* **2016**, *90*, 54–82. [CrossRef]
10. Song, D.; Song, T.; Paik, U.; Lyu, G.; Jung, Y.-G.; Choi, B.-G.; Kim, I.-S.; Zhang, J. Crack-growth behavior in thermal barrier coatings with cyclic thermal exposure. *Coatings* **2019**, *9*, 365. [CrossRef]
11. Ahmadian, S.; Jordan, E.H. Explanation of the effect of rapid cycling on oxidation, rumpling, microcracking and lifetime of air plasma sprayed thermal barrier coatings. *Surf. Coat. Technol.* **2014**, *244*, 109–116. [CrossRef]
12. Dong, H.; Yang, G.-J.; Cai, H.-N.; Li, C.-X.; Li, C.-J. Propagation feature of cracks in plasma-sprayed YSZ coatings under gradient thermal cycling. *Ceram. Int.* **2015**, *41*, 3481–3489. [CrossRef]
13. Ahmadian, S.; Thistle, C.; Jordan, E.H.; Hsueh, C.H. Experimental and Finite Element Study of an Air Plasma Sprayed Thermal Barrier Coating under Fixed Cycle Duration at Various Temperatures. *J. Am. Ceram. Soc.* **2013**, *96*, 3210–3217. [CrossRef]
14. Huang, J.; Wang, W.; Yu, J.; Wu, L.; Feng, Z. Effect of Particle Size on the Micro-cracking of Plasma-Sprayed YSZ Coatings During Thermal Cycle Testing. *J. Therm. Spray Technol.* **2017**, *26*, 755–763. [CrossRef]
15. Viswanathan, V.; Dwivedi, G.; Sampath, S. Engineered Multilayer Thermal Barrier Coatings for Enhanced Durability and Functional Performance. *J. Am. Ceram. Soc.* **2014**, *97*, 2770–2778. [CrossRef]
16. Viswanathan, V.; Dwivedi, G.; Sampath, S. Multilayer, multimaterial thermal barrier coating systems: Design, synthesis, and performance assessment. *J. Am. Ceram. Soc.* **2015**, *98*, 1769–1777. [CrossRef]
17. Li, C.J.; Li, Y.; Yang, G.J.; Li, C.X. Evolution of Lamellar Interface Cracks During Isothermal Cyclic Test of Plasma-Sprayed 8YSZ Coating with a Columnar-Structured YSZ Interlayer. *J. Therm. Spray Technol.* **2013**, *22*, 1374–1382. [CrossRef]
18. Huang, J.; Wang, W.; Lu, X.; Liu, S.; Li, C. Influence of lamellar interface morphology on cracking resistance of plasma-sprayed YSZ coatings. *Coatings* **2018**, *8*, 187. [CrossRef]
19. Dwivedi, G.; Viswanathan, V.; Sampath, S.; Shyam, A.; Lara-Curzio, E. Fracture Toughness of Plasma-Sprayed Thermal Barrier Ceramics: Influence of Processing, Microstructure, and Thermal Aging. *J. Am. Ceram. Soc.* **2014**, *97*, 2736–2744. [CrossRef]
20. Mauer, G.; Du, L.; Vaßen, R. Atmospheric Plasma Spraying of Single Phase Lanthanum Zirconate Thermal Barrier Coatings with Optimized Porosity. *Coatings* **2016**, *6*, 49. [CrossRef]
21. Huang, J.B.; Wang, W.Z.; Li, Y.J.; Fang, H.J.; Ye, D.D.; Zhang, X.C.; Tu, S.T. A novel strategy to control the microstructure of plasma-sprayed YSZ thermal barrier coatings. *Surf. Coat. Technol.* **2020**, *402*, 126304. [CrossRef]
22. Huang, J.B.; Wang, W.Z.; Li, Y.J.; Fang, H.J.; Ye, D.D.; Zhang, X.C.; Tu, S.T. Novel-structured plasma-sprayed thermal barrier coatings with low thermal conductivity, high sintering resistance and high durability—ScienceDirect. *Ceram. Int.* **2020**, *47*, 5156–5167. [CrossRef]
23. Bakan, E.; Vaßen, R. Ceramic Top Coats of Plasma-Sprayed Thermal Barrier Coatings: Materials, Processes, and Properties. *J. Therm. Spray Technol.* **2017**, *26*, 992–1010. [CrossRef]
24. Odhiambo, J.G.; Li, W.; Zhao, Y.; Li, C. Porosity and its significance in plasma-sprayed coatings. *Coatings* **2019**, *9*, 460. [CrossRef]
25. Medřický, J.; Curry, N.; Pala, Z.; Vilemova, M.; Chraska, T.; Johansson, J.; Markocsan, N. Optimization of High Porosity Thermal Barrier Coatings Generated with a Porosity Former. *J. Therm. Spray Technol.* **2015**, *24*, 622–628. [CrossRef]
26. Huang, J.; Wang, W.; Li, Y.; Fang, H.; Ye, D.; Zhang, X.; Tu, S. Improve durability of plasma-splayed thermal barrier coatings by decreasing sintering-induced stiffening in ceramic coatings. *J. Eur. Ceram. Soc.* **2019**, *40*, 1433–1442. [CrossRef]
27. Cheng, B.; Yang, N.; Zhang, Q.; Zhang, Y.-M.; Chen, L.; Yang, G.-J.; Li, C.-X.; Li, C.-J. Sintering induced the failure behavior of dense vertically crack and lamellar structured TBCs with equivalent thermal insulation performance. *Ceram. Int.* **2017**, *43*, 15459–15465. [CrossRef]
28. Zou, Z.; Xing, C.; He, L.; Shan, X.; Luo, L.; Zhao, X.; Guo, F.; Xiao, P. A Highly Strain and Damage Tolerant Thermal Barrier Coating Fabricated by Electro: Prayed Zirconia Hollow Spheres. *J. Am. Ceram. Soc.* **2018**, *101*, 4375–4386. [CrossRef]
29. Bakan, E.; Mack, D.E.; Mauer, G.; Mücke, R.; Vaßen, R. Porosity–Property Relationships of Plasma-Sprayed Gd2Zr2O7/YSZ Thermal Barrier Coatings. *J. Am. Ceram. Soc.* **2015**, *98*, 2647–2654. [CrossRef]
30. Lin, X.; Zeng, Y.; Lee, S.W.; Ding, C. Characterization of alumina–3 wt.% titania coating prepared by plasma spraying of nanostructured powders. *J. Am. Ceram. Soc.* **2004**, *24*, 627–634. [CrossRef]
31. Huang, J.; Wang, W.; Lu, X.; Hu, D.; Feng, Z.; Guo, T. Effect of Particle Size on the Thermal Shock Resistance of Plasma-Sprayed YSZ Coatings. *Coatings* **2017**, *7*, 150. [CrossRef]
32. Li, G.R.; Li, C.X.; Li, C.J.; Yang, G.J. Sintering characteristics of plasma-sprayed TBCs: Experimental analysis and an overall modelling. *Ceram. Int.* **2018**, *44*, 2982–2990. [CrossRef]
33. Yan, J.; Wang, X.; Chen, K.; Lee, K.N. Sintering modeling of thermal barrier coatings at elevated temperatures: A review of recent advances. *Coatings* **2021**, *11*, 1214. [CrossRef]
34. Markocsan, N.; Nylen, P.; Wigren, J.; Li, X.; Tricoire, A. Effect of thermal aging on microstructure and functional properties of zirconia-base thermal barrier coatings. *J. Therm. Spray Technol.* **2009**, *18*, 201–208. [CrossRef]
35. Erdogan, G.; Ustel, F.; Bobzin, K.; Öte, M.; Linke, T.F.; Zhao, L. Influence of long time post annealing on thermal stability and thermophysical properties of plasma sprayed $La_2Zr_2O_7$ coatings. *J. Alloys Compd.* **2017**, *695*, 2549–2555. [CrossRef]
36. Liu, T.; Chen, X.; Yang, G.-J.; Li, C.-J. Properties evolution of plasma-sprayed La 2 Zr 2 O 7 coating induced by pore structure evolution during thermal exposure. *Ceram. Int.* **2016**, *42*, 15485–15492. [CrossRef]

37. Nakamura, T.; Qian, G.; Berndt, C.C. Effects of Pores on Mechanical Properties of Plasma-Sprayed Ceramic Coatings. *J. Am. Ceram. Soc.* **2000**, *83*, 578–584. [CrossRef]
38. Białas, M. Finite element analysis of stress distribution in thermal barrier coatings. *Surf. Coat. Technol.* **2008**, *202*, 6002–6010. [CrossRef]
39. Jiang, J.; Xu, B.; Wang, W.; Adjei, R.A.; Zhao, X.; Liu, Y. Finite element analysis of the effects of thermally grown oxide thickness and interface asperity on the cracking behavior between the thermally grown oxide and the bond coat. *J. Eng. Gas Turbines Power* **2017**, *139*, 022504. [CrossRef]
40. Rad, M.R.; Farrahi, G.; Azadi, M.; Ghodrati, M. Stress analysis of thermal barrier coating system subjected to out-of-phase thermo-mechanical loadings considering roughness and porosity effect. *Surf. Coat. Technol.* **2015**, *262*, 77–86.

coatings

MDPI

Editorial

Advanced Cold-Spraying Technology

Wen Sun [1,*], Adrian Wei-Yee Tan [2,*], Xin Chu [1,*] and Jian Huang [1,*]

1   National Engineering Laboratory of Modern Materials Surface Engineering Technology,
    Guangdong Provin-cial Key Laboratory of Modern Surface Engineering Technology, Institute of New
    Materials, Guangdong Academy of Sciences, Guangzhou 510650, China
2   School of Engineering, University of Southampton Malaysia, Iskandar Puteri 79100, Malaysia
*   Correspondence: sunw0013@e.ntu.edu.sg (W.S.); adrian.tan@soton.ac.uk (A.W.-Y.T.);
    xin.chu@hotmail.com (X.C.); huangjian@gdinm.com (J.H.)

Citation: Sun, W.; Tan, A.W.-Y.; Chu,
X.; Huang, J. Advanced
Cold-Spraying Technology. *Coatings*
2022, *12*, 1986. https://doi.org/
10.3390/coatings12121986

Received: 5 December 2022
Accepted: 13 December 2022
Published: 18 December 2022

Publisher's Note: MDPI stays neutral
with regard to jurisdictional claims in
published maps and institutional affiliations.

Cold-spraying technology is a solid-state, powder-based coating deposition and addi-tive manufacturing (AM) technology, which utilises a high-pressure gas stream to accelerate micron-sized particles through a de-Laval nozzle for supersonic speed and impact on sub-strates and to generate dense, high-quality deposits. Cold-spraying technology has many unique features (e.g., high deposition rate, high adhesion strength; solid-state bonding at low temperatures) contrary to other fusion-based techniques, thus opening up new opportunities in various industrial markets. After over thirty years of rapid development, cold-spraying technology has been successfully applied in surface repair, surface enhance-ment, functional coating, and AM in many fields, including the aerospace, weapon, energy and power, electronics, and medical equipment industries.

With the developments of fundamental understanding and practical applications in cold-spraying technology, this technology has attracted increasing attention around the world, pushing this technology to a more 'advanced' level. The supersonic projectile be-haviours of micron-sized particles in cold-spraying allow investigations into fundamental material phenomena under extreme conditions, e.g., "size effects" in mechanics, dynamic recrystallisation, and phase transformation. In addition, the advanced hybrid cold-spraying process incorporates in-situ or post-mortem treatment techniques, which could resolve many inherent drawbacks of this technology (e.g., lack of ductility and insufficient cohe-sion) and further promote the deployment of cold-spraying technology. Moreover, the integration of artificial intelligence and deep learning technology in cold-spraying helps to realize better toolpath and process optimization, as well as process control for industrial production campaigns.

The advancement of cold-spraying technology still faces many challenges that need to be addressed, such as those related to (1) feedstock materials, (2) bonding mechanisms, (3) numerical and modelling issues, (4) composite coatings, (5) AM, (6) process control, (7) equipment, and (8) carbon emissions.

The stringent selection and availability of the feedstock material properties directly impact the advancement of cold-sprayed coatings. Metal and alloyed powders are primarily utilised as feedstock for the cold-spraying process because plastic deformation is required to form the coating during the deposition process. Some unique cermet powder can also be deposited using the cold-spraying process [1]. On the other hand, ceramic powders have poor ductility and can be rarely deposited by cold-spraying. The metallic powders used in the cold-spraying process can be manufactured by atomisation (gas or water) or through a crushing or electrolytic process. Atomised powders are more spherical and regular, while crushed powders are blocky with sharp edges and are irregular. Electrolytic powders usually have dendritic structures. The difference in shape and regularity affects the particle speed and influences the coating quality due to different deformation characteristics; however, this would differ from material to material.

Several bonding mechanisms are being proposed for cold-spraying due to the complex particle–particle/carrier gas interactions, which can influence the research direction of the community. Studies have shown that successful bonding requires the sprayed particles to have a specific amount of energy, a critical velocity, and an adequate temperature [2]. This will lead to high-strain-rate deformation and breakage of oxide films, coupled with microscopic protrusions, grain refinement and localised heating at the particle impact interfaces (particle–particle/substrate interfaces), resulting in possible metallurgical bonding [3–5]. According to Assadi et al. [2], when the particle impact velocity surpasses a critical velocity, adiabatic shear instabilities cause bonding at the particle impact surfaces. The adiabatic shear instabilities are shown to be accompanied by interfacial jetting of plastically deformed material that contributes to the removal of oxide films and production of clean contact surfaces that promote bonding [6]. However, Hassani et al. [7,8] demonstrated that adiabatic softening and adiabatic shear instability are not prerequisites for hydrodynamic jetting, but an interaction of strong pressure waves with the free surface at the particle edges that affects the bonding.

Modelling has been widely researched to understand the impact process of the particles. The approximations of the deformation mechanism, critical velocity, and stresses have been evaluated by numerical methods such as the Eulerian, Lagrangian, smoothed particle hydrodynamics, molecular dynamics and coupled Eulerian–Lagrangian methods [9–11]. These models allow investigations into the distribution of pressure, stress, strain, and temperature during particle impact [6,12–15]. However, some material phenomena are not considered in these models, for example, the role of cracks, stress relaxation, interactions between splats (impact onto an uneven surface or previously deposited layer), phase transformations, and microstructural changes.

Cold-spraying is a strategic method used to deposit composite coatings with comparable or improved properties compared to other methods. There is considerable freedom to combine powders of various properties to create a composite coating with phase retention and specific functional requirements [16]. Several composite structures that have been deposited include metal–metal, metal–ceramic, and metal–intermetallic composite coatings [17,18]. Some critical investigations into the tailoring of cold-sprayed composites focus on the following aspects: (1) the size, concentration, and distribution optimisation of multi-materials, (2) effect of process parameters on composite coating microstructure and properties and (3) control of the degree of phase transformation that may occur at the bonding interfaces between the multi-materials.

Cold-spraying is integrated with AM to build 3D components that have the opportunity to revolutionise the manufacturing industry, such as mass production, fabrication, and restoration of engineering components [19]. The key advantages of cold-spraying additive manufacturing (CSAM) compared to other AM techniques are rapid building times, higher process flexibility, unrestricted product size and suitability for repairing and restoring damaged components. Additionally, CSAM is helpful in producing highly reflective metals, such as Cu and Al, which are very challenging to produce using laser-based AM methods. However, there are several drawbacks for CSAM in its industrial applications that needs attention, such as the limited resolution of the printed parts due to the large size of the nozzle outlet diameter, semi-finished components with a rough surface that requires post-machining and some inherent defects caused by difficulties in regulating process parameters, which result in poorer properties in their finished condition.

Understanding the effect of cold-spraying process controls is required to determine the optimal input parameters for a range of coating/substrate material combinations. The coating properties are affected by feedstock properties, particle velocity, gas temperature, stand-off distance, gas pressure, process gas, spray angle, traverse speed, substrate conditions (temperature, hardness, roughness and material) and many other factors [4,20]. Each parameter has a critical linkage with each other, and every change provides a certain level of trade-offs. Deep insights into each parameter are needed to fully advance cold-spraying technology.

The selection and capability of cold-spraying equipment affect, to a certain extent, the potential to produce the required coating quality. The cold-spraying equipment can be categorised as low-pressure cold-spraying (LPCS, 5–10 bar) and high-pressure cold-spraying (HPCS, >10 bar) [21]. The LPCS is suitable for depositing low-melting-point and low-yield-strength materials, for example, aluminium alloys, babbitt alloys, magnesium alloys [22–25], etc. As for the HPCS system, it has higher potential to deposit materials that are difficult to deform, such as steel, titanium [26], Inconel, etc. [27–32]. Several companies have successfully developed commercial cold-spraying systems to meet the increasing demand for this technology, such as VRC from the USA, Impact Innovations from Germany, Plasma Giken from Japan, Centerline from Canada and Dycomet from Russia.

Advanced cold-spraying technology should also become a greener process, which could involve gas and feedstock recycling in the future. This is because the requirements for carbon emission limits in different industry sectors are becoming increasingly stringent. The cold-spraying process does not produce toxic or harmful fumes or by-products because of chemically inert gas propellants such as nitrogen or helium. However, the resources used to prepare/extract these chemically inert gases do pose some impacts on the environment. Powders used for the cold-spraying process are also complicated to produce, and this is still coupled with inevitable powder wastage during the spraying process, including rebounded powders and during overspray (spraying sequence not on the target). In addition, cold-spraying repair and remanufacturing applications should be further developed to reduce material waste. Cold-spraying can restore damaged metal parts and components to their original state without decommissioning the part, component, or structure. This technology can significantly reduce the cost and environmental impact of producing additional components to replace the decommissioned ones [33].

In conclusion, the advancement of cold-spraying technology will benefit numerous industries. The challenges need to be effectively and efficiently solved by strong collaboration between industry and academia, in order to accelerate the development of the fundamentals and applications of this technology.

Funding: This work is funded by the Science and Technology Program of Guangzhou (No. 202201011080), the Guangdong Academy of Sciences Special Fund for Comprehensive Industrial Technology Innovation Center Building (No. 2022GDASZH-2022010107), and the University of Southampton Malaysia Seed Fund (UoSM/SF2021/5).

Conflicts of Interest: The authors declare no conflict of interest.

References

1. Huang, R.; Fukanuma, H. 6-Future trends in cold spray techniques. In *Future Development of Thermal Spray Coatings*; Espallargas, N., Ed.; Woodhead Publishing: Sawston, UK, 2015; pp. 143–162.
2. Assadi, H.; Gärtner, F.; Stoltenhoff, T.; Kreye, H. Bonding mechanism in cold gas spraying. *Acta Mater.* **2003**, *51*, 4379–4394. [CrossRef]
3. Kumar, S.; Bae, G.; Lee, C. Influence of substrate roughness on bonding mechanism in cold spray. *Surf. Coat. Technol.* **2016**, *304*, 592–605. [CrossRef]
4. Singh, S.; Raman, R.K.S.; Berndt, C.C.; Singh, H. Influence of cold spray parameters on bonding mechanisms: A review. *Metals* **2021**, *11*, 2016. [CrossRef]
5. Lek, J.Y.; Bhowmik, A.; Tan, A.W.-Y.; Sun, W.; Song, X.; Zhai, W.; Buenconsejo, P.J.; Li, F.; Liu, E.; Lam, Y.M.; et al. Understanding the microstructural evolution of cold sprayed Ti-6Al-4V coatings on Ti-6Al-4V substrates. *Appl. Surf. Sci.* **2018**, *459*, 492–504. [CrossRef]
6. Grujicic, M.; Zhao, C.; DeRosset, W.; Helfritch, D. Adiabatic shear instability based mechanism for particles/substrate bonding in the cold-gas dynamic-spray process. *Mater. Des.* **2004**, *25*, 681–688. [CrossRef]
7. Hassani-Gangaraj, M.; Veysset, D.; Champagne, V.K.; Nelson, K.A.; Schuh, C.A. Adiabatic shear instability is not necessary for adhesion in cold spray. *Acta Mater.* **2018**, *158*, 430–439. [CrossRef]
8. Guo, D.; Kazasidis, M.; Hawkins, A.; Fan, N.; Leclerc, Z.; MacDonald, D.; Nastic, A.; Nikbakht, R.; Ortiz-Fernandez, R.; Rahmati, S.; et al. Cold spray: Over 30 years of development toward a hot future. *J. Therm. Spray Technol.* **2022**, *31*, 866–907. [CrossRef]
9. Fardan, A.; Berndt, C.C.; Ahmed, R. Numerical modelling of particle impact and residual stresses in cold sprayed coatings: A review. *Surf. Coat. Technol.* **2021**, *409*, 126835. [CrossRef]

10. Song, X.; Everaerts, J.; Zhai, W.; Zheng, H.; Tan, A.W.Y.; Sun, W.; Li, F.; Marinescu, I.; Liu, E.; Korsunsky, A.M. Residual stresses in single particle splat of metal cold spray process—Numerical simulation and direct measurement. *Mater. Lett.* **2018**, *230*, 152–156. [CrossRef]

11. Song, X.; Ng, K.L.; Chea, J.M.-K.; Sun, W.; Tan, A.W.-Y.; Zhai, W.; Li, F.; Marinescu, I.; Liu, E. Coupled Eulerian-Lagrangian (CEL) simulation of multiple particle impact during metal cold spray process for coating porosity prediction. *Surf. Coat. Technol.* **2020**, *385*, 125433. [CrossRef]

12. Schmidt, T.; Gärtner, F.; Assadi, H.; Kreye, H. Development of a generalised parameter window for cold spray deposition. *Acta Mater.* **2006**, *54*, 729–742. [CrossRef]

13. Sun, W.; Tan, A.W.-Y.; Wu, K.; Yin, S.; Yang, X.; Marinescu, I.; Liu, E. Post-process treatments on supersonic cold sprayed coatings: A review. *Coatings* **2020**, *10*, 123. [CrossRef]

14. Li, W.; Zhang, D.; Huang, C.; Yin, S.; Yu, M.; Wang, F.; Liao, H. Modelling of impact behaviour of cold spray particles. *Surf. Eng.* **2014**, *30*, 299–308. [CrossRef]

15. Sun, W.; Tan, A.W.-Y.; Bhowmik, A.; Marinescu, I.; Song, X.; Zhai, W.; Li, F.; Liu, E. Deposition characteristics of cold sprayed Inconel 718 particles on Inconel 718 substrates with different surface conditions. *Mater. Sci. Eng. A* **2018**, *720*, 75–84. [CrossRef]

16. Chu, X.; Che, H.; Teng, C.; Vo, P.; Yue, S. Understanding particle–particle interactions from deposition efficiencies in cold spray of mixed Fe/316L powders with different particle size combinations. *J Therm. Spray Technol.* **2019**, *29*, 413–422. [CrossRef]

17. Li, W.; Assadi, H.; Gaertner, F.; Yin, S. A review of advanced composite and nanostructured coatings by solid-state cold spraying process. *Crit. Rev. Solid State Mater. Sci.* **2019**, *44*, 109–156. [CrossRef]

18. Tan, A.W.-Y.; Lek, J.Y.; Sun, W.; Bhowmik, A.; Marinescu, I.; Buenconsejo, P.J.; Dong, Z.; Liu, E. Microstructure, mechanical and tribological properties of cold sprayed Ti6Al4V–CoCr composite coatings. *Compos. Part B Eng.* **2020**, *202*, 108280. [CrossRef]

19. Prashar, G.; Vasudev, H. A comprehensive review on sustainable cold spray additive manufacturing: State of the art, challenges and future challenges. *J. Clean. Prod.* **2021**, *310*, 127606. [CrossRef]

20. Tan, A.W.-Y.; Lek, J.Y.; Sun, W.; Bhowmik, A.; Marinescu, I.; Song, X.; Zhai, W.; Li, F.; Dong, Z.; Boothroyd, C.B.; et al. Influence of particle velocity when propelled using N2 or N2-He mixed gas on the properties of cold-sprayed Ti6Al4V coatings. *Coatings* **2018**, *8*, 327. [CrossRef]

21. Sun, W.; Chu, X.; Lan, H.; Huang, R.; Huang, J.; Xie, Y.; Huang, J.; Huang, G. Current implementation status of cold spray technology: A short review. *J. Therm. Spray Technol.* **2022**, *31*, 848–865. [CrossRef]

22. Tillmann, W.; Hagen, L.; Kensy, M.; Abdulgader, M.; Paulus, M. Microstructural and tribological characteristics of Sn-Sb-Cu-based composite coatings deposited by cold spraying. *J. Therm. Spray Technol.* **2020**, *29*, 1027–1039. [CrossRef]

23. DeForce, B.S.; Eden, T.J.; Potter, J.K. Cold spray Al-5% Mg coatings for the corrosion protection of magnesium alloys. *J. Therm. Spray Technol.* **2011**, *20*, 1352–1358. [CrossRef]

24. Azizpour, M.J.; Majd, S.N.H.M. Babbitt casting and babbitt spraying processes. *Int. J. Mech. Aerosp. Ind. Mechatron. Manuf. Eng.* **2011**, *5*, 1628–1630.

25. Chen, Z.H.; Sun, X.F.; Huang, Y.L. A brief discussion about nickel aluminum bronze propeller failure modes and its repair methods. *Key Eng. Mater.* **2017**, *723*, 125–129. [CrossRef]

26. Tan, A.W.-Y.; Sun, W.; Bhowmik, A.; Lek, J.Y.; Marinescu, I.; Li, F.; Khun, N.W.; Dong, Z.; Liu, E. Effect of coating thickness on microstructure, mechanical properties and fracture behaviour of cold sprayed Ti6Al4V coatings on Ti6Al4V substrates. *Surf. Coat. Technol.* **2018**, *349*, 303 317. [CrossRef]

27. Sun, W.; Bhowmik, A.; Tan, A.W.-Y.; Li, R.; Xue, F.; Marinescu, I.; Liu, E. Improving microstructural and mechanical characteristics of cold-sprayed Inconel 718 deposits via local induction heat treatment. *J. Alloys Comp.* **2019**, *797*, 1268–1279. [CrossRef]

28. Sun, W.; Chu, X.; Huang, J.; Lan, H.; Tan, A.W.-Y.; Huang, R.; Liu, E. Solution and double aging treatments of cold sprayed Inconel 718 coatings. *Coatings* **2022**, *12*, 347. [CrossRef]

29. Wu, K.; Sun, W.; Tan, A.W.-Y.; Marinescu, I.; Liu, E.; Zhou, W. An investigation into microstructure, tribological and mechanical properties of cold sprayed Inconel 625 coatings. *Surf. Coat. Technol.* **2021**, *424*, 127660. [CrossRef]

30. Wu, K.; Chee, S.W.; Sun, W.; Tan, A.W.-Y.; Tan, S.C.; Liu, E.; Zhou, W. Inconel 713C coating by cold spray for surface enhancement of Inconel 718. *Metals* **2021**, *11*, 2048. [CrossRef]

31. Sun, W.; Tan, A.W.-Y.; King, D.J.Y.; Khun, N.W.; Bhowmik, A.; Marinescu, I.; Liu, E. Tribological behavior of cold sprayed Inconel 718 coatings at room and elevated temperatures. *Surf. Coat. Technol.* **2020**, *385*, 125386. [CrossRef]

32. Wu, K.; Sun, W.; Tan, A.W.-Y.; Tan, S.C.; Liu, E.; Zhou, W. High temperature oxidation and oxychlorination behaviors of cold sprayed Inconel 718 deposits at 700 °C. *Corros. Sci.* **2022**, *207*, 110536. [CrossRef]

33. Tan, A.W.-Y.; Tham, N.Y.S.; Chua, Y.S.; Wu, K.; Sun, W.; Liu, E.; Tan, S.C.; Zhou, W. Cold spray of nickel-based alloy coating on cast iron for restoration and surface enhancement. *Coatings* **2022**, *12*, 765. [CrossRef]

MDPI

Review

Research and Development on Cold-Sprayed MAX Phase Coatings

Weiwei Zhang, Shibo Li *, Xuejin Zhang and Xu Chen

Center of Materials Science and Engineering, School of Mechanical and Electronic Control Engineering, Beijing Jiaotong University, Beijing 100044, China; 22110428@bjtu.edu.cn (W.Z.); 22121382@bjtu.edu.cn (X.Z.); 22126130@bjtu.edu.cn (X.C.)
* Correspondence: shbli1@bjtu.edu.cn; Tel.: +86-10-51685554

Abstract: Cold spraying is an attractive solid-state processing technique in which micron-sized solid particles are accelerated towards a substrate at high velocities and relatively low temperatures to produce a coating through deformation and bonding mechanisms. Metal, ceramic, and polymer powders can be deposited to form functional coatings via cold spraying. MAX phase coatings deposited via cold spraying exhibit several advantages over thermal spraying, avoiding tensile residual stresses, oxidation, undesirable chemical reactions and phase decomposition. This paper presents a review of recent progress on the cold-sprayed MAX phase coatings. Factors influencing the formation of coatings are summarized and discussions on the corresponding bonding mechanisms are provided. Current limitations and future investigations in cold-sprayed MAX coatings are also listed to facilitate the industrial application of MAX phase coatings.

Keywords: cold spray; MAX phases; coatings; microstructure; properties; applications

Citation: Zhang, W.; Li, S.; Zhang, X.; Chen, X. Research and Development on Cold-Sprayed MAX Phase Coatings. *Coatings* **2023**, *13*, 869. https://doi.org/10.3390/coatings 13050869

Academic Editor: Cecilia Bartuli

Received: 14 April 2023
Revised: 29 April 2023
Accepted: 3 May 2023
Published: 5 May 2023

1. Introduction

MAX phases (M is an early transition metal, A is an A-group element, and X is carbon, nitrogen or boron) are ternary-layered carbides, nitrides, and borides [1–4]. MAX phases have a hexagonal crystal structure consisting of MX slabs interleaved with a single "A" layer. The M-X bonds in the MX slabs are strong covalent bonds, but the bonds between the MX and A layers are relatively weak. This special layered-structure endows MAX materials with a combination of the attractive properties of ceramics and metals, such as self-lubrication, self-healing, machinability, high thermal and electrical conductivities, superior resistance to corrosion and oxidization, and nonsusceptibility to thermal shock [1–5]. To date, MAX materials have been successfully developed into products such as heating elements, gas burner nozzles, electrical contacts, pantographs, and thread bolts (Figure 1).

In addition to the above applications, MAX phases can be used as protective coatings against oxidation, corrosion, and friction. So far, MAX phase coatings, with thicknesses ranging from tens to hundreds of micrometers, have been mainly prepared via the feasible or cost-effective techniques of thermal spraying and cold spraying.

Thermal spraying techniques use thermal energy to melt and soften particles, which are then sprayed on a substrate using process gases to form a coating. According to the different heat sources used, thermal spraying techniques are divided into plasma spraying techniques, electric arc spraying techniques, and high-velocity oxy-fuel (HVOF) spraying. Among the above techniques, the HVOF technique is generally used to prepare MAX phase coatings due to its employment of a median heat energy (2100–3000 °C) [6–11]. For example, Frodelius et al. [6] prepared a Ti_2AlC MAX coating with a thickness greater than 100 μm on a stainless-steel surface using the HVOF technique. The results showed that the Ti_2AlC coating adhered to the substrate well. Chen et al. [7] fabricated a Cr_2AlC MAX coating with a thickness greater than 200 μm using supersonic flame spraying. Thermally-sprayed MAX phase coatings have a dense microstructure and offer good adhesion to a substrate.

However, the HVOF technique requires the use of high temperatures (2100–3000 °C), which causes the oxidation and even decomposition of MAX phases. The oxidation or decomposition of MAX phases results in impurities with a high content in the resultant coatings, thereby deteriorating the functional performance of MAX coatings compared to bulk MAX phases. In addition, tensile stresses are inevitably generated in thermally-sprayed MAX coatings, inducing the crack formation and even the peeling of coatings. Therefore, achieving large-scale MAX coatings and retaining the original composition and properties of MAX phases through thermal spraying techniques remains challenging.

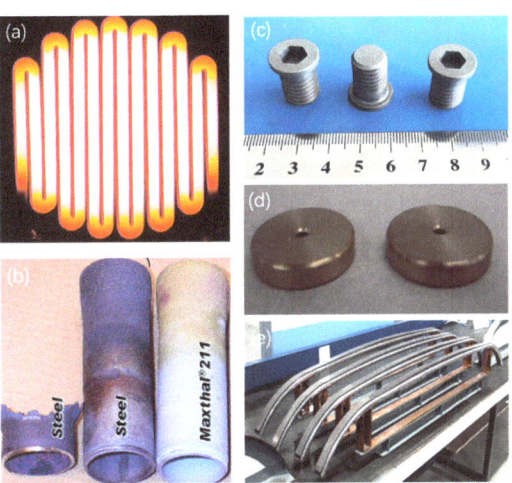

Figure 1. MAX products. (**a**) Maxthal 211 heater glowing at 1723 K (courtesy 3-ONE-2, LLC), (**b**) gas burner nozzles made of Maxthal 211 and 353MA steel after one year in furnace at 1773 K (courtesy 3-ONE-2, LLC and Kanthal), (**c**) MAX thread bolts used in corrosive environments, (**d**) MAX/Cu vacuum contact materials, and (**e**) MAX/Cu pantographs.

Cold spraying is a relatively new solid-state processing technique in which micron-size solid particles are accelerated to high velocities (500–1000 m/s) towards a substrate with a lower amount of thermal energy (<1000 °C) to produce a coating through complex deformation and bonding mechanisms [12]. To date, not only metal powders, but also ceramic and polymer powders have been successfully deposited on the surfaces of substrates (metals, ceramics and polymers) [12–19]. Cold spraying presents unique advantages, such as the avoidance of powder melting and oxidation, high density, high bond strength, and compressive residual stresses [12].

MAX phase coatings prepared via cold spraying have been extensively studied. The successful cold spray deposition of Ti_3AlC_2, Ti_3SiC_2, Cr_2AlC, and Ti_2AlC, as representative MAX coatings, has been reported [20–22]. In contrast to thermal spraying, cold spraying offers attractive properties with respect to the preparation of MAX coatings due to the following typical advantages: (1) MAX particles experience little oxidation and no decomposition upon low-temperature cold spraying, and the resulting coatings retain the original composition and properties of the employed powders; (2) MAX coatings are dense and their bonding strength is high under high-energy impacts; (3) compressive residual stresses rather than tensile stresses are generated in MAX coatings, which is beneficial to a coating's integrity. Cold-sprayed composite coatings containing MAX phases and metals exhibit improved abradable damage tolerance, improved thermal cycling and durability, tailored thermal expansion, and self-lubrication. These coatings are now being used in a number of important applications in turbine engine components, including turbine blades, the vane and the tip of an airfoil, and piston rings [23,24].

This work presents a review of the cold spray deposition of MAX phase coatings. This review principally focuses on the microstructure, mechanical properties, and tribological behaviors of cold-sprayed Ti_3AlC_2, Ti_3SiC_2, Cr_2AlC, and Ti_2AlC MAX coatings. The current limitations and outlooks for future applications in MAX coating research are presented.

2. Influencing Factors of Cold-Sprayed MAX Phase Coatings

Some important factors of cold spraying are interrelated and greatly influence the deposition process, quality, and properties of the formed coatings. These major influencing factors include powder characteristics, the driving gas, and the substrate. The influences of such factors on the formation of MAX coatings are discussed as follows.

2.1. Powder Characteristics

In the cold-spraying process, the characteristics of the feedstock powder employed, such as particle size, size distribution, and particle morphology (or shape), all influence the resultant coatings.

2.1.1. Particle Size

Feedstock powders are fed into the gas flow and accelerated to high velocity prior to impacting the substrate. A fine powder is easily accelerated to a high velocity and will ease the formation of a denser, thicker coating. However, the price of these powders is a function of their particle size; thus the smaller the powder size, the higher the cost. In addition, if the particles are too small, the bow shock effect due to the presence of a substrate in the spray stream will significantly decelerate their impact speed [12]. Therefore, the particle size of metal powders is usually less than 50 µm for use in cold spraying [12,25]. However, the above particle size requirement is not suitable for ceramic particles because ceramic powders are brittle and thus experience little plastic deformation when impacting a substrate.

Great efforts have been devoted to preparing cold-sprayed MAX phase coatings. The adopted MAX particle sizes are in the range of 0.5–70 µm. Generally, fine MAX particles can achieve a high velocity with which to impact a substrate, thus rendering the coatings thicker and denser. Elsenberg and coworkers [26] prepared different sizes of Ti_3SiC_2 MAX powders and discussed the effects of particle size on particle velocity and coating thickness. The initial state of the as-received Ti_3SiC_2 powder consists of large particles with an equiaxed shape and smaller ones with a flake-like shape, and has a particle size of D50 = 6.9 µm (Figure 2a). Subsequent high-energy milling led to particle size reduction and morphological change. The D50 particle diameters decreased from 6.2 to 0.5 µm, and their morphology changed from platelet-like to equiaxed shapes after milling for different durations (Figure 2b–e). The authors found that particle size has a profound influence on particle velocity, while other process conditions such as gas flow and pressure have minor influences on particle velocity. High velocities greater than 350 m/s were only achieved for particle sizes smaller than 2 µm. However, the impact velocity of fine particles, especially with sizes smaller than 1 µm, is strongly affected by the bow shock [26,27]. Via optimization of spray conditions, the coating thickness continuously increased with a decreasing mean particle size. At a gas pressure level of 1.1 bar, the best degree of layer buildup, with a thickness of about 500 nm, was observed for the finest powder (D50 = 0.5 µm). Based on their results, it was concluded that a suitable particle size should be optimized to deposit a desirable MAX coating. The use of powder that is too fine may result in a suboptimal coating.

To avoid inducing the bow shock effect on the fine powder and realize the deposition of ceramic coatings, an effective approach is employing agglomerated particles, which are micro-sized but composed of submicron or nanosized grains. For example, Bai et al. [28] prepared spherical Ti_2AlC agglomerated particles with a size of 5–150 µm using spray-drying technology. We have also prepared Ti_3AlC_2 and Cr_2AlC_2 MAX agglomerated particles. The agglomerated Ti_3AlC_2 particles are roughly spherical in shape, and about

75 µm in diameter (Figure 3a). Each agglomerated particle is composed of fine, flake-like Ti$_3$AlC$_2$ grains. In addition, spherical Cr$_2$AlC$_2$ particles consisting of small grains with sizes of less than 5 µm have been successfully prepared using spray-drying technology (Figure 3b). These spherical particles have good flowability during cold spraying. If necessary, these agglomerated particles or grains can be further refined via ball milling until they meet the requirements of good cold-sprayed coatings.

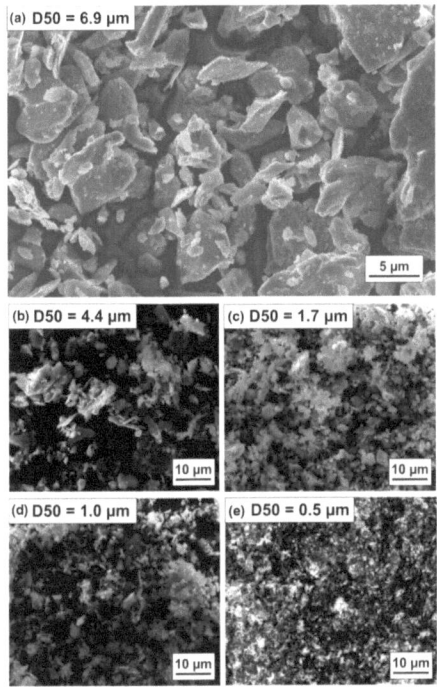

Figure 2. SEM images of (**a**) the as-received and (**b**–**e**) ball-milled Ti$_3$SiC$_2$ powders with different sizes used for cold spraying. Adapted with permission from Ref. [26]. Copyright 2021 Springer Nature.

| (**a**) | (**b**) |

Figure 3. SEM images of (**a**) agglomerated Ti$_3$AlC$_2$ particles and (**b**) agglomerated Cr$_2$AlC particles.

2.1.2. Particle Size Distribution

MAX phase particles with a single particle size distribution are not suitable for the formation of cold-sprayed coatings; instead, particles comprising a mixture of coarse and fine particles or grains are preferred for the preparation of such coatings. Rech et al. [29] reported that a Ti$_2$AlC powder with particle sizes of 25~40 µm was used to deposit a coating on an Al substrate via cold spraying. A 50 µm thick Ti$_2$AlC MAX phase coating was obtained. Gutzmann et al. [30] prepared a 100 μm thick Ti_2AlC coating via the cold spraying of Ti_2AlC powder with a D50 particle size of 34.3 μm. However, cracks and internal delamination were detected in the coatings due to the limited plastic deformation capacity of Ti_2A1C. Many experiments demonstrated that ceramic coatings without cracks and delamination can be achieved via the cold spraying of micron agglomerated-particles containing submicron or nanosized grains. Zang et al. [31] prepared 6 μm-agglomerated Ti_2A1C particles via the hydrothermal treatment of ball-milled Ti_2A1C powder with sizes of less than 0.3 μm. A 100 μm thick and dense Ti_2A1C coating without cracks and delamination was successfully obtained on a Zr-4 substrate by cold spraying the above agglomerated particles at a gas temperature of 400 °C.

2.1.3. Particle Morphology

Besides particle size, particle morphology also influences the flowability, the ability to be cold-sprayed, and ultimately, the coating's porosity [12]. In general, spherical particles offer a better flowability than irregular particles. However, previous studies demonstrated that metal particles with an irregular shape can be sprayed at a higher velocity than spherical particles because irregular particles have high drag coefficients [12,32–34]. Thus, irregular particles can be deposited with higher deposition efficiency than spherical particles but may not lead to a lower degree of porosity in the coating since irregular powders are more difficult to pack. The effects of the spherical and irregular morphologies of metal and ceramic powders on a coating's microstructure are complicated. Shockley [35] investigated the influence of Al_2O_3 ceramic particle's morphology on coating formation and found that the deposition efficiency (DE) of the irregular Al_2O_3 particles was slightly higher than that of the spherical ones. However, the coating containing spherical Al_2O_3 particles had improved tribological properties compared to a coating consisting of angular ones. To date, most studies have focused on the cold spray deposition of MAX coatings with irregular particles. The effect of particle morphology on MAX coatings' microstructure and performance should be investigated in the future.

2.2. Driving Gases

In the cold-spraying process, the nature, velocity, pressure, and temperature of the driving gases employed have profound influences on the formed coatings. Three types of gases, namely, air, nitrogen (N_2), and helium (He), are used as driving gases in cold spraying. The latter two gases are the preferred driving gases due to their inertness. He is the best candidate because it has the lowest molecular weight and can be used to reach the highest velocity among the three gases, but its high cost limits its application in cold spraying. Therefore, N_2 is used as the driving gas in cold spraying. Driving gases must be able to be sprayed at high velocities to accelerate feedstock particles with a high enough level of kinetic energy upon impacting a substrate in order to produce a desirable coating. Particles traveling at an insufficient velocity may bounce off the substrate. The impact velocity of particles is generally in the range of 200–1200 m/s. The gas temperature employed can not only increase the gas velocity, but also soften the impact of particles to improve their plastic deformation properties. The driving gas can be heated using an electric heater, for which temperatures range from room temperature to 1200 °C. In addition, high gas pressure provides high drag force for particle acceleration, and affects the gas temperature in the cold-spraying system [36].

During the cold spraying of MAX phase coatings, the pressure and temperature of the N_2 driving gas are 3.4–5 MPa and 500–1100 °C, respectively. MAX phase particles can reach a sufficiently high impact velocity with an increasing gas pressure and temperature. Table 1 summarizes the experimental parameters adopted for the cold spray deposition of MAX coatings. Cold-sprayed MAX phase coatings demonstrate the following advantages over thermally-sprayed MAX coatings: no phase transformation, no coating oxidation, and low porosity even at a process temperature of 1100 °C. Elsenberg et al. [21] used Ti_3SiC_2, Ti_2AlC, and Cr_2AlC MAX phase particles to prepare coatings, and found that Ti_3SiC_2 is unsuitable for cold-spraying due to its brittle characteristics, whereas Ti_2AlC and Cr_2AlC could be cold-sprayed on metal substrates. A Ti_3SiC_2 coating is highly non-uniform after cold spraying at a gas pressure of 5 MPa and a gas temperature of 1000 °C (Figure 4a). However, Ti_2AlC and Cr_2AlC MAX coatings can be obtained under the same conditions (Figure 4b,c). Ti_2AlC MAX particles can be easily cold-sprayed into a thicker coating with a size ranging from 300 to 520 μm (Figure 4b). Even at a low temperature of 400 °C, a dense Ti_2AlC coating with a thickness of 100 μm was achieved via the cold spraying of agglomerated Ti_2AlC particles containing submicron grains while using compressed air as the driving gas [31].

Table 1. Experimental parameters for cold spray deposition of MAX phase coatings.

Coating Material	Particle Size (μm)	Substrate Material	Driving Gas	Gas Pressure (MPa)	Gas Temperature (°C)	Particle Speed (m/s)	Stand of Distance (mm)	Thickness (μm)	Remarks	Ref.
Ti_3AlC_2	20–40	Ti4Al6V	N_2	3.5–5	600–1000	723–780	30	50	No oxidation at 1000 °C; internal cracks in coatings.	[20]
Ti_3SiC_2	42 (D50)	304 SS Cu	N_2	4–5	800–1100	699–801	60	-	No oxidation; non-uniform Ti_3SiC_2 coating.	[21]
Ti_2AlC	<62	Stainless steel	N_2	3.7	-	-	20	55	Continuous transversal cracks in Ti_2AlC coating.	[10]
	11 (D50)	304 SS Cu	N_2	5	1000	717–802	60	300–530	No oxidation; low porosity; lateral cracks in coatings.	[21]
	<20	Zr alloy	N_2	3.5	600	-	-	90	No oxidation or phase transformation; low porosity; no delamination in coating.	[22]
	5–150	TiAl alloy	Air	1–5	100–1000	-	-	5–300	No oxidation; low porosity.	[23]
	5.9 (D50)	Zr alloy	Air	2–2.5	400	-	20	100	No oxidation; low porosity.	[31]
	34 (D50)	Cu Stainless steel	N_2	4	600 1000	-	60	110 155	No oxidation or phase transformation; low porosity; cracks and internal delamination in coatings.	[30]
	25–40	Al alloy Stainless steel	N_2	3.4 3.9	500–800	-	20	50 80	No oxidation; a continuous transversal crack in coating.	[29]
	<20	Inconel 625	-	-	-	-	-	70	No processing details; voids and microcracks in coating.	[37]
	5–50	Zr alloy	N_2 He	-	500 800	-	26	25–30	Non-uniform coating; microcracks in coating.	[38]
Cr_2AlC	9 (D50)	304 SS Cu	N_2	5	1000	733–849	60	200–320	No oxidation; low porosity; lateral cracks in coatings.	[21]
	7.6 (D50)	Stainless Steel	N_2	4	650–950	-	60	40–97	No oxidation; low porosity; cracks in coating.	[39]

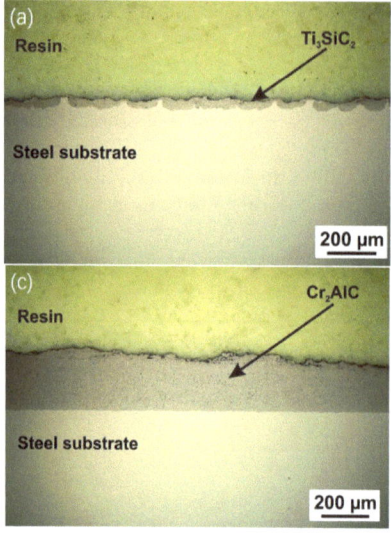
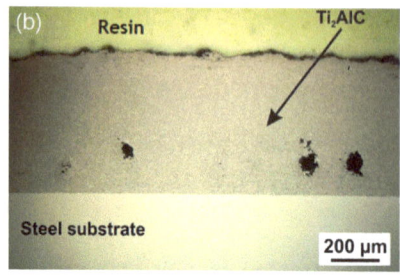

Figure 4. Optical micrographs of ten-layer coatings of (**a**) Ti$_3$SiC$_2$, (**b**) Ti$_2$AlC, and (**c**) Cr$_2$AlC after being cold-sprayed on steel substrate with a gas pressure of 5 MPa, a gas temperature of 1000 °C and a traversal speed of 100 mm/s. Adapted with permission from Ref. [21]. Copyright 2020 Springer Nature.

2.3. Substrates

So far, cold-sprayed MAX coatings on metal substrates such as Ti, Cu, Zr, Al, Inconel, and stainless steel have been achieved [10,20–22,37–40]. Through plastic deformation, metal substrates mechanically interlock with MAX particles, thus resulting in good adhesion between MAX coatings and metal substrates. However, there is no information on the formation of MAX coatings on ceramic substrates. Cold-sprayed metal particles adhere to the surface of ceramic substrates via the plastic deformation of the particle itself. MAX phases denote ceramic materials with little plastic deformation upon impacting the substrates. It has been reported that cold-sprayed oxide ceramic coatings can be realized on glass or ceramic brittle substrates if the agglomerated particles contain nanostructured crystals [18,41–43]. For example, a dense TiO$_2$ coating with a uniform thickness of several tens of micrometers on an indium tin oxide glass substrate has been achieved via the cold spraying of 25 nm TiO$_2$ particles [43]. Good adhesion between a brittle substrate and nanoparticles may benefit mechanical interlocking. Using these results as an inspiration, MAX coatings on ceramic substrates may be achieved via the cold spraying of nanostructured MAX particles.

3. Bonding Mechanisms in Cold-Sprayed MAX Phase Coatings

Understanding the bonding mechanisms involved in cold spraying helps clarify how particles are bonded together and how coatings are built up on the substrate. In this section, microstructural evolution, bonding strength, and residual stresses are discussed to explain the bonding mechanisms in the cold-sprayed MAX coatings.

3.1. Microstructural Evolution

During the cold spraying of conventional metal powders, the presence of high temperatures results in the thermal softening of the metal particles, which improves their plastic deformation on the substrate toward the formation of dense coatings. Elsenberg and coworkers [21] investigated the impact morphologies of Ti$_3$SiC$_2$, Ti$_2$AlC, and Cr$_2$AlC MAX powders on 304 steel via cold spraying with a pressure of 4 MPa and a gas temperature of 1000 °C. They found that Ti$_3$SiC$_2$ is the most prone to fracture, in contrast, Ti$_2$AlC and Cr_2AlC particles show greater plastic deformation at 1000 °C. Ti_2AlC particles' morphologies are shown in Figure 5. At the aforementioned high temperature, most particles adhered to the substrate (marked with "i" and "iii" in Figure 5a). The larger adhering particles are fractured, presenting a nanolamellar structure and cracks (marked with "v" Figure 5b,c). Some smaller particles rebounded and left empty craters (marked with "ii" in Figure 5a,b) in the substrate. Gutzmann and coworkers [30] also reported that most of the impacting Ti_2A1C particles rebounded and left empty craters in the substrate at 600 °C but that increasing the temperature up to 1000 °C improved adhesion. Secondary impacting particles (marked with "iv" in Figure 5b,c) adhered to the primary ones, for which deformation features appeared via slipping along lamellas (marked with "vi" in Figure 5d). A viscous-like deformation effect also occurred along a lamellar failure (marked with "vii" in Figure 5d). Similar impacting morphologies were also observed upon the cold spraying of Cr_2AlC particles [21]. The thermal softening of Ti_2AlC and Cr_2AlC particles at a process temperature of 1000 °C enables good bonding between MAX phase coatings and metal substrates.

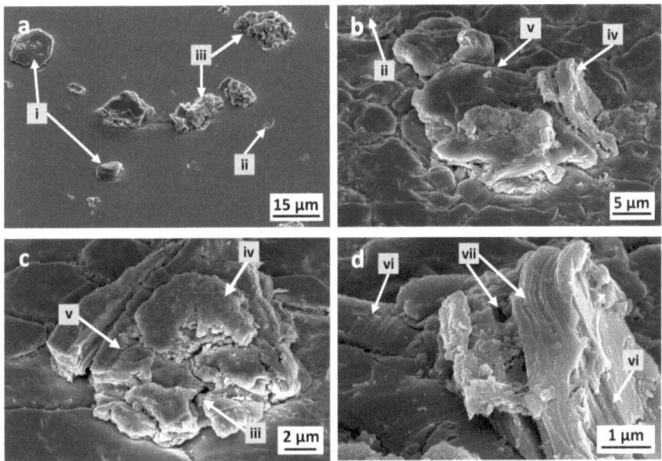

Figure 5. Impact morphologies of Ti_2AlC MAX particles on 304 steel substrate after being cold sprayed under conditions of 5 MPa/1000 °C. SEM images with different magnifications (**a–d**). The inserts describe the following: i— adhesion of complete, flattened particles, ii—empty impact crater with ASI of steel substrate, iii—fractured particles, iv—secondary particle impacts, v—cracks, vi—deformation on laminae, and vii—viscous-like flow at laminae. Adapted with permission from Ref. [21]. Copyright 2020 Springer Nature.

In cold-sprayed MAX phase coatings, no oxidation or decomposition occurs; thus, the initial composition and structure of the MAX phases are preserved. The thickness of MAX coatings can be tunable, ranging from twenty to five hundred micrometers (see Table 1). The prepared MAX phase coatings are dense and have low porosity. For example, dense, thick Ti_2AlC (Figure 6a) and Cr_2AlC (Figure 6b) on steel were achieved via spraying under conditions of 5 MPa/1000 °C and 4 MPa/950 °C, respectively [21,39]. These coatings were homogeneous, but pores and lateral cracks were detected in the coatings, which deteriorated the mechanical properties of the MAX coatings. However, temperature has a profound influence on porosity and crack concentration; accordingly, the number of pores and cracks decreased with an increasing processing temperature. Porosity decreased from 12.4 vol.% to 9.1 vol.% as temperature increased from 650 to 950 °C [39]. The microstructures shown in Figures 4 and 5 further demonstrate that no cracks or delamination were detected at the interfaces of Ti_2AlC/steel and Cr_2AlC/steel, suggesting that there was good adhesion between the MAX coatings and the steel substrate.

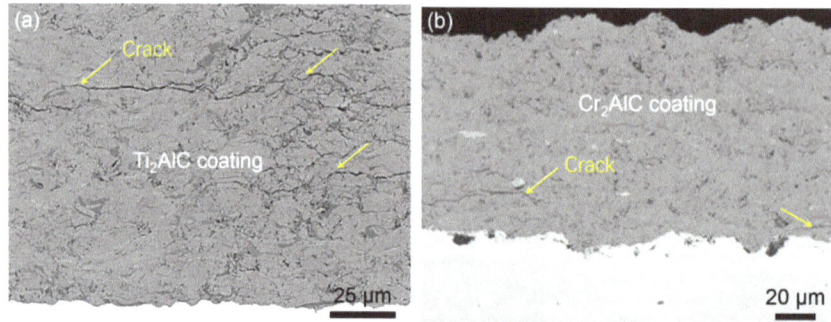

Figure 6. Cross-sectional SEM micrographs of (**a**) Ti$_2$AlC coating on steel sprayed under conditions of 5 MPa/1000 °C. Adapted with permission from Ref. [21]. Copyright 2020 Springer Nature. (**b**) Cr$_2$AlC coating on steel sprayed with 4 MPa/950 °C. Cracks are marked with arrows. Adapted with permission from Ref. [39]. Copyright 2018 Elsevier Ltd.

3.2. Bonding Strength of Cold-Sprayed MAX Phase Coatings

Adhesion is a fundamental property in a coating/substrate system. The adhesion of cold-sprayed MAX coatings is determined via the bonding of particles to the substrate surface. So far, few studies have focused on the bonding strength between MAX coatings and substrates. Zang et al. [31] reported that a bonding strength of 44 MPa was achieved in a Ti$_2$AlC coating/Zr-4 alloy substrate system prepared via the cold spraying of 0.6 µm Ti$_2$AlC particles. Bonding strength was measured using a tensile adhesion test according to the ASTM C633 standard [44], in which tensile stresses are applied to a coated system consisting of a coated sample glued to another sample. The bonding strength of 44 MPa is even higher than that of 40 MPa of an Al coating/Al alloy substrate [45], 10–20 MPa of Cu coating/steel [46], and 34 MPa of a Ti coating/Al alloy substrate [47] achieved using the same test method.

3.3. Residual Stresses in Cold-Sprayed MAX Phase Coatings

Residual stresses are typically compressive stresses occurring on the surface of cold-sprayed coatings due to the peening effect and plastic deformation through the continuous, high-velocity impact of particles. The presence of compressive residual stresses on a coating's surface is beneficial for increasing fatigue strength, resistance to stress corrosion cracking, and bending strength [12]. Go et al. [39] prepared Cr$_2$AlC coatings via cold spraying, and, using X-ray diffraction method, determined that residual stresses were compressive in the coatings. The values of compressive residual stresses increased with increasing process temperature and coating thickness. For example, the compressive residual stresses changed from −200 MPa in the 39.8 µm coating prepared at 650 °C to −310 MPa in the 97.4 µm coatings deposited at 950 °C. However, if the residual stresses are large, they can cause cracking and local delamination in the coatings, or even partial layer spallation [26].

4. Mechanical Properties and Tribological Behaviors of Cold-Sprayed MAX Phase Coatings

The reliability of a coating depends on its mechanical properties. The mechanical properties of a cold-sprayed coating are determined by various factors such as microstructure, the adhesion of the coating to the substrate, and residual stress. A dense, fine, and homogeneous microstructure of a cold-sprayed coating is beneficial to the improvement of its mechanical properties. In this section, the hardness and tribological behaviors of MAX coatings will be discussed.

4.1. Hardness of Cold-Sprayed MAX Phase Coatings

Hardness is a measure of a material's resistance to localized plastic deformation. It is important to predict the behavior of a coating. The Vickers microhardness test is used to measure a coating's hardness. However, depending on the influence of the substrate, the nano-hardness test is also adopted to measure the hardness of thin coatings.

The micro- and nano-hardness values of different MAX phase coatings, together with those of MAX bulks for comparison, are listed in Table 2. During the cold-spraying process, the high velocity impact of powder induces a refined microstructure, the corresponding cold-working effect, and residual compressive stresses in a coating, leading to an increment in the coating's hardness. The microhardness values of MAX phase coatings are close to or slightly lower than those of MAX bulks. The presence of pores and cracks and a lack of cohesive strength at the particle-particle boundaries negatively impact the hardness of MAX coatings.

Table 2. Hardness of MAX bulks and coatings.

MAX Phase		GRAIN Size (μm)	Microhardness (GPa)	Nanohardness (GPa)	Ref.
Ti_3SiC_2	Bulk	3–200	2–6	7.3	[48–50]
	Coating	42 (D50)	3.75	7.9	[21]
Ti_3AlC_2	Bulk	25	3.5	-	[51]
	Coating	20-40	-	-	[20]
Ti_2AlC	Bulk	25–50	3.3-4.5	8.2	[52–54]
		11 (D50)	3.68	7.89	[21]
	Coating	25–40	-	10.1	[29]
		<20	-	7–8	[37]
		<20	-	11.8	[40]
Cr_2AlC	Bulk	2–35	3.5–6.4	-	[55–57]
	Coating	9 (D50)	5.73	11.3	[21]

Note: hardness values of MAX coatings are converted into values with a unit of "GPa".

The nanohardness values of MAX phase coatings are greatly higher than the corresponding microhardness values (Table 2). This feature can be attributed to the fact that the indented areas are significantly reduced when the indentation load decreases. When the nanoindentation load is low enough, only small indentations will be formed within a single flattened particle. For example, a microhardness of 3.68 GPa was achieved in the Ti_2AlC coating tested using a Vickers indenter under a load of 1 N, but a nano-hardness value of 7.89 GPa was obtained for the same coating when tested with a Berkovich indenter under a load of 0.01 N [21]. In another study, the nano-hardness value changed from 7 to 15.8 GPa for a Ti_2AlC coating as the load was decreased from 8 N to 0.007 N [37]. This phenomenon is commonly referred to as the indentation size effect (ISE). One of the possible mechanisms of the ISE is that the deformation under the indenter occurs in a discrete manner rather than a continuous one [58].

MAX phases have a hexagonal crystal structure; thus, their nano-hardness value is anisotropic, as the basal planes are parallel or perpendicular with respect to the surface. Concerning a Ti_3SiC_2 single crystal with grain sizes of 50–200 μm, Kooi et al. [50] reported that a nano-hardness value of 7.3 GPa was achieved when the basal planes were perpendicular to the surface, which is higher than the value of 4.9 GPa achieved when the basal planes were parallel to the surface, which was due to the greater degree of plastic deformation and the larger pile-up around the indentations for the parallel orientation compared to the perpendicular orientation.

It has been reported that gas temperature, pressure, and particle velocity offer beneficial contributions to the incrementation in the hardness of metal coatings [59,60]. However, there has been little research on the hardness evolution of MAX phase coatings as a function of gas temperature/pressure and particle velocity so far. Therefore, a systematical study on the relationship between MAX coating hardness and gas temperature/pressure and particle velocity is required.

4.2. Tribological Behaviors of Cold-Sprayed MAX Phase Coatings

Wear-resistant coatings on different metallic substrates have been developed to protect metallic components used in different industrial applications. One of the most attractive properties of MAX phases is their self-lubrication [2]. Therefore, a lubricative MAX film or coating can greatly protect substrates from damage by vastly increasing wear resistance and reducing surface friction.

Maier and coworkers [22] prepared a cold-sprayed Ti_2AlC coating on a Zircaloy (zirconium alloy) substrate and investigated its wear resistance. A pin-on-disk wear test was conducted using a 3 mm diameter alumina ball under a 0.02 kg applied load, and a scratch test was performed with a Vickers diamond-tipped scribe under a constant 50 N load. The wear test showed that the wear resistance of the Ti_2AlC coatings was significantly superior to that of the Zircaloy substrate and that the wear track depth decreased from 12 to 1 µm after the cold spray deposition of the Ti_2AlC coating, suggesting enhanced wear resistance. The scratch test demonstrated that scratching did not cause any spallation or delamination of the Ti_2AlC coating from the substrate, indicating good adhesion of the coating to the substrate [22].

Loganathan et al. [37] deposited a Ti_2AlC coating on an Inconel (a nickel-chromium-based superalloy) substrate, which demonstrated enhanced wear resistance both at room temperature and a high temperature of 600 °C. A ball-on-disc wear test was performed using a tribometer, for which an alumina ball of 3 mm in diameter was used as the counter material and a normal load of 5 N and a speed of 100 RPM were applied in dry air both at room temperature (25 °C) and 600 °C. The wear test showed that the average coefficient of friction (COF) of the coating was 0.767 at room temperature but 0.603 at 600 °C. The wear depth decreased from 25 m at room temperature to 12 µm at 600 °C, and the calculated wear rates also changed from 4.7×10^{-7} mm^3/Nmm at room temperature to 2.84×10^{-7} mm^3/Nmm at 600 °C, constituting a 40% decrease. The lower COF and wear rate at 600 °C resulted from the fact that a mostly uniform and continuous oxide tribofilm consisting of TiO_2 and Al_2O_3, which are relatively lubricious at high temperature, formed at 600 °C and provided a lubricious effect. The wear mechanism of the cold-sprayed Ti_2AlC coating changed from brittle at room temperature to ductile at a high temperature. The above results suggested that cold-sprayed Ti_2AlC coatings on metal substrates have great potential applications at high temperatures up to 600 °C.

5. Summary and Outlook

This review has presented the recent progress on the cold spraying of MAX phase coatings. The influencing factors on the formation, bonding mechanisms, mechanical properties, and wear resistance of MAX phase coatings have been discussed. It is clear that cold spraying is a valid technique for developing MAX phase coatings. This work provides some basic information for the future preparation and research of cold-sprayed MAX phase coatings and accelerates the practical applications of such coatings. However, there are still some issues that have not been comprehensively explained, and future work is still needed to obtain such knowledge. Future directions and research gaps are listed as follows:

(1)   Powder size and morphology

Irregular MAX phase particles with sizes of 0.5–70 µm have been used to prepare MAX coatings via cold spraying. The influences of particle size (especially in a nano scale) and morphology (irregular and spherical shapes) on the microstructure and properties of cold-sprayed MAX coatings have received less attention and should thus be further investigated.

(2)   Interface characterization

MAX phase coatings can be cold-sprayed on metal substrates, for which good adhesion is achieved. However, the absence of interfacial characterization renders the nature of the bonding mechanisms between MAX coatings and metal substrates ambiguous. Interfacial features influence bond strength. One of the most readily identifiable interfacial features is interfacial melting. Interfacial melting occurs during the cold spraying of some metal particles on metal substrates, wherein the high energy impact of cold spray particles can render the particles or the substrate molten. Large splashes and droplets can be seen to radiate from bonded particles and rebound sites, and nanosized, spheroidized droplets can also be detected at particle-particle interfaces, contributing to high bond strength [61–63]. As MAX particles impact metal substrates with high velocity, does interfacial melting occur between the coating and the metal substrate or at particle-particle interfaces? In addition, are there interfacial amorphous layers between MAX coatings and metal substrates due to strain-rate-induced amorphization? At the bonded interface, microstructural inhomogeneity must also be characterized via electron microscopy.

(3)   Bonding mechanisms

Cold spray bonding is a combination of mechanical interlocking and metallurgical bonding. Recrystallization, atomic diffusion, and interfacial melting contribute to the metallurgical bonding mechanisms for cold-sprayed metal coatings. At present, mechanical interlocking is considered to be the predominant bonding mechanism between MAX phase coatings and metal substrates. The metallurgical bonding mechanism for cold-sprayed MAX phase coatings has been discussed to a lesser extent. The detailed bonding mechanisms that affect particle-to-substrate adhesion and particle-to-particle adhesion are still unclear. Research on bonding, especially in the MAX phase particles, is required.

(4)   Expand the types of MAX phase coatings and substrates.

So far, only four kinds of cold-sprayed MAX phases (Ti_3AlC_2, Ti_3SiC_2, Cr_2AlC, and Ti_2AlC) coatings have been deposited on metal substrates. In the MAX phase family, about one hundred types of compounds with different functional properties have been synthesized. It is necessary to develop different cold-sprayed MAX phase coatings with tailored properties for specific applications. In addition, several metal substrates with cold-sprayed MAX phase coatings have been investigated. In the future, the cold spraying of MAX phase coatings on metal, ceramic, and polymer substrates is necessary in order to meet practical application requirements.

(5)   Computational simulation of cold-sprayed MAX phase coatings.

Computational simulation plays an important role in the development of coatings and the understanding of cold spray bonding. Great efforts have been devoted to the computational simulation of cold-sprayed coatings on different substrates. Computational simulation is also required to analyze the influencing factors with respect to the quality of MAX phase coatings and to understand their bonding mechanisms.

(6)   Performance of cold-sprayed MAX phase coatings

The mechanical and wear resistance properties of cold-sprayed MAX phase coatings have been investigated. Future work on high-temperature oxidation resistance, corrosion resistance, electrical properties, and other properties of MAX phase protective coatings, and on the relationship between influencing factors and the performance of cold-sprayed MAX phase coatings, is required.

(7)   The post-heat-treatment is essential for coating consolidation and microstructure modification.

The influence of heat treatment on the microstructure and properties of MAX phase coatings has not been assessed, and this is required for further investigation.

Author Contributions: Investigation, W.Z., X.Z. and X.C.; Supervision, S.L.; Writing—Original Draft Preparation, W.Z.; Writing—Review and Editing, S.L.; All authors have read and agreed to the published version of the manuscript.

Funding: This work was supported by the National Natural Science Foundation of China under Grant No. 52275171, and the Pre-Research Program of the National 14th Five-Year Plan (No. 80923010304).

Institutional Review Board Statement: Not applicable.

Informed Consent Statement: Not applicable.

Data Availability Statement: Not applicable.

Conflicts of Interest: The authors declare no conflict of interest.

References

1. Zhou, A.; Liu, Y.; Li, S.; Wang, X.; Ying, G.; Xia, Q.; Zhang, P. From Structural Ceramics to 2D Materials with Multi-Applications: A Review on the Development from MAX Phases to MXenes. *J. Adv. Ceram.* **2021**, *10*, 1194–1242. [CrossRef]
2. Barsoum, M.W. The Mn+1AXn phases: A new class of solids. *Prog. Solid State Chem.* **2000**, *28*, 201–281. [CrossRef]
3. Sun, Z.M. Progress in research and development on MAX phases: A family of layered ternary compounds. *Int. Mater. Rev.* **2011**, *56*, 143–166. [CrossRef]
4. Qin, Y.; Zhou, Y.; Fan, L.; Feng, Q.; Grasso, S.; Hu, C. Synthesis and Characterization of Ternary Layered Nb$_2$SB Ceramics Fabricated by Spark Plasma Sintering. *J. Alloy. Compd.* **2021**, *878*, 160344. [CrossRef]
5. Hu, W.; Huang, Z.; Wang, Y.; Li, X.; Zhai, H.; Zhou, Y.; Chen, L. Layered ternary MAX phases and their MX particulate derivative reinforced metal matrix composite: A review. *J. Alloy. Compd.* **2021**, *856*, 157313. [CrossRef]
6. Frodelius, J.; Sonestedt, M.; Bjorklund, S.; Palmquist, J.P.; Stiller, K.; Hogberg, H.; Hultman, L. Ti$_2$AlC coatings deposited by high velocity oxy-fuel spraying. *Surf. Coat. Technol.* **2008**, *202*, 5976–5981. [CrossRef]
7. Chen, Y.; Chu, M.; Wang, L.; Shi, Z.; Wang, X.; Zhang, B. Microstructure and performance of Cr$_2$AlC coatings deposited by HVOF spraying. *Chin. J. Rare Met.* **2012**, *36*, 568–573.
8. Jiang, J.; Fasth, A.; Nylen, P.; Choi, W.B. Microindentation and inverse analysis to characterize elastic-plastic properties for thermal sprayed Ti$_2$AlC and NiCoCrAlY. *J. Therm. Spray Technol.* **2009**, *18*, 194–200. [CrossRef]
9. Sonestedt, M.; Frodelius, J.; Palmquist, J.P.; Hogberg, H.; Hultman, L.; Stiller, K. Microstructure of high velocity oxy-fuel sprayed Ti$_2$AlC coatings. *J. Mater. Sci.* **2010**, *45*, 2760–2769. [CrossRef]
10. Markocsan, N.; Manitsas, D.; Jiang, J.; Bjorklund, S. MAX-phase coatings produced by thermal spraying. *J. Superhard Mater.* **2017**, *39*, 355–364. [CrossRef]
11. Zhang, Z.; Lim, S.H.; Chai, J.; Lai, D.M.Y.; Lim, P.C.; Cheong, A.K.H.; Wang, S.; Jin, H.; Pan, J. Kerosene-fuelled high velocity oxy-fuel (HVOF) spray of Ti$_2$AlC MAX phase powders. *J. Alloy. Compd.* **2018**, *735*, 377–385. [CrossRef]
12. Villafuerte, J. *Modern Cold Spray-Materials, Process and Application*, 1st ed.; Springer International Publishing: Cham, Switzerland, 2015.
13. Poza, P.; Garrido-Maneiro, M.A. Cold-sprayed coatings: Microstructure, mechanical properties, and wear behaviour. *Prog. Mater. Sci.* **2022**, *123*, 100839. [CrossRef]
14. Lupoi, R.; O'Neill, W. Deposition of metallic coatings on polymer surfaces using cold spray. *Surf. Coat. Technol.* **2010**, *205*, 2167–2173. [CrossRef]
15. Khalkhali, Z.; Xie, W.; Champagne, V.K.; Lee, J.H.; Rothstein, J.P. A comparison of cold spray technique to single particle micro-ballistic impacts for the deposition of polymer particles on polymer substrates. *Surf. Coat. Technol.* **2018**, *351*, 99–107. [CrossRef]
16. Bush, T.B.; Khalkhali, Z.; Champagne, V.; Schmidt, D.P.; Rothstein, J.P. Optimization of cold spray deposition of high-density polyethylene powders. *J. Therm. Spray Technol.* **2017**, *26*, 1548–1564. [CrossRef]
17. Kliemann, J.O.; Gutzmann, H.; Gärtner, F.; Hübner, H.; Borchers, C.; Klassen, T. Formation of cold-sprayed titanium dioxide layers on metal surface. *J. Therm. Spray Technol.* **2011**, *20*, 292–298. [CrossRef]
18. Lee, H.Y.; Yu, Y.H.; Lee, Y.C.; Hong, Y.P.; Ko, K.H. Thin film coatings of WO$_3$ by cold gas dynamic spray: A technical note. *J. Therm. Spray Technol.* **2005**, *14*, 183–186. [CrossRef]
19. Seo, D.; Sayar, D.; Ogawa, K. SiO$_2$ and MoSi$_2$ formation on Inconel 625 surface via SiC coating deposited by cold spray. *Surf. Coat. Technol.* **2012**, *206*, 2851–2858. [CrossRef]
20. Ion, A.; Sallot, P.; Badea, V.; Duport, P.; Popescu, C.; Denoirjean, A. The dual character of MAX phase nano-layered structure highlighted by supersonic particles deposition. *Coatings* **2021**, *11*, 1038. [CrossRef]
21. Elsenberg, A.; Busato, M.; Gärtner, F.; List, A.; Bruera, A.; Bolelli, G.; Lusvarghi, L.; Klassen, T. Influence of MAX-phase deformability on coating formation by cold spraying. *J. Therm. Spray Technol.* **2021**, *30*, 617–642. [CrossRef]
22. Maier, B.R.; Garcia-Diaz, B.L.; Hauch, B.; Olson, L.C.; Sindelar, R.L.; Sridharan, K. Cold spray deposition of Ti$_2$AlC coatings for improved nuclear fuel cladding. *J. Nucl. Mater.* **2015**, *466*, 712–717. [CrossRef]

23. Amini, S.; Strock, C.W.; Sharon, J.A.; Klecka, M.A.; Nardi, A.T. Cold Spray Manufacturing of MAXMET Composites. U.S. Patent 10000851, 19 June 2018.

24. Zhao, Y.; Zhou, L.; Hua, Y.; Liu, C.; Tian, C.; Wu, L. Composite Material for Piston Ring Coating, Piston Ring Coating and Preparation Method Thereof. CN Patent 102517577A, 12 January 2012.

25. Li, W.Y.; Li, C.J. Optimization of spray conditions in cold spraying based on the numerical analysis of particle velocity. *T. Nonferr. Met. Soc.* **2004**, *14*, 43–48.

26. Elsenberg, A.; Gärtner, F.; Klassen, T. Aerosol deposition of Ti_3SiC_2-MAX-phase coatings. *J. Therm. Spray Technol.* **2021**, *30*, 1121–1135. [CrossRef]

27. Park, H.; Kwon, H.; Kim, Y.; Lee, C. Computational research on factors affecting particle velocity in a vacuum kinetic spray process. *J. Therm. Spray Technol.* **2019**, *28*, 1945–1958. [CrossRef]

28. Bai, C.; Xie, X.; Cui, Y.; Li, T.; Xiong, T.; Yang, R. Method for Preparing Ti2AlC Phase Ceramic Coating through Cold Spraying In-Situ Formation. CN Patent 201610210730.3, 26 October 2016.

29. Rech, S.; Surpi, A.; Vezzù, S.; Patelli, A.; Trentin, A.; Glor, J.; Frodelius, J.; Hultman, L.; Eklund, P. Cold-spray deposition of Ti_2AlC coatings. *Vacuum* **2013**, *94*, 69–73. [CrossRef]

30. Gutzmann, H.; Gartner, F.; Hoche, D.; Blawert, C.; Klassen, T. Cold spraying of Ti_2AlC max-phase coatings. *J. Therm. Spray Technol.* **2013**, *22*, 406–412. [CrossRef]

31. Zang, X.; Xie, X.; Shen, Y.; Wang, J.; Bai, C.; Xiong. T. Influences of agglomeration powder preparation on deposition of cold-sprayed Ti_2AlC coating. *Chn. Surf. Eng.* **2018**, *31*, 140–147.

32. Wong, W.; Vo, P.; Irissou, E.; Ryabinin, A.N.; Legoux, J.G.; Yue, S. Effect of particle morphology and size distribution on cold-sprayed pure titanium coatings. *J. Therm. Spray Technol.* **2013**, *22*, 1140–1153. [CrossRef]

33. Munagala, V.N.V.; Chakrabarty, R.; Song, J.; Chromik, R.R. Effect of metal powder properties on the deposition characteristics of cold-sprayed Ti6Al4V-TiC coatings: An experimental and finite element study. *Surf. Interfaces* **2021**, *25*, 101208. [CrossRef]

34. Yamada, M.; Isago, H.; Nakano, H.; Fukumoto, M. Cold spraying of TiO_2 photocatalyst coating with nitrogen process gas. *J. Therm. Spray Technol.* **2010**, *19*, 1218–1223. [CrossRef]

35. Shockley, J.M.; Descartes, S.; Voc, P.; Irissou, E.; Chromik, R.R. The influence of Al_2O_3 particle morphology on the coating formation and dry sliding wear behavior of cold sprayed Al-Al_2O_3 composites. *Surf. Coat. Technol.* **2015**, *270*, 324–333. [CrossRef]

36. Schmidt, T.; Assadi, H.; Gärtner, F.; Richter, H.; Stoltenhoff, T.; Kreye, H.; Klassen, T. From particle acceleration to impact and bonding in cold spraying. *J. Therm. Spray Technol.* **2009**, *18*, 794–808. [CrossRef]

37. Loganathan, A.; Sahu, A.; Rudolf, C.; Zhang, C.; Rengifo, S.; Laha, T.; Boesl, B.; Agarwal, A. Multi-scale tribological and nanomechanical behavior of cold sprayed Ti_2AlC MAX phase coating. *Surf. Coat. Technol.* **2018**, *334*, 384–393. [CrossRef]

38. Garcia-Diaz, B.; Olson, L.; Verst, C.; Sindelar, R.; Hoffman, E.; Hauch, B.; Sridharan, K. MAX phase coatings for accident tolerant nuclear fuel. *Trans. Am. Nucl. Soc.* **2014**, *110*, 994–996.

39. Go, T.; Sohn, Y.J.; Mauer, G.; Vaßen, R.; Gonzalez-Julian, J. Cold spray deposition of Cr_2AlC MAX phase for coatings and bondcoat layers. *J. Eur. Ceram. Soc.* **2019**, *39*, 860–867. [CrossRef]

40. Gigax, J.G.; Kennas, M.; Kim, H.; Wang, T.; Maier, B.R.; Yeom, H.; Johnson, G.O.; Sridharan, K.; Shao, L. Radiation response of Ti_2AlC MAX phase coated Zircaloy-4 for accident tolerant fuel cladding. *J. Nucl. Mater.* **2019**, *523*, 26–32. [CrossRef]

41. Lee, H.Y.; Yu, Y.H.; Lee, Y.C.; Hong, Y.P.; Ko, K.H. Interfacial studies between cold-sprayed WO_3, Y_2O_3 films and Si substrate. *Appl. Surf. Sci.* **2004**, *227*, 244–249. [CrossRef]

42. Toibah, A.R.; Sato, M.; Yamada, M.; Fukumoto, M. Cold-Sprayed TiO_2 coatings from nanostructured ceramic agglomerated powders. *Mater. Manuf. Process.* **2016**, *31*, 1527–1534. [CrossRef]

43. Fan, S.Q.; Yang, G.J.; Li, C.J.; Liu, G.J.; Li, C.X.; Zhang, L.Z. Characterization of microstructure of nano-TiO_2 coating deposited by vacuum cold spraying. *J. Therm. Spray Technol.* **2006**, *15*, 513–517. [CrossRef]

44. *ASTM C633-13*; Standard Test Method for Adhesion or Cohesion Strength of Thermal Spray Coatings. R&B, Inc.: Colmar, PA, USA, 2021.

45. Irissou, E.; Legoux, J.G.; Arsenault, B.; Moreau, C. Investigation of Al-Al_2O_3 cold spray coating formation and properties. *J. Therm. Spray Technol.* **2007**, *16*, 661–668. [CrossRef]

46. Gärtner, F.; Stoltenhoff, T.; Schmidt, T.; Kreye, H. The cold spray process and its potential for industrial applications. *J. Therm. Spray Technol.* **2006**, *15*, 223–232. [CrossRef]

47. Cinca, N.; Barbosa, M.; Dosta, S.; Guilemany, J.M. Study of Ti deposition onto Al alloy by cold gas spraying. *Surf. Coat. Technol.* **2010**, *205*, 1096–1102. [CrossRef]

48. Li, S.B.; Xie, J.X.; Zhang, L.T.; Cheng, L.F. Synthesis and some properties of Ti_3SiC_2 by hot pressing of Ti, Si and C powders. part II: Mechanical and other properties of Ti_3SiC_2. *Mater. Sci. Techol.* **2005**, *21*, 1054–1058. [CrossRef]

49. Low, I.M. Vickers contact damage of micro-layered Ti_3SiC_2. *J. Eur. Ceram. Soc.* **1998**, *18*, 709–713. [CrossRef]

50. Kooi, B.J.; Poppen, R.J.; Carvalho, N.J.M.; De Hosson, J.T.M.; Barsoum, M.W. Ti_3SiC_2: A damage tolerant ceramic studied with nanoindentations and transmission electron microscopy. *Acta Mater.* **2003**, *51*, 2859–2872. [CrossRef]

51. Tzenov, N.V.; Barsoum, M.W. Synthesis and characterization of Ti_3AlC_2. *J. Am. Ceram. Soc.* **2000**, *83*, 825–832. [CrossRef]

52. Li, S.B.; Hu, S.J.; Hee, A.C.; Zhao, Y. Surface modification of a Ti_2AlC soft ceramic by plasma nitriding treatment. *Surf. Coat. Technol.* **2015**, *281*, 164–168. [CrossRef]

53. Barsoum, M.W.; Ali, M.; El-Raghy, T. Processing and characterization of Ti_2AlC, Ti_2AlN, and $Ti_2AlC_{0.5}N_{0.5}$. *Met. Mater. Trans. A* **2000**, *31*, 1857–1865. [CrossRef]
54. Singh, J.; Wani, M.F.; Banday, S.; Shekhar, C.; Singh, G. Nano scratch and nanoindentation: An approach to understand the tribological behavior of MAX phase material Ti_2AlC. *IOP Conf. Ser. Mater. Sci. Eng.* **2019**, *561*, 012111. [CrossRef]
55. Li, S.B.; Yu, W.B.; Zhai, H.X.; Song, G.M.; Sloof, W.G.; van der Zwaag, S. Mechanical properties of low temperature synthesized dense and fine grained Cr_2AlC ceramics. *J. Eur. Ceram. Soc.* **2011**, *31*, 217–224. [CrossRef]
56. Tian, W.; Wang, P.; Zhang, G.; Kan, Y.; Li, Y.; Yan, D. Synthesis and thermal and electrical properties of bulk Cr_2AlC. *Scr. Mater.* **2006**, *54*, 841–846. [CrossRef]
57. Tian, W.; Wang, P.; Zhang, G.; Kan, Y.; Li, Y. Mechanical properties of Cr_2AlC ceramics. *J. Am. Ceram. Soc.* **2007**, *90*, 1663–1666. [CrossRef]
58. Bull, S.J.; Page, T.F.; Yoffe, E.H. An explanation of the indentation size effect in ceramics. *Phil. Mag. Lett.* **1989**, *59*, 281–288. [CrossRef]
59. Chavan, N.M.; Kiran, B.; Jyothirmayi, A.; Sudharshan, P.P.; Sundararajan, G. The corrosion behavior of cold sprayed zinc coatings on mild steel substrate. *J. Therm. Spray Technol.* **2013**, *22*, 463–470. [CrossRef]
60. Luo, X.T.; Li, C.X.; Shang, F.L.; Yang, G.J.; Wang, Y.Y.; Li, C.J. High velocity impact induced microstructure evolution during deposition of cold spray coatings: A review. *Surf. Coat. Technol.* **2014**, *254*, 11–20. [CrossRef]
61. King, P.C.; Bae, G.; Zahiri, S.; Jahedi, M.; Lee, C. An experimental and finite element study of cold spray copper impact onto two aluminum substrates. *J. Therm. Spray Technol.* **2010**, *19*, 620–634. [CrossRef]
62. Li, W.Y.; Zhang, C.; Guo, X.; Li, C.J.; Liao, H.; Coddet, C. Study on impact fusion at particle interfaces and its effect on coating microstructure in cold spraying. *Appl. Surf. Sci.* **2007**, *254*, 517–526. [CrossRef]
63. Zhang, D.; Shipway, P.H.; McCartney, D.G. Cold gas dynamic spraying of aluminium: The role of substrate characteristics in deposit formation. *J. Therm. Spray Technol.* **2005**, *14*, 109–116. [CrossRef]

Article

Influence of Powder Plasticity on Bonding Strength of Cold-Sprayed Copper Coating

Fu-Jun Wei [1,2,*], Bang-Yen Chou [2], Kuan-Zong Fung [1], Shu-Yi Tsai [3] and Chung-Wei Yang [4]

1   Department of Materials Science and Engineering, National Cheng Kung University, Tainan 70101, Taiwan
2   Department of Research and Development, Plus Metal Tech Co., Ltd., Tainan 71247, Taiwan
3   Hierarchical Green-Energy Materials (Hi-GEM) Research Center, National Cheng Kung University, Tainan 70101, Taiwan
4   Department of Materials Science and Engineering, National Formosa University, Yunlin 63201, Taiwan
*   Correspondence: this.is.fjwei@gmail.com

Abstract: When cold spraying is performed at a velocity equivalent to or greater than a specific material-dependent critical velocity, powders suffer intensive plastic deformation and localized heating of interacting surfaces. The thermomechanical reaction of the sprayed powder upon impacting the substrate material triggers thermally dependent metallurgical bonding and/or mechanical interlocking mechanisms. In this study, three Cu feedstocks, fabricated through electrolysis (EP), gas-assisted water atomization (WA), and inert gas atomization (GA), were characterized and annealed before cold spraying. The electron back-scattered diffraction technique was used to analyze the grain structure and plastic microstrain within the powders and coatings. The plastic microstrains that originally existed in the Cu powders were released after 30 min of annealing at 500 °C. The influence of plastic deformation behavior (associated with the grain structure and plastic microstrain of powder feedstocks) on the bonding strength of the cold-sprayed Cu coatings on AA6061 aluminum alloy substrates was examined. The results indicate that EP powder with an asymmetric dendrite morphology was not conducive to the intensive plastic deformation that may cause recrystallized twin grains to form after cold spraying. Furthermore, the homogeneous microstructure of the spherical Cu feedstocks, which may be induced by strain release as recrystallized twin grains and low-angle boundary grain growth through annealing, caused the cold-sprayed Cu coating to have high ductility and low hardness. The findings reveal the low strain hardening and residual stress in the cold-sprayed coating—characteristics regarded as providing key advantages for the bonding strength of the coating.

Keywords: cold spray; plastic deformation; strain; recrystallization; bonding strength

Citation: Wei, F.-J.; Chou, B.-Y.; Fung, K.-Z.; Tsai, S.-Y.; Yang, C.-W. Influence of Powder Plasticity on Bonding Strength of Cold-Sprayed Copper Coating. *Coatings* **2022**, *12*, 1197. https://doi.org/10.3390/coatings12081197

Academic Editor: Yasutaka Ando

Received: 22 July 2022
Accepted: 15 August 2022
Published: 17 August 2022

Publisher's Note: MDPI stays neutral with regard to jurisdictional claims in published maps and institutional affiliations.

1. Introduction

Cold spraying is a technology that is more than two decades old, and it has attracted considerable attention in both academia and industry. In contrast to magnetron sputtering deposition with a thickness of several micrometers [1], cold spray deposition can attain a thickness up to several centimeters [2]. Cold spraying is a rapid kinetic deposition process in which feedstock powders are accelerated to high velocities ranging from 300 to 1200 m/s in a supersonic jet of compressed inert gas, which is preheated to a temperature exceeding 300 °C [2,3]. Upon impacting a substrate, sprayed powders deform in a solid state at temperatures below their melting point. Cold spraying is thus highly suitable for preparing oxidation-sensitive coatings [4]. At a velocity that is equivalent to or greater than a specific material-dependent critical velocity, powders suffer intensive plastic deformation and localized heating of the interacting surfaces because of the aforementioned impact [5–7]. Adiabatic shear instability (i.e., thermal mechanical reaction at an impact interface), which results from the high-strain-rate deformation processes of the powder sample, is believed to cause a localized increase in temperature. The increasing temperature may lead to the formation of metallurgical bonding and dynamic recrystallization, and it may also cause viscous flow, which leads to the formation of a metal jet comprising powder and substrate materials and contributes to mechanical interlocking [6,8–10]. The formation of a metal jet is proposed to be advantageous for the fracturing of the oxide layers that cover a powder and a substrate, as well as a means of enabling true metal to metal contact to be established [3,11].

The bonding of cold-sprayed powder with substrate is highly influenced by powder's plastic deformation, and consequently, various studies have investigated the bonding mechanism by depositing various powder materials with different properties onto different substrates [3,5–7,11–18]. Kim et al. [16] studied bonding mechanisms by examining the recrystallization phenomenon associated with a cold-sprayed titanium coating onto three different substrates (i.e., titanium, aluminum, and zirconia), and various dynamic recrystallization behaviors and interface features were reported. Meng et al. [17] proposed layered powder/substrate and crater powder models for material systems in which the powder and substrate exhibit considerably different deformability (i.e., soft powder (Al)/hard substrate (Cu) and hard powder (Cu)/soft substrate (Al) systems, respectively) in numerical simulations and have metallurgical and mechanical interlocking bonding mechanisms, respectively. Hussain et al. [18] reported that performing various substrate treatments prior to cold spraying a Cu coating exerts various effects on the dominant bonding mechanism. These effects can be attributed to metallurgical and/or mechanical interlocking mechanisms. Hussain et al. further discovered that mechanical interlocking usually accounts for a large proportion of total bonding strength; however, metallurgical bonding only contributed substantially when the AA6082 aluminum alloy substrate was polished and annealed prior to cold spraying.

Four methods are used to commercially produce metal powder for industrial applications, namely, the chemical method, the electrolytic method, the mechanical method, and atomization. Metal powders produced through these four methods differ in terms of properties such as powder size distribution, morphology, oxidation, and flow characteristics [19]. Commercial Cu feedstocks are commonly fabricated through electrolysis (EP) and atomization. Cold-sprayed Cu coatings have potential applications in power electronics [20] and pin fin array heat sinks [21]. In our previous study, three commercially available Cu feedstocks, fabricated through EP, gas-assisted water atomization (WA), and inert gas atomization (GA) processes, were characterized and cold sprayed as coatings. The original features of the powder plasticity of these three Cu powders, as demonstrated through their grain structure and plastic microstrain distribution, caused different thermomechanical reactions upon their impact with AA6061 aluminum alloy substrates and thus resulted in various coating microstructures [22]. The factors that influence the powder plasticity of Cu upon impact with a substrate include the velocity [5,15,23], temperature [8,24–26], morphology [27–29], and strain state [22] of the powder feedstock. The relationship of velocity, temperature, and morphology of Cu feedstocks with the coating microstructure and thus the bonding mechanisms has been reported in aforementioned studies. However, the microstrain of the Cu feedstocks, which significantly determine the plasticity, and its influence on the microstructure and bonding strength of cold-sprayed coating are rarely studied. In the present study, the three Cu feedstocks, which were fabricated through EP, gas-assisted WA, and inert GA processes, were characterized before and after annealing, and then cold sprayed as coatings. The influences of various thermomechanical reactions (originating from the different forms of dynamic recrystallization (plastic deformation) of the powder feedstocks upon impact) on the bonding strength of the cold-sprayed coatings on AA6061 aluminum alloy substrates are discussed herein, accompanied by the demonstration of microstrain both in Cu feedstocks and cold-sprayed coatings.

2. Materials and Methods

2.1. Cu Feedstocks and Annealing

The Cu powders used in the present study were fabricated through EP (JX Nippon Mining & Metals, Tokyo, Japan), gas-assisted WA (Fukuda Metal Foil & Powder, Kyoto, Japan), and inert GA (Thintech Materials Technology, Kaohsiung, Taiwan). The powder size distributions of the feedstocks were determined through laser diffraction (Malvern Mastersizer 2000, Malvern, UK). The morphologies of the Cu powders were inspected through scanning electron microscopy (SEM; Hitachi S3000N, Tokyo, Japan). The Cu powders were annealed for 30 min at 500 °C in a vacuum before cold spraying process. The purpose of this procedure was to further elucidate the differences in the thermomechanical properties of the EP, WA, and GA powders and the effects of annealing-induced microstructural changes on the bonding strength of cold-sprayed coatings.

2.2. Cold-Sprayed Cu Coatings

The PCS-1000 (Plasma Giken, Osato-gun, Japan) cold spray system was used in the present study. The Cu powders were introduced through N_2 carrier gas into the high-pressure chamber of a converging–diverging de Laval-type nozzle and accelerated in a supersonic stream of N_2 inert gas, which was preheated to 600 °C and controlled at 5 MPa. In the present study, the combined effects of velocity and temperature on the plasticity of the Cu powders were assumed to be equivalent under a given set of cold spray parameters. We adopted cold spray parameters sufficient for accelerating Cu feedstocks above the critical velocity of 570 m/s [5,15,23]. The substrate used in the cold spraying experiment was an AA6061 aluminum alloy substrate with the dimensions of 100 (length) × 100 (width) × 3 (thickness) mm^3. The standoff distance was set to 30 mm. The original and annealed Cu powders were cold sprayed as coatings onto the AA6061 substrates. The cold-sprayed coatings produced from the original EP, WA, and GA powders were denoted as the OEP, OWA, and OGA coatings, respectively; the cold-sprayed coatings produced from annealed EP, WA, and GA powders were denoted as the AEP, AWA, and AGA coatings, respectively. The cross-section specimens for measuring Vickers hardness (HV0.1) were cut from the cold-sprayed Cu-coated AA6061 substrates and mounted in thermosetting resin (Buehler KonductoMet, Lake Bluff, IL, USA), after which they were ground and polished. The hardness of a cold-sprayed Cu coating was measured using the Mitutoyo Hardness Testing Machine HM, Kawasaki, Japan.

The bonding strength of the cold-sprayed Cu coating was measured according to the standard test method of ASTM C633 using a tensile test machine. Discs of cold-sprayed Cu-coated AA6061 substrate with a 1 in diameter and 3 mm thickness were cut using a wire saw from 100 (length) × 100 (width) × 3 (thickness) mm^3 plates that have Cu coatings thereon with a nominal thickness of 150 μm. To prepare tensile specimens, two steel rods (each rod having a 1 in diameter and 55 mm length) were grit-blasted by Al_2O_3 particles to roughen and clean the end surfaces. Each cold-sprayed Cu-coated AA6061 disc was affixed between the clean-end surfaces of the two rods by using a heat-cured epoxy (3M Scotch-Weld 2214, St. Paul, MN, USA). The adhesion and curing of the components were conducted under pressure in a fixture at 190 °C for 2 h. The completed specimens were then subjected to the tensile test at a constant crosshead speed of 0.02 mm/s until failure. The bonding strength measurements for each type of testing material and treatment were performed on eight specimens ($n = 8$), and the bonding strength data were reported as means and standard deviations.

2.3. Microstructure Analysis

The electron back-scattered diffraction (EBSD) technique was applied to analyze the grain structure and plastic microstrain within the powders and coatings. Field-emission SEM (FESEM; Zeiss Supra 55 equipped with an Oxford Nordlys EBSD detector, Jena, Germany) was used to obtain secondary electron images and EBSD data. The scan step sizes of 15 or 30 nm were used for EBSD measurements, and the electron beam conditions of

20 kV and 10 nA were applied for the FESEM-based analysis. The cross-section specimens for the EBSD analysis were cut from cold-sprayed Cu-coated AA6061 substrates and mounted in KonductoMet thermosetting resin. Similarly, the Cu powder specimens for the EBSD analysis were first mixed with KonductoMet resin powder and then mounted after undergoing a thermosetting process. The mounted specimens were grounded with SiC paper with a grit up to 1500 and then polished using a 1 μm MicroPolish II suspension (Buehler, Lake County, IL, USA) and 0.25 μm MasterPolish suspension (Buehler, Lake County, IL, USA). The specimens were further polished using a Leica EM TIC 3X Ion Beam Milling System (Wetzlar, Germany) with a beam size of 0.8 mm, voltage level of 8 kV, and current level of 3 mA. For FESEM EBSD analysis, all the mounted specimens were tilted at 70° and kept at a working distance of 12 or 14 mm.

3. Results

3.1. Microstructures and Mechanical Properties of Cu Powders

3.1.1. Morphology

The EP, WA, and GA powders were fabricated through EP, gas-assisted WA, and inert GA, respectively. The powder size distributions and morphologies are reported to have effects on the velocity and temperature of powder feedstocks upon impact substrates [8,15,26–28]. The results for the powder size distributions of these three Cu powders are listed in Table 1, revealing that the WA powder had the smallest powder size. SEM-attained morphologies of the EP, WA, and GA powders are presented in Figure 1a–c, respectively. The morphology of the EP powder (Figure 1a) exhibits a dendritic structure and irregular shape; by contrast, the WA (Figure 1b) and GA (Figure 1c) powders have near-spherical and spherical shapes, respectively.

Table 1. Properties of original and annealed electrolyzed powder (EP), gas-assisted water-atomized (WA) powder, and gas-atomized (GA) powder: morphology, powder size distributions, and percentages of misorientations within grains when angles are >1.5° in kernel average misorientation (KAM).

Powders	EP		WA		GA	
	Original	Annealed	Original	Annealed	Original	Annealed
Morphology	Dendritic		Near-spherical		Spherical	
Powder size $D_{10}/D_{50}/D_{90}$ (μm)	10/30/67	–	7/11/20	–	17/40/79	–
% KAM (>1.5°)	2.3%	0%	0.5%	0.3%	2.3%	0%

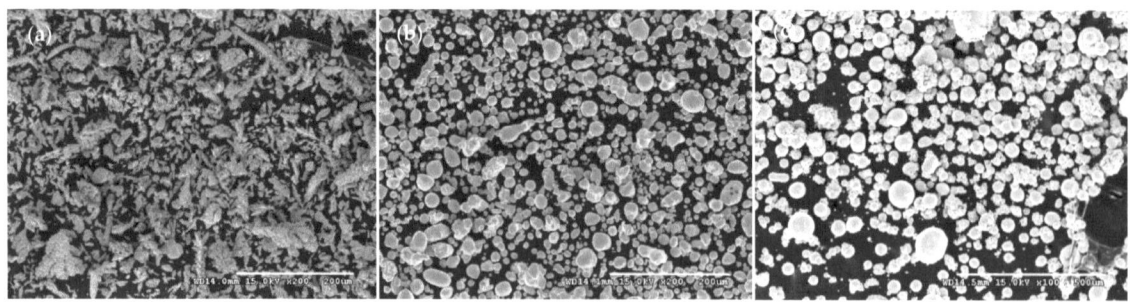

Figure 1. Scanning electron microscopy morphologies of (**a**) electrolyzed, (**b**) gas-assisted water-atomized, and (**c**) gas-atomized Cu powders.

3.1.2. Grain Structure and Strain State

The plasticity of powder can be characterized by the grain structure and strain state. The grain structure and plastic microstrain within the powders before and after annealing were analyzed through EBSD, and the results were presented as grain boundary (GB) and kernel average misorientation (KAM) data. The GBs were resolved and classified as the oriented angles of >15°, 5°–15°, and 1°–5°. The secondary electron (SE) images of the original EP, WA, and GA powders mounted in thermosetting resins are shown in Figure 2a,c,e, respectively, and the corresponding areas analyzed by subsequent EBSD are indicated by circular lines. The KAM + GB maps of the original EP, WA, and GA powders are illustrated in Figure 2b,d,f, respectively. The analysis results reveal that the individual particles of EP, WA, and GA powders are polycrystalline and most GBs have angles greater than 5°. The local crystallographic misorientation or subgrain features within grains can be recognized as the plastic microstrain [30]. The KAM technique was used to reveal the plastic microstrain of the Cu powders represented by local crystallographic misorientation. A scan step size of 15 nm was used to acquire sufficient data to perform a KAM analysis. Plastic microstrain refers to the deformation state of a powder's grain, and it can influence the microstructure of a cold-sprayed coating that has suffered intensive plastic deformation. In the KAM legend in Figure 2, the green area indicates high local misorientation with the angles 0.8°–2.6° relative to the blue area with angles of <0.8° (see the bottom section of Figure 2b,d,f). KAM values of >1.5° were summarized and normalized to the total KAM, and the resultant data are listed as proportional percentages in Table 1. The high angles of local misorientation within the grains of an individual powder indicate the presence of severe strain. All the EP, WA, and GA powders exhibited inherited strain from manufacturing processes; nevertheless, for the percentages of the local misorientation within grains with angles of >1.5°, those of the EP (2.3%) and GA (2.3%) powders were similar, and that of the WA powder (0.5%) was the smallest (Table 1). The results were also consistent with our previous study [22].

The annealed Cu powders were also analyzed by EBSD. The SE images and the corresponding areas analyzed through the EBSD KAM + GB mapping of the annealed EP, WA, and GA powders are presented in Figure 3a–f, respectively. The recrystallization and grain growth phenomena of the annealed Cu powders were assessed by comparing the KAM + GB maps of the annealed Cu powders with those of the original powders (see Figure 2b,d,f and Figure 3b,d,f). The KAM + GB maps presented in Figure 3b,d,f reveal that the local misorientations with high angles of 0.8–2.6° (the green area) almost disappeared. These maps also reveal that the strain release in the grains, as indicated by the blue area (<0.8°) and the presence of low-angle (1–5°) grain boundaries. The aforementioned phenomena are not observed in Figure 2b,d,f, in contrast to the annealing effect. The recrystallization of Cu into a twin-grain structure has been reported by other studies [31–33], and this phenomenon is also observed in the annealed EP and GA powders (Figure 3b,f) examined in the present study. However, the grains of the annealed WA powder were not recrystallized into a twin-grain structure but into equiaxed grains instead. The amount of local misorientation with angles of >1.5° within the grains of the annealed EP, WA, and GA powders was also calculated and is presented as percentage values in accordance with the respective KAM legends in Figure 3b,d,f. Table 1 also lists the analysis results, revealing that the changes in the >1.5° misorientation (presented as percentage values) within the grains of the EP, WA, and GA powders before and after annealing were 2.3% to 0%, 0.5% to 0.3%, and 2.3% to 0%, respectively. The plastic microstrain remaining in the original Cu grains (Figure 2b,d,f) was released after annealing through recrystallization and grain growth (Figure 3b,d,f). The twin boundaries exhibited by the annealed EP and GA powders may influence powder plasticity [33].

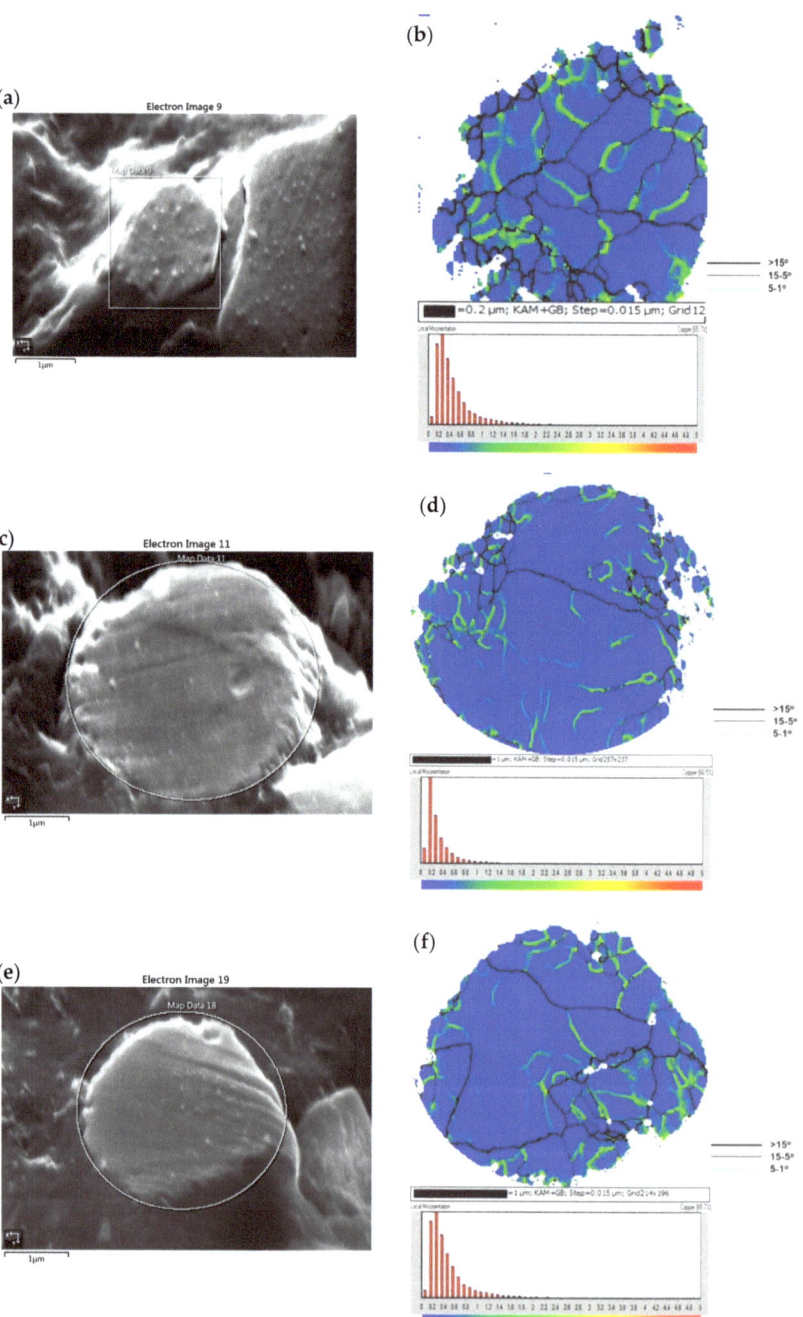

Figure 2. Secondary electron images and corresponding kernel average misorientation (KAM) + grain boundary (GB) maps of original Cu powders produced through various processes: (**a**,**b**), electrolyzed powders; (**c**,**d**), gas-assisted water-atomized powders; (**e**,**f**), gas-atomized powders.

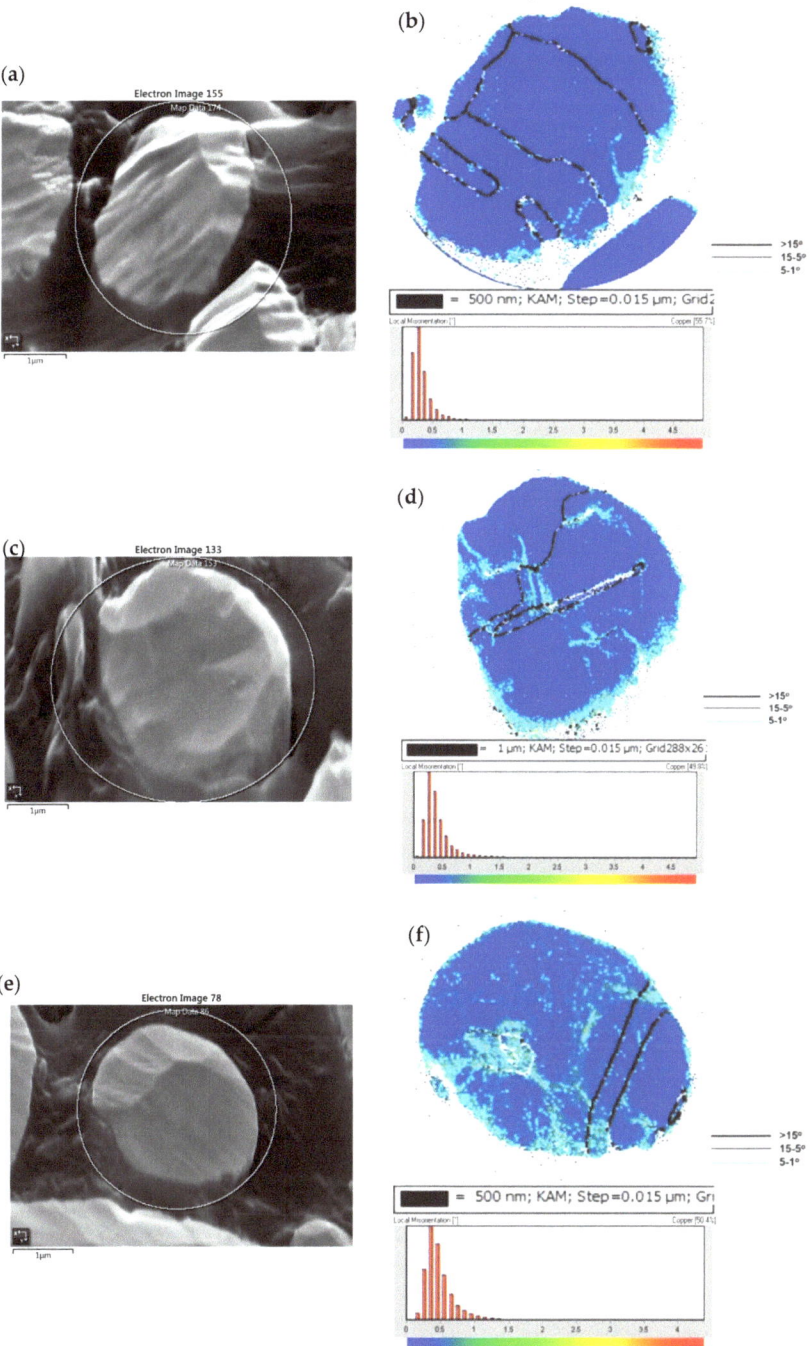

Figure 3. Secondary electron images and corresponding kernel average misorientation (KAM) + grain boundary (GB) maps of annealed Cu powders produced through various processes: (**a**,**b**), electrolyzed powder; (**c**,**d**), gas-assisted water-atomized powder; (**e**,**f**), gas-atomized powders.

3.2. Microstructures and Mechanical Properties of Cold-Sprayed Coatings

3.2.1. Hardness and Bonding Strength

The measured Vickers hardness (HV0.1) values of the cold-sprayed Cu coatings, which can reveal the deformation hardening effect during cold spray deposition, made from original and annealed feedstock powders are listed in Table 2. All the coatings made from annealed powders (i.e., the AEP, AWA, and AGA coatings) have lower hardness values than those made from original powders (i.e., the OEP, OWA, and OGA coatings). The lower hardness and thus lower deformation hardening of the AEP, AWA, and AGA coatings may be due to the improved plasticity of the annealed EP, WA, and GA powders, achieved through strain release and grain growth in addition to the appearance of low-angle (1°–5°) GBs. Table 2 also reveals that the bonding strength levels of the AEP, AWA, and AGA coatings were higher than those of the OEP, OWA, and OGA coatings. It represents that the AGA coating had the most significant improvement in bonding strength with a value of 45.6 ± 3.2 MPa and an increment ratio of 1.98 compared with the OGA coating.

Table 2. Properties of cold-sprayed coatings produced from original electrolyzed powder (OEP), gas-assisted water-atomized powder (OWA), and gas-atomized powder (OGA), and from respective annealed powders (AEP, AWA, and AGA): Vicker's hardness (HV0.1), bonding strength, and percentages of misorientations within grains when angles are >1.5° in kernel average misorientation (KAM) legend.

Coatings	OEP	AEP	OWA	AWA	OGA	AGA
Hardness (HV0.1)	154	120	156	119	136	105
Bonding strength (MPa)	8.0 ± 0.7	11.3 ± 1.8	12.0 ± 1.5	21.3 ± 4.4	23.0 ± 2.4	45.6 ± 3.2
% KAM (>1.5°)	1.2%	0.7%	3.0%	0.5%	1.5%	0%

3.2.2. Grain Structure and Strain State

The grain structure and plastic strain distribution of the cold-sprayed coating located close to the coating–substrate interface are presented as EBSD Euler contrast and KAM + GB maps, respectively. For EBSD measurements, the normal direction of the polished plane of a cross-section specimen is represented as the z-axis of the EBSD coordinate system. Accordingly, the impact direction of the cold spray is parallel to the y-axis. The Euler contrast and KAM + GB maps of the cold-sprayed OEP, OWA, and OGA coatings are presented in Figure 4a–f, respectively. The Euler contrast maps reveal clear grain structures, and the recrystallization of Cu near the coating/substrate interface is observed in the OEP, OWA, and OGA coatings, which are shown in Figure 4a,c,e, respectively. Notably, in the OEP (Figure 4a) and OGA (Figure 4e) coatings, the recrystallization of Cu is characterized as a twin GB; this phenomenon is not observed in the OWA coating (Figure 4c). The Cu powder produced through gas-assisted WA and the subsequent cold-sprayed Cu coating appear to be preferably recrystallized as equiaxed grains. Furthermore, the OGA coating (Figure 4e) appears to have a greater tendency to develop recrystallized twin grains than the OEP coating, which exhibits more new grains that are recrystallized as equiaxed grains instead of twin grains (Figure 4a). According to the KAM + GB maps of the cold-sprayed OEP, OWA, and OGA coatings, as shown in Figure 4b,d,f, respectively, the percentages of local misorientations with angles of >1.5° within the grains were calculated, and the results are listed in Table 2. As indicated in Table 2, the OEP and OGA coatings exhibited more strain release relative to the OWA coating because the percentages of misorientations with angles of >1.5° within the grains in the OEP (1.2%) and OGA (1.5%) coatings were approximately equivalent and smaller than those in the OWA coating (3.0%). Even though the cold-sprayed OEP and OGA coatings had a similar strain release, the hardness of the OGA coating was less than that of the OEP coating (Table 2). The plasticity of Cu can be improved by the occurrence of deformation twins [33]. The OGA coating had a greater tendency to have recrystallized twin grains than did the OEP coating, which appears to explain the lower hardness of the OGA coating relative to the OEP coating. Among the OEP, OWA, and OGA coatings, the cold-sprayed OWA coating had the highest percentage of misorientations, with angles of >1.5°, but did not exhibit a considerable increase in hardness. This phenomenon may be due to the lower level of strain that accumulated in the WA powder (0.5% in Table 1), which resulted in fewer tangled dislocations after the cold spraying process for the WA powder.

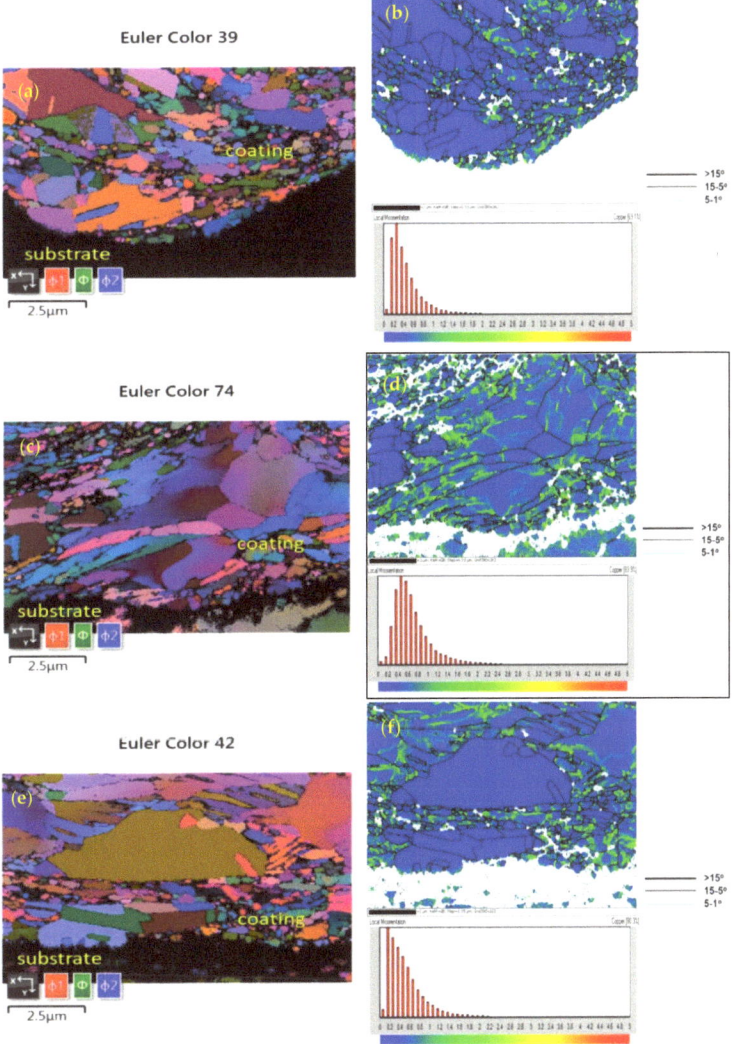

Figure 4. Euler contrast and kernel average misorientation (KAM) + grain boundary (GB) maps of cold-sprayed OEP (**a,b**), OWA (**c,d**), and OGA coatings (**e,f**), recorded at the coating–substrate interface.

The EBSD Euler contrast and KAM + GB maps of the cold-sprayed AEP, AWA, and AGA coatings made from the annealed EP, WA, and GA powders are presented in Figure 5a–f, respectively. The percentages of misorientations with angles of >1.5° within the grains of the AEP, AWA, and AGA coatings were also calculated on the basis of the

KAM + GB maps in Figure 5b,d,f, respectively, and the results are listed in Table 2. Notably, the Euler contrast maps in Figure 5a,c,e reveal that the cold-sprayed AEP, AWA, and AGA coatings generally exhibit well-recrystallized grains and strain release even after intensive deformation upon impact; furthermore, this phenomenon is most prominent in the AGA coating. The percentages of misorientations with angles of >1.5° within the grains in the AEP (0.7%), AWA (0.5%), and AGA coatings (0%) were low (Table 2); this finding is consistent with the occurrence of well-recrystallized grains in the Euler contrast maps of the AEP (Figure 5a), AWA (Figure 5c), and AGA (Figure 5e) coatings and the lower hardness of the AEP, AWA, and AGA coatings relative to the OEP, OWA, and OGA coatings (Table 2).

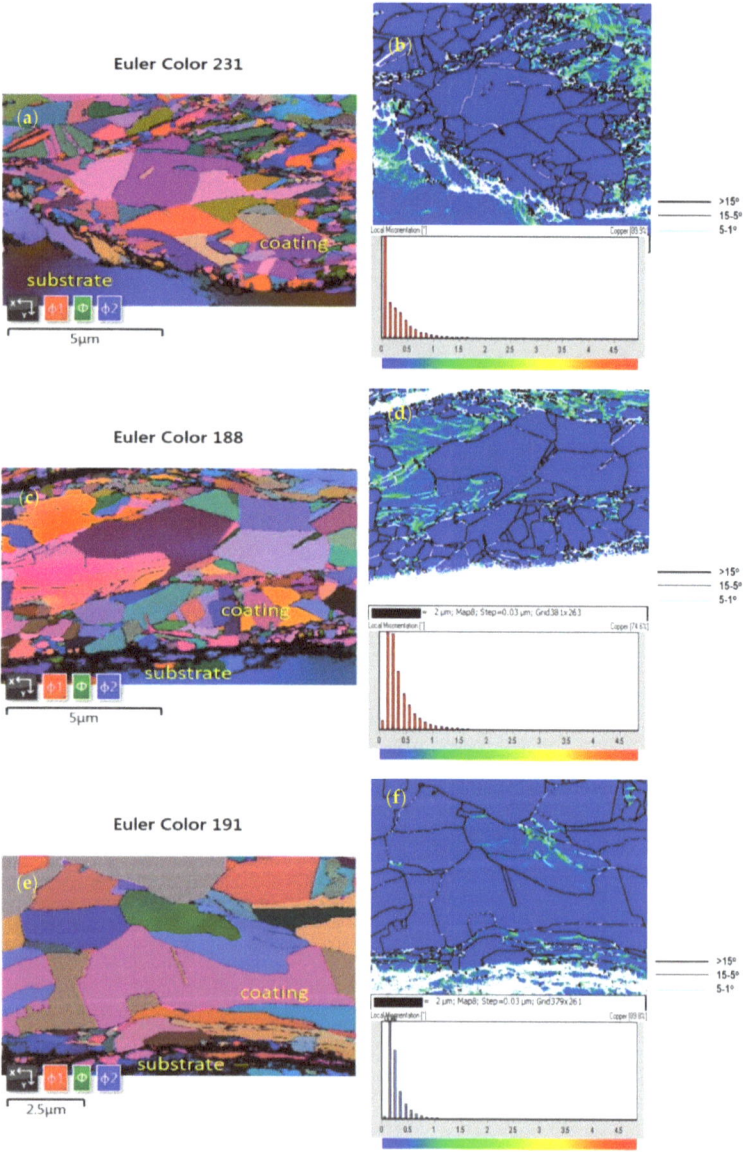

Figure 5. Euler contrast and kernel average misorientation (KAM) + grain boundary (GB) maps of cold-sprayed AEP (**a,b**), AWA (**c,d**), and AGA coatings (**e,f**), recorded at the coating–substrate interface.

4. Discussion

The specimen preparation for EBSD analysis was carried out carefully to minimize any residual surface deformation introduced during grinding and polishing, which can affect background correction and band detection. Etchants which preferentially attack grain boundaries were also avoided. In addition to traditional SiC grinding paper and polish suspension for specimen preparation, ion beam milling was adopted to ensure the quality of EBSD patterns in the present study [34].

4.1. Powder Plasticity

Generally, small powders are easily accelerated to high velocities [14]. However, the critical velocity of a powder greatly depends on powder temperature and it increases with decreasing powder temperature [8,26]. T. Schmidt et al. reported that the larger Cu powder (50 μm) has significantly higher powder temperature than the smaller one (5 μm), with a value of 140 °C in numerical simulation [14]. The significant effect of powder temperature (critical velocity) on the plastic deformation of powder upon impact is also simulated and reported in some studies [24,25]. Consequently, in the present study, the combined effects of velocity and temperature on the plasticity of the Cu powders were assumed to be equivalent under a given set of cold spray parameters. Notably, the property changes in the original powders after undergoing annealing were revealed to have distinctive effects on the microstructures of cold-sprayed coatings. The annealing treatment of the EP, WA, and GA powders were demonstrated to have effects on strain release and low GB angle with respect to the original powders because of recrystallization and grain growth. The angles of the GBs of the original powders were greater than those of the annealed powders even though the microstrains were low, as evident in the original WA powder (0.5%), and the consequent microstrains in the OEP (1.2%), OWA (3.0%), and OGA (1.5%) coatings were higher than those in the AEP (0.7%), AWA (0.5%), and AGA (0%) coatings after cold spraying. Therefore, the strain release and low GB angle of the annealed powders were inferred to be responsible for the formation of recrystallized ductile coatings after cold spraying. Furthermore, the hardness of Cu (approximately 350 HV) is typically higher than that of AA6061 (approximately 100 HV). Several studies have discussed the bonding mechanism of cold-sprayed Cu coating on Al substrate. A study conducted a numerical simulation and reported that the penetration of hard Cu powder into soft Al substrate contributes the bonding mechanism of mechanical interlocking in addition to plastic recrystallization [17], which is regarded as the foundation for the triggering of a metallurgical bonding mechanism [6,8–10]. Schmidt et al. [14] discussed the key role of powder mechanical properties in predicting powder bonding during cold spraying, and they proposed the taking of hardness measurements at the cross-section of a powder for use in predictions. Hussain et al. [18] demonstrated that the mechanical interlocking of cold-sprayed Cu coating and AA6082 substrate materials accounts for a large proportion of total bonding strength. The higher plasticity of the annealed EP, WA, and GA powders (relative to the original powders) because of their strain release and low GB angle may help the viscous flow of Cu powder to penetrate more deeply into the AA6061 substrate under a given set of cold spray parameters. Therefore, a viscous flow leads to the formation of a metal jet consisting of Cu and AA6061 materials, and thus, the mechanical interlocking may be more effective for the cold spraying of annealed Cu powder. In the present study, the bonding strength levels of the cold-sprayed AEP, AWA, and AGA coatings made from the annealed EP, WA, and GA powders, respectively, were generally higher than those of the OEP, OWA, and OGA coatings made from the original powders (Table 2). No evidence of a metal jet consisting of Cu and AA6061 located close to the coating–substrate interfaces of the AEP, AWA, and AGA coatings was revealed through the microstructural EBSD investigation, and this is a topic that requires further clarification. However, the improved powder plasticity that resulted from the strain release and the low GB angle achieved through annealing appears to be beneficial for both metallurgical and mechanical interlocking bonding mechanisms.

The bonding strength of the OEP coating was considerably less than that of the OGA coating (Table 2) even when the microstrains were similar to those of the original powders (2.3% for both powders) and the coatings (1.2% for OEP coating and 1.5% for OGA coating) after cold spraying. In addition to the microstrain factor, the factors influencing powder plasticity upon the impact of a powder with a substrate include velocity [15], temperature [8,26], and morphology [27,28]. In the present study, the combined effects of velocity [15,27,28] and temperature [8,26] on the plasticity of the Cu powders were assumed to be equivalent under a given set of cold spray parameters. Therefore, the morphology of the original EP powder with a dendritic structure was inferred to be responsible for the lower bonding strength of the OEP coating than OGA coating. In one study [27], Cu powders produced through EP and GA were used to prepare cold-sprayed coatings, and the irregular dendritic structure of the EP powder in that study was reported as the reason for the Archimedes porosity of the resulting EP coating being higher than that of the resulting GA coating. Accordingly, the asymmetric dendritic morphology of EP powder was inferred not to be conducive to the intensive plastic deformation that determines the bonding strength of cold-sprayed coating. This reasoning can also be applied to explain the difference in the bonding strength levels of the AEP and AGA coatings.

In our previous study [22], the OWA coating was the only one that exhibited a jet-forming coating structure under metallographic microscopy; it consequently had the highest microstrain out of all tested coatings. The WA powder, which had the smallest powder size among the EP, WA, and GA powders, should have the lowest powder temperature among the three powders as it streams out of the cold spray nozzle [14]. This phenomenon increases the critical velocity of the WA powder [8,26] and limits plastic deformation and recrystallization upon impact. By contrast, the strain release and ductility of the OGA coating (Table 2), which has a 1.5% microstrain and a recrystallized twin-grain structure, result in the coating having a higher bonding strength relative to the OWA coating. Thus, the well-recrystallized structure of a cold-sprayed Cu coating is inferred to have the effect of enhancing bonding strength; however, having less recrystallization and the formation of a residual jet-forming structure do not have the same effect, even though a jet-forming structure is usually regarded as a representation of intensive plastic deformation [5–7].

4.2. Recrystallized Twin Grains

Notably, recrystallized twin grains were observed in the OEP and OGA coatings and the annealed EP and GA powders. Under the similar microstrain conditions of OEP (1.2%) and OGA (1.5%) coatings, it can be recognized that twin grains were more well-recrystallized in the OGA coating (Figure 4e) than in the OEP coating (Figure 4a). Similar results were also shown in our previous study [22]. Twin grains that directly recrystallize through a rotational mechanism because of high-strain-rate deformation during or shortly after impact are arranged in characteristic parallel arrays, whereas twin grains that are formed through migration during the subsequent annealing that occurs in the cooling period following adiabatic strain heating are not arranged in the aforementioned pattern [35]. In the present study, the twins in the OGA coating (Figure 4e) did not have prominent parallel arrays [22]; however, the occurrence of migration twins was more prominent in the OGA coating (Figure 4e) than in the OEP coating (Figure 4a).

The microstrains in the annealed EP and GA powders were 0% and released as recrystallized twin grains, but those in the annealed WA powder were 0.3% and recrystallized as equiaxed grains. The twin grains in the annealed EP and GA powders are expected to be beneficial for plastic deformation upon impact during cold spraying. The higher bonding strength of the AGA coating relative to the AEP and AWA coatings demonstrated that the influencing factors of the powder plasticity include not only strain release and the angles of low GBs but also having a spherical shape and recrystallized twin grains. Table 2 reveals that the cold-sprayed Cu coatings prepared from annealed powders generally have lower levels of hardness relative to those made from the original powders. The greater strain release and ductility of the AEP, AWA, and AGA coatings relative to the OEP, OWA, and

OGA coatings indicate that dynamic recrystallization is essential for improving bonding strength. The homogeneous microstructure of the spherical Cu feedstocks may be induced by strain release, recrystallized twin grains, and grain growth (low-angle boundaries) through annealing such that cold-sprayed Cu coatings exhibit improved ductility and reduced hardness. This finding indicates that the cold-sprayed coatings have less strain hardening and residual stress, which are regarded as key advantages for bonding strength.

5. Conclusions

Three copper feedstocks fabricated through EP, gas-assisted WA, and inert GA processes were characterized and annealed before cold spraying. The thermomechanical reactions of sprayed powders with AA6061 substrates and their effects on the bonding strength of coatings were assessed by examining grain structures and the microstrain of powders and corresponding coatings. The following conclusions are summarized based on our results and discussion:

1. The plasticity of Cu powder is improved through strain release and low GB angle after annealing treatment at 500 °C for 30 min. This plasticity plays a role in increasing the bonding strength of cold-sprayed coatings, characterized by an excellent dynamic recrystallization structure and a bonding strength increment ratio up to 1.98.
2. The Cu powders with an asymmetric dendritic morphology are detrimental for intensive plastic deformation, which results in the lowest bonding strength (8.0 ± 0.7 MPa) of cold-sprayed coating.
3. Recrystallized twin grains are beneficial for plasticity, and they can help increase the bonding strength of cold-sprayed coatings.

Author Contributions: Conceptualization, experiment, writing—original draft preparation, F.-J.W.; Methodology, investigation, B.-Y.C.; Supervision, K.-Z.F.; Writing—review and editing, S.-Y.T. and C.-W.Y. All authors have read and agreed to the published version of the manuscript.

Funding: This research received no external funding.

Institutional Review Board Statement: Not applicable.

Informed Consent Statement: Not applicable.

Data Availability Statement: No new data were created or analyzed in this study. Data sharing is not applicable to this article.

Conflicts of Interest: The authors declare no conflict of interest.

References

1. Mohammadpour, E.; Liew, W.Y.H.; Radevski, N.; Lee, S.; Mondinos, N.; Altarawneh, M.; Minakshi, M.; Amri, A.; Rowles, M.R.; Lim, H.N.; et al. High temperature (up to 1200 °C) thermal-mechanical stability of Si and Ni doped CrN framework coatings. *J. Mater. Res. Technol.* **2021**, *14*, 2406–2419. [CrossRef]
2. Papyrin, A.N.; Kosarev, V.F.; Klinkov, S.; Alkhimov, A.P.; Fomin, V. *Cold Spray Technology*; Elsevier: Amsterdam, The Netherlands, 2007.
3. Alkhimov, A.P.; Papyrin, A.N.; Kosarev, V.F.; Nesterovich, N.I.; Shuspanov, M.M. Gas-Dynamic Spraying Method for Applying a Coating. U.S. Patent No. 5,302,414, 12 April 1994.
4. Voyer, J.; Stoltenhoff, T.; Kreye, H. Development of Cold Sprayed Coatings. In *Thermal Spray 2003: Advancing the Science and Applying the Technology, Proceedings of the 2003 International Thermal Spray Conference, Orlando, FL, USA, 5–8 May 2003*; Marple, B.R., Moreau, C., Eds.; ASM International: Almere, The Netherlands, 2003; pp. 71–78.
5. Stoltenhoff, T.; Kreye, H.; Richter, H.J. An analysis of the cold spray process and its coatings. *J. Therm. Spray Technol.* **2002**, *11*, 542–550. [CrossRef]
6. Assadi, H.; Gärtner, F.; Stoltenhoff, T.; Kreye, H. Bonding mechanism in cold gas spraying. *Acta Mater.* **2003**, *51*, 4379–4394. [CrossRef]
7. Schmidt, T.; Gärtner, F.; Assadi, H.; Kreye, H. Development of a generalized parameter window for cold spray deposition. *Acta Mater.* **2006**, *54*, 729–742. [CrossRef]
8. Grujicic, M.; Zhao, C.L.; DeRosset, W.S.; Helfritch, D. Adiabatic shear instability based mechanism for particles/substrate bonding in the cold-gas dynamic-spray process. *Mater. Des.* **2004**, *25*, 681–688. [CrossRef]

9. Dykhuizen, R.C.; Smith, M.F.; Gilmore, D.L.; Neiser, R.A.; Jiang, X.; Sampath, S. Impact of high velocity cold spray particles. *J. Therm. Spray Technol.* **1999**, *8*, 559–564. [CrossRef]
10. Li, W.Y.; Liao, H.L.; Li, C.J.; Bang, H.S.; Coddet, C. Numerical simulation of deformation behavior of Al particles impacting on Al substrate and effect of surface oxide films on interfacial bonding in cold spraying. *Appl. Surf. Sci.* **2007**, *253*, 5084–5091. [CrossRef]
11. Champagne, V.K. *The Cold Spray Materials Deposition Process: Fundamentals and Applications*; CRC: Cambridge, UK, 2007.
12. Gilmore, D.L.; Dykhuizen, R.C.; Neiser, R.A.; Smith, M.F.; Roemer, T.J. Particle velocity and deposition efficiency in the cold spray process. *J. Therm. Spray Technol.* **1999**, *8*, 576–582. [CrossRef]
13. Gärtner, F.; Stoltenhoff, T.; Schmidt, T.; Kreye, H. The cold spray process and its potential for industrial applications. *J. Therm. Spray Technol.* **2006**, *15*, 223–232. [CrossRef]
14. Schmidt, T.; Assadi, H.; Gärtner, F.; Richter, H.; Stoltenhoff, T.; Kreye, H.; Klassen, T. From particle acceleration to impact and bonding in cold spraying. *J. Therm. Spray Technol.* **2009**, *18*, 794–808. [CrossRef]
15. Huang, R.; Ma, W.; Fukanuma, H. Development of ultra-strong adhesive strength coatings using cold spray. *Surf. Coat. Technol.* **2014**, *258*, 832–841. [CrossRef]
16. Kim, K.; Watanabe, M.; Kuroda, S. Bonding mechanisms of thermally softened metallic powder particles and substrates impacted at high velocity. *Surf. Coat. Technol.* **2010**, *204*, 2175–2180. [CrossRef]
17. Meng, F.; Hu, D.; Gao, Y.; Yue, S.; Song, J. Cold-spray bonding mechanisms and deposition efficiency prediction for particle/substrate with distinct deformability. *Mater. Design* **2016**, *109*, 503–510. [CrossRef]
18. Hussain, T.; McCartney, D.G.; Shipway, P.H.; Zhang, D. Bonding mechanisms in cold spraying: The contributions of metallurgical and mechanical components. *J. Therm. Spray Technol.* **2009**, *18*, 364–379. [CrossRef]
19. Huppmann, W.J.; Dalal, K. *Metallographic Atlas of Powder Metallurgy*; Verlag Schmid Gmbh: Freiburg, Germany, 1986.
20. Kosarev, V.F.; Klinkov, S.V.; Melamed, B.M.; Nepochatov, Y.K.; Ryashin, N.S.; Shikalov, V.S. Cold spraying for power electronics: Deposition of thick topologically patterned copper layers on ceramics. In Proceedings of the International Conference on the Methods of Aerophysical Research, Novosibirsk, Russia, 13–19 August 2018; p. 030047.
21. Perry, J.; Richer, P.; Jodoin, B.; Matte, E. Pin fin array heat sinks by cold spray additive manufacturing: Economics of powder recycling. In Proceedings of the International Thermal Spray Conference, Orlando, FL, USA, 7–9 May 2018; ASM International: Almere, The Netherlands, 2018.
22. Wei, F.J.; Chou, B.Y.; Tsai, S.Y.; Fung, K.Z. Thermomechanical properties of cold-sprayed copper coatings from differently fabricated powders. *Surf. Coat. Technol.* **2022**, *434*, 128128. [CrossRef]
23. Jakupi, P.; Keech, P.G.; Barker, I.; Ramamurthy, S.; Jacklin, R.L.; Shoesmith, D.W.; Moser, D.E. Characterization of commercially cold sprayed copper coatings and determination of the effects of impacting copper powder velocities. *J. Nucl. Mater.* **2015**, *466*, 1–11. [CrossRef]
24. Yu, M.; Li, W.Y.; Wang, F.F.; Suo, X.K.; Liao, H.L. Effect of particle and substrate preheating on particle deformation behavior in cold spraying. *Surf. Coat. Technol.* **2013**, *220*, 174–178. [CrossRef]
25. Yin, S.; Wang, X.; Suo, X.; Liao, H.; Guo, Z.; Li, W.; Coddet, C. Deposition behavior of thermally softened copper particles in cold spraying. *Acta Mater.* **2013**, *61*, 5105–5118. [CrossRef]
26. Fukanuma, H.; Ohno, N.; Sun, B.; Huang, R. In-flight particle velocity measurements with DPV-2000 in cold spray. *Surf. Coat. Technol.* **2006**, *201*, 1935–1941. [CrossRef]
27. Li, Y.J.; Luo, X.T.; Li, C.J. Dependency of deposition behavior, microstructure and properties of cold sprayed Cu on morphology and porosity of the powder. *Surf. Coat. Technol.* **2017**, *328*, 304–312. [CrossRef]
28. Li, Y.J.; Luo, X.T.; Rashid, H.; Li, C.J. A new approach to prepare fully dense Cu with high conductivities and anti-corrosion performance by cold spray. *J. Alloy. Comp.* **2018**, *740*, 406–413. [CrossRef]
29. Luo, X.T.; Ge, Y.; Xie, Y.; Wei, Y.; Huang, R.; Ma, N.; Ramachandran, C.S.; Li, C.J. Dynamic evolution of oxide scale on the surfaces of feed stock particles from cracking and segmenting to peel-off while cold spraying copper powder having a high oxygen content. *J. Mater. Sci. Technol.* **2021**, *67*, 105–115. [CrossRef]
30. Hsu, W.C.; Chang, L.; Kao, P.W. Study of potential recrystallization nuclei in the cold-rolled microstructure of an electrical steel by electron backscatter diffraction. *Mater. Sci. Eng.* **2019**, *580*, 012034. [CrossRef]
31. Borchers, C.; Gärtner, F.; Stoltenhoff, T.; Assadi, H.; Kreye, H. Microstructural and macroscopic properties of cold sprayed copper coatings. *J. Appl. Phys.* **2003**, *93*, 10064–10070. [CrossRef]
32. Feng, Y.; Li, W.; Guo, C.; Gong, M.; Yang, K. Mechanical property improvement induced by nanoscaled deformation twins in cold-sprayed Cu coatings. *Mater. Sci. Eng. A* **2018**, *727*, 119–122. [CrossRef]
33. Wang, Y.B.; Sui, M.L.; Ma, E. In situ observation of twin boundary migration in copper with nanoscale twins during tensile deformation. *Philos. Mag. Lett.* **2007**, *87*, 935–942. [CrossRef]
34. Nowell, M.M.; Witt, R.A.; True, B. EBSD sample preparation: Techniques, tips, and tricks. *Microsc. Microanal.* **2005**, *11* (Suppl. 2), 504–505. [CrossRef]
35. Murr, L.E.; Niou, C.-S.; Pappu, S.; Rivas, J.M.; Quinones, S.A. LEDS in ultra-high strain-rate deformation. *Phys. Status Solidi A* **1995**, *149*, 253–274. [CrossRef]

Review

A Review of Advances in Cold Spray Additive Manufacturing

Rodolpho Fernando Vaz *, Andrea Garfias, Vicente Albaladejo, Javier Sanchez and Irene Garcia Cano

Thermal Spray Centre CPT, Universitat de Barcelona, 08028 Barcelona, Spain
* Correspondence: rvaz@cptub.eu

Abstract: Cold Spray Additive Manufacturing (CSAM) produces freeform parts by accelerating powder particles at supersonic speed which, impacting against a substrate material, trigger a process to consolidate the CSAM part by bonding mechanisms. The literature has presented scholars' efforts to improve CSAM materials' quality, properties, and possibilities of use. This work is a review of the CSAM advances in the last decade, considering new materials, process parameters optimization, post-treatments, and hybrid processing. The literature considered includes articles, books, standards, and patents, which were selected by their relevance to the CSAM theme. In addition, this work contributes to compiling important information from the literature and presents how CSAM has advanced quickly in diverse sectors and applications. Another approach presented is the academic contributions by a bibliometric review, showing the most relevant contributors, authors, institutions, and countries during the last decade for CSAM research. Finally, this work presents a trend for the future of CSAM, its challenges, and barriers to be overcome.

Keywords: cold spray; additive manufacturing; 3D-printing; geometries; properties; innovation

1. Introduction

Additive Manufacturing (AM) has been an industrial revolution in recent decades, starting with producing polymeric parts and advancing to metallic components. Many alloys and methods have been studied, some more industrially mature and others in a developing stage. The definition of Additive Manufacturing (AM) given by ISO/ASTM 52,900:2015 standard [1] is the "process of joining materials to make parts from 3D model data, usually layer upon layer, as opposed to subtractive manufacturing and formative manufacturing methodologies". Other nomenclatures have been used worldwide as synonyms for AM, such as 3D printing, additive fabrication, rapid prototyping, and others. AM has been used to build prototypes, manufacture the final products, or even repair damaged components, innovating the global manufacturing industry [2–6]. Many companies have invested in developing new AM techniques and materials, optimizing the process parameters, reducing costs, and making the AM a competitive piece of technology [7,8]. Different sectors have benefited from using AM [9,10], such as medical [11–16], aerospace [17–20], automotive [21–23], supply chain [6,24–26], and others. Compared to the traditional subtractive manufacturing techniques, AM is characterized by being less wasteful, enhancing resource efficiencies, and changes in the design and production phases. Kozoir [27] presents the effectiveness of optimizing AM processing parameters to reduce the mass of models, keeping the desired mechanical properties. AM also extends the product life cycle by repairing high-cost parts, and reconfigures the value chains to be shorter, collaborative, and offer remarkable sustainability benefits [6,28]. In this way, AM offers clear benefits from the viewpoint of sustainability [29–31].

The commercial use of AM emerged for polymers in the 1980s, introducing Stereolithography (SL), which involves curing a photosensitive liquid polymer by a laser beam [32,33]. An evolution in equipment changed the raw material to the powder form, using Selective Laser Sintering (SLS) to fuse this powder [34]. Other classes of AM for polymers are Material Jetting (MJ) [35,36], Binder Jetting (BJ) [37,38], Material Extrusion

Citation: Vaz, R.F.; Garfias, A.; Albaladejo, V.; Sanchez, J.; Cano, I.G. A Review of Advances in Cold Spray Additive Manufacturing. *Coatings* **2023**, *13*, 267. https://doi.org/10.3390/coatings13020267

Academic Editor: Mohammadreza Daroonparvar

Received: 31 December 2022
Revised: 19 January 2023
Accepted: 20 January 2023
Published: 23 January 2023

(ME) [39,40], and Sheet Lamination or Laminated Object Manufacturing (LOM) [41]. The techniques consolidated for polymers have been successfully applied for other materials also, such as BJ for ceramics and metals [37,42,43], LOM for metals [44,45], and ME for composites [40,46]. Various processes are available for metal AM processing for the most different alloys and applications. The selection or choice of the adequate process depends on the part's geometry, complexity, mechanical properties, and other factors [47,48].

The metal AM processes differ from the heat source and metal feeding method or type. Some options are the laser process, Selective Laser Melting (SLM) or Sintering (SLS), Direct Metal Laser Melting (DMLM) or Sintering (DMLS), or Laser Metal Fusion (LMF), besides the Electron Beam Melting (EBM) process [49–51]. These are methods which are applied to the parts that need low or no machining post-processing or are used directly as end-use products. Other processes are presented in the literature but are not capable of producing complex geometries, such as Gas Tungsten Arc Welding (GTAW) [52–54], Gas Metal Arc Welding (GMAW) or Wire Arc Additive Manufacturing (WAAM) [55–58], Plasma Arc Welding (PAW) [57,59–61], Friction Stir Energy Manufacturing (FSAM) [62,63], and Ultrasonic Additive Manufacturing (UAM) [64,65]. Examples of AM by welding processes that demand post-machining are repairing long fatigue cracks in hydro powerplant runners [66] or repairing eroded gas turbine blades [67].

Cold Spray (CS) is a thermal spray process designed for coatings that has extended its use to produce freeform parts [28,68–70]. CS produces harder microstructures than other AM processes, as studied by Gamon et al. [71], who present CSAM-ed Inconel 625 with 600 HV. On the other hand, WAAM, SLM, EBM, DMLM, and BJ resulted in less than 300 HV. Figure 1 presents the AM technology maturity, evidencing the actual industrial use of the laser processes, SLM and DMLM, as WAAM. The prediction is to use CSAM industrially in a short time, less than two years, but a long development journey for FSAM and UAM [72]. This work aims to present the trodden path by CS as an AM technique and the foreseen way to consolidate and diffuse CSAM in the industry. Figure 2 shows examples of AM-made products employing different strategies.

This paper presents and discusses the evolution and advances of CSAM critically, following the scheme shown in Figure 3.

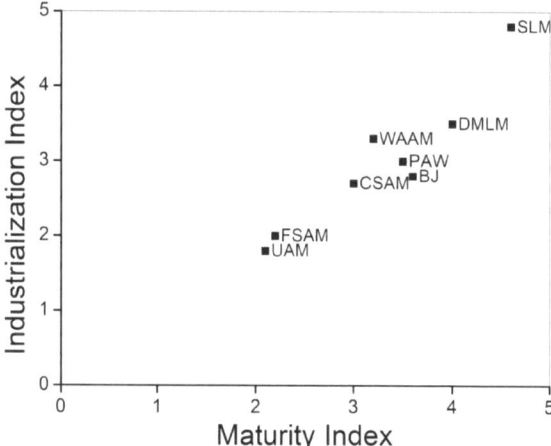

Figure 1. AM maturity index for producing metallic parts.

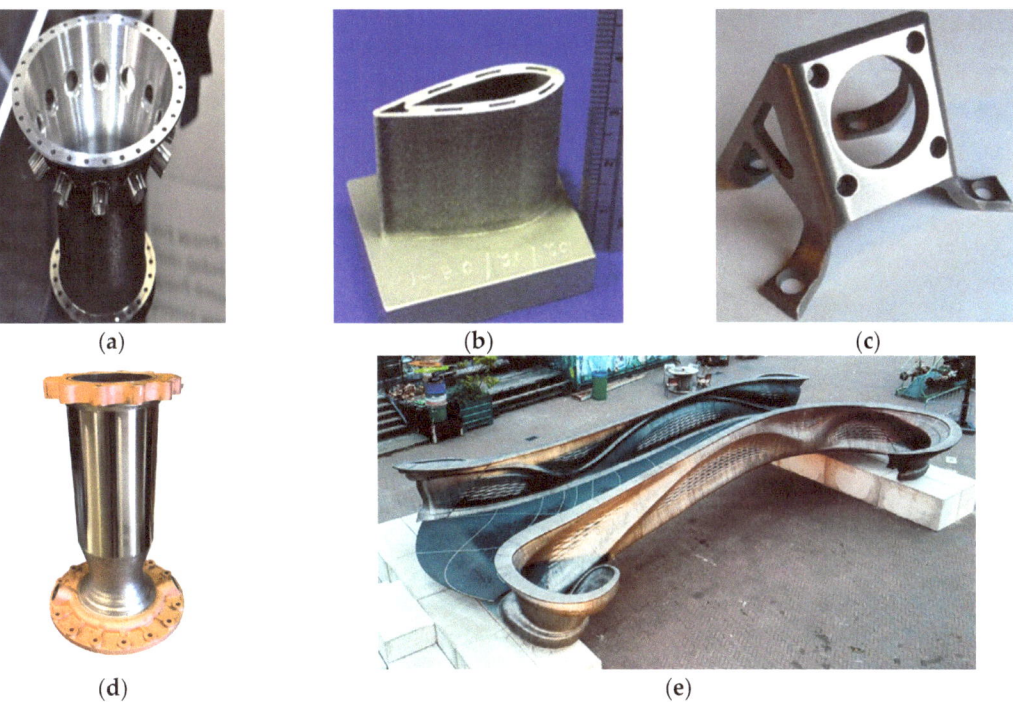

(a)   (b)   (c)

(d)   (e)

Figure 2. Metal AM parts. (**a**) DMLM-ed stainless steel gas turbine housing. Reprinted with permission from Ref. [18], Elsevier, 2017. (**b**) DMLM-ed Ti6Al4V airfoil [18], (**c**) CSAM-ed Ti bracket. Reprinted with permission from Ref. [2], Elsevier, 2018. (**d**) CSAM-ed and DMLM-ed bimetallic thrust chamber. Reprinted with permission from Ref. [73], NASA, 2021. (**e**) WAAM-ed stainless steel bridge in Amsterdam. Reprinted with permission from Ref. [74], Elsevier, 2019. Unit: mm.

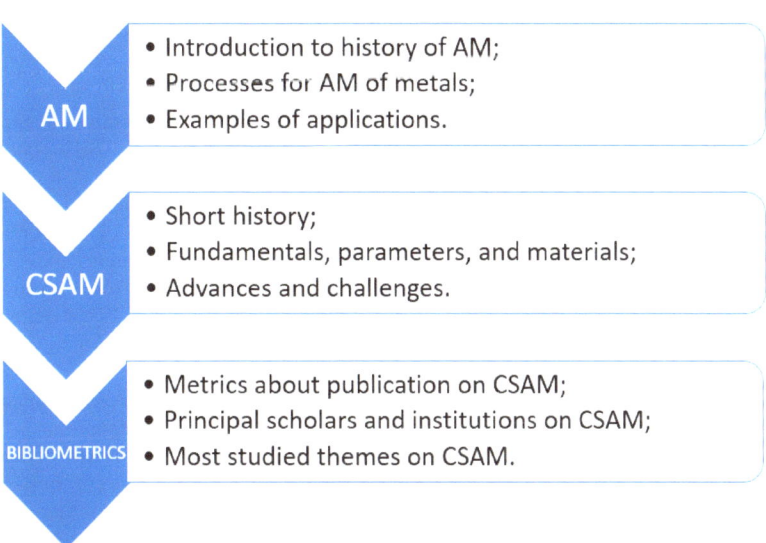

AM
- Introduction to history of AM;
- Processes for AM of metals;
- Examples of applications.

CSAM
- Short history;
- Fundamentals, parameters, and materials;
- Advances and challenges.

BIBLIOMETRICS
- Metrics about publication on CSAM;
- Principal scholars and institutions on CSAM;
- Most studied themes on CSAM.

Figure 3. Flowchart of the topics presented in the work.

2. Cold Spray Process

This section describes the CS process, its fundamentals, principles, parameters, and their selection, which is essential to understanding and placing CS as an AM process. CS is a thermal spray process investigated and presented by many authors as an alternative to producing AM freeform parts. It has severe differences from the laser, welding, and other thermal spray processes since CS does not change the properties of the feedstock powder by heating or melting during the AM part fabrication because the powder is kept below its recrystallization temperature during the spraying time [75–77]. However, CSAM produces parts with a very high density, >99%, due to the very high velocity imposed on the particles, reaching supersonic velocity values [78–80]. Therefore, the correct selection of feedstock powder, deposition parameters, and strategy are fundamental for achieving high Deposition Efficiency (DE) and good CSAM performance [81–84]. CS also prevents materials oxidizing during the deposition due to the relatively low temperature that the material absorbs at the spraying time [85]. In addition, CS avoids other harmful effects seen in other AM or thermal spray processes, such as evaporation, melting, recrystallization, tensile residual stresses, debonding, and gas releasing, besides the ability to deposit high-reflective metals such as Cu and Al [28,83,86,87]. A great CSAM advantage is the possibility of the deposition of dissimilar materials, e.g., a sandwich-like structure of Cu and Al [88], which is not feasible by welding.

Historically, CS has been presented in the literature by different names: "kinetic energy metallization", "kinetic spraying", "solid-state deposition", or "high-velocity powder deposition", and others [78,89]. Its principle and physics of operation were studied during the XX century, with the operational evolution starting in the 1980s. Still, its commercial development started just in the early 2000s [86,89,90], increasing its expansion from the R & D sector to the industry since then, and with a prediction of widespread industrial use in less than two years [91]. CS is the thermal spraying process to which a large number of studies and publications have been devoted over recent years, presenting its principles and physics, but, nowadays, emphasizing industrial or real applications of the technique and mainly its use in the AM field [2,28,81,92–97]. The monetary benefits are imperative to select CSAM as an industrial production technology. A comparison among the metal AM technologies was presented by Munsh et al. [91], and CSAM was highlighted as the lowest cost per volume fabricated and the highest deposition rate, reaching kg·h^{-1} [79,81]. Besides the component at hand, the advantages of AM over traditional or subtractive fabrication processes include the redesign potential of the whole system, which is not easily measurable [91].

CSAM produces a coating or bulk component generated by a solid-state cohesion during the powders' impact on a substrate. The working gas is previously heated in a chamber, reaching high pressure, flowing through a de Laval or similar convergent–divergent nozzle, accelerating it to supersonic velocities, and dragging the feedstock powders. [68,78,98]. The working gas pressure classifies CS, as presented schematically in Figure 4. Low-Pressure Cold Spray (LPCS) operates under 1 MPa, and High-Pressure Cold Spray (HPCS) uses higher pressure levels. A Medium-Pressure Cold Spray (MPCS) is a commercially available system, Titomic D623. LPCS is limited to a few materials and can be portable or manually operated, accrediting it for in-field operation and repair services. At the same time, HPCS is the CSAM used for many materials, but has heavier and more equipment than LPCS, employing a bigger gun, heat exchanger, energy source, robot, and acoustic enclosure (soundproof booth) for the operation, because the noise usually exceeds 100 dB [2,10,99,100]. This change in gas pressure and equipment configuration influences the sprayed particle velocity since the high velocity of particles is a consequence of high gas pressure and the nozzle design [101–105]. Another difference between the LPCS and HPCS is the powder feeding; for the first one, the particles are dragged by the working gas in the nozzle directly, using a downstream mode. On the other hand, HPCS uses an upstream injection mode, and the powder feeder is connected to a feeding gas line, which improves the powder flowability, and increases the range of powders which are CS sprayable [90,106].

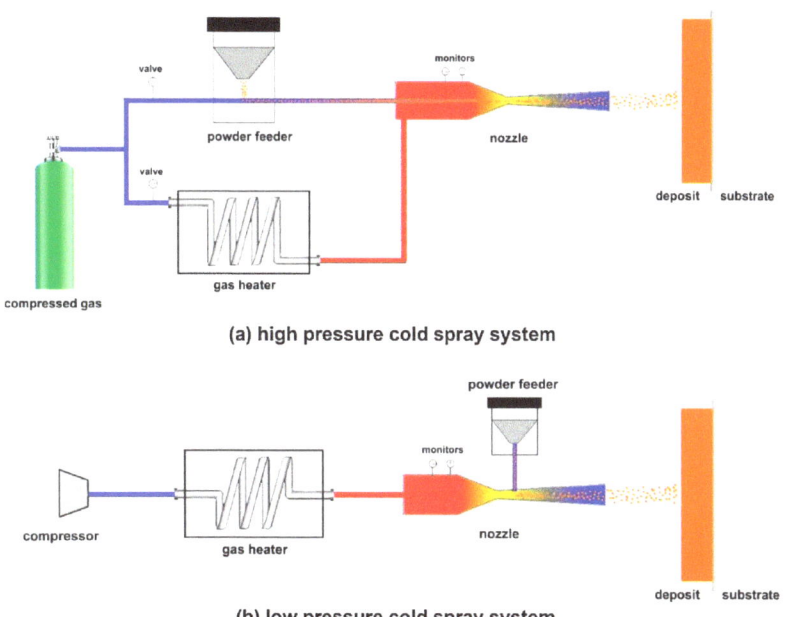

(a) high pressure cold spray system

(b) low pressure cold spray system

Figure 4. LPCS and HPCS schemes. Reprinted with permission from Ref. [2], Elsevier, 2018.

The bonding mechanism of the solid-state particles to a substrate still has to be understood entirely. However, it is believed that their high energy at the impact disrupts the oxide films on the particle and substrate surfaces, pressing their atomic structures into intimate contact with each other under short high interfacial pressures and temperatures. This mechanism is called Adiabatic Shear Instability (ASI) [107,108]. It supports the success in coating ductile materials, such as Cu and Al, and flops in spraying brittle materials, such as ceramics or carbides [78]. At the impact, most of the kinetic energy from the in-flight particles is converted into heat or the plastic deformation of the substrate and the particle, which can produce strain, ultimately shear instability, and jetting. With an increase in local temperature, thermal softening alters the capacity of the material to transmit shear forces, and eventually, the softening process dominates over strain hardening [109–111]. Hassani-Gangaraj et al. [112] show the jetting happening with or without the material having a thermal softening capacity, proposing that CS jetting is formed as a result of strong pressure waves in the particles, expanding the particle edges. This mechanism is related to hydrodynamic processes that promote jetting, such as liquid droplet impacting, shaped-charge jetting, and explosion welding. The critical velocity (V_{cr}) of particle for the bonding was mathematically related to the bulk speed of sound, which was minutely commented by Assadi et al. [113], who refuted those conclusions and sustained the ASI as the strongest and the primary bonding mechanism for CS-ed particle. Chen et al. [114] also proposed a low-velocity impact-induced metal bonding, in which the conventionally accepted metal jetting and melting may not be prerequisites for solid-state impact-induced bonding.

2.1. Cold Spray Parameters

The properties of CSAM-ed parts, such as density, porosity, adhesion, or hardness, depend on the CS spraying parameters, which have to be set to spray the particles in a specific velocity range or deposition window [79,80,85,103]. A velocity of a particle below a $V_{critical}$ or V_{cr} value does not promote the particle bonding, and an excessive velocity,

$V_{erosion}$ or V_{er}, results in the erosion of the substrate instead of a deposition consolidation. This ideal velocity depends on the particles' properties and substrate materials [80,115,116]. Table 1 lists the V_{cr} and V_{er} for the most CSAM-ed materials.

Table 1. Window of deposition for CS.

Material	$V_{critical}$ [m·s^{-1}]	$V_{erosion}$ [m·s^{-1}]	Ref.
Al	625	1250	[79,117]
316L	550	1500	[79,117–119]
Cu	570	1000	[79,120]
Ti	700	1750	[79,117,121]
Ti6Al4V	750	2500	[122]
Ni	570		[117,120]
Inconel 718	600	1700	[123,124]

Process parameters optimization is based on particular applications and equipment, working gases, substrate and feedstock materials' characteristics, and others. Typically, these parameters include the gas type, temperature, pressure, nozzle geometry, throat size, and deposition robot strategy. In addition, a critical point is the feedstock powder material itself, particle size distribution, shape, and particle attributes, such as oxide skins and mechanical properties, which influence the ability to form a compacted layer [78,80,83–85,99].

For the CSAM, the working or main gas commonly used is N_2 or He, or N_2/He mixtures, but for LPCS and MPCS, compressed air is a low-cost option also [2,90,125,126]. N_2 has a lower cost than He and, due to the high consumption of the working gas, it is the choice for the main gas. For CS using He instead of N_2, the particles are propelled with a higher velocity due to He's higher atomic mass [126–130], e.g., CS-ed 316L (particle size 28 μm) with He reaches 750 m·s^{-1}, but less than 500 m·s^{-1} with N_2 as the working gas [131]. The CS working gas temperature is set up to high values in the CS gun heating chamber, e.g., 1100 °C for spraying 316L [132]; however, after passing through the nozzle, the gas expands, reducing the density and temperature [131,133]. Lee et al. [134] presented a CFD gas flow simulation in which a CS gas heating chamber at 1200 K and 20 bar resulted in less than 800 K in the CS gas jet, but a velocity higher than 1300 m·s^{-1}. Considering the heat transfer inertia from the gas to the particle and the short time of exposition, the temperature of the particles is much lower than 800 K, maintaining the sprayed particles below their recrystallization temperature. It influences the properties of the sprayed material, such as the particles' cohesion, adhesion, strength, and others. For example, for Ti coatings, the cohesion measured by TCT (Tubular Coating Tensile) [135] had a linear relation with the gas temperature [121], and higher cohesion corroborates a material with a lower porosity, higher strength, and DE. However, by selecting a high gas temperature, the cold work and hardness in the material are dwindled by partial recovery and recrystallization phenomena [136,137].

The Standoff Distance (SD) is how far the substrate surface is from the gun nozzle exit. This distance has an optimum value, where the velocity of particles reaches the peak, impacting the substrate with the highest energy possible. A relation presented in the literature as a reference for an excellent SD is seven times the gas jet diameter. Further, the pressure reduces drastically [86], e.g., for a 3 mm gas jet diameter, the SD should be 21 mm. Turbulences, the oscillation of the gas jet, and the irregular distribution of the particles impacting the substrate are also seen to increase the SD, which reduces the DE [138], as confirmed experimentally for CS-ed Al, Cu, and Ti [139]. The adherence of CSAM-ed Ti6Al4V on the steel substrate increased by optimizing the SD parameter, reaching the best value of 50 mm without delamination [140], showing that the relation of an SD seven times the gas jet diameter proposed by Kosarev et al. [86] is just a starting point for parameter optimization and not a rule.

The robot path and velocity influence the characteristics and properties of the CS-ed material; the step between the sprayed single tracks has to be optimized to guarantee good adherence and produce a flat and smooth deposit surface because an insufficient overlapping distance results in a wave surface [138]. Therefore, rotating is one of the most applied strategies for CSAM, building up the part by coating a rotating pipe-like substrate, resulting in parts with symmetry, such as the one presented in Figure 2d, after the post-machining process. This strategy promotes the good adhesion and cohesion of particles but limits the geometries feasible to the symmetrical ones. The use of alternate directions, Figure 5b with the CS laden-jet particles in the Z-direction, increased the material's isotropy when compared to the traditional strategy, Figure 5a with the CS laden-jet particles in the Z-direction, for CS-ed Cu thick parts [141]. Compression tests in the X- and Y-direction indicated different crack propagation paths for the bidirectional strategy, revealing that the robot path influences the preferential direction for crack propagation [142]. The robot path also may change the angle of the impact of particles, drastically affecting the DE and material microstructure. For CSAM, the robot path has a crucial function since the part sidewalls grow up and follow an angle, which has to be rectified to the designed and desired inclination. An adequate robot programme can spray on the inclined sidewall with a jet angle that corrects it, improving DE [92,143–145].

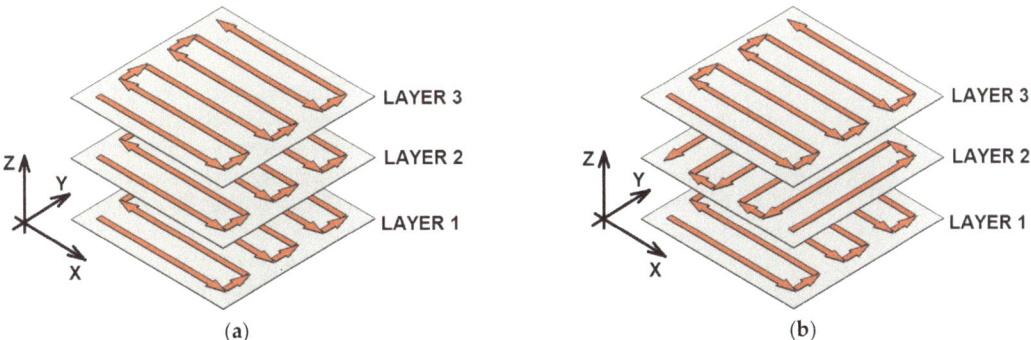

Figure 5. Different robot path strategies for CS-ed deposition. (**a**) Traditional or bidirectional, and (**b**) cross-hatching.

CS almost always uses conventional powders as feedstock materials developed for Air Plasma Spray (APS), High-Velocity Oxy-Fuel (HVOF), or laser processes in a spherical and finer particle size range at best. Various techniques are available to produce metallic powders, which are chosen by the chemical composition, characteristics, and/or properties required for the powder [146]. For the CS, the particles' metallurgical, morphological, and physico-chemical characteristics influence the spraying success and material performance [83]. Since CS does not promote recrystallization during the deposition, a deposit with a refined microstructure is obtained by selecting a small grain-size feedstock powder. It improves the mechanical properties; however, a larger gain size promotes more ductility to the particle. Using HT to reach the ideal powder microstructure was an alternative presented by Poirier et al. [147] for CS-ed H13 tool steel and by Story and Brewer [148] for aluminum alloys, resulting in a DE increase from 35 to 60% and from 70 to 90% to Al7075 and Al6061, respectively. Silvello et al. [84] summarized the relationship between powder characteristics, CS process parameters, and the CS-ed material properties by modeling and the experimental results. Table 2 presents coefficients for the model proposed using modeFRONTIER software, in which negative values represent inverse input/output relationships. It is noticed that the particle diameter and hardness influence the CS-ed material characteristics, highlighting the porosity, which is responsible for some CS drawbacks, such as a short fatigue life.

Table 2. Correlation behavior among the different input/output for CS [84].

Input/ Output	Particle Diameter	Particle Hardness	Gas Pressure	Gas Temperature	Particle Velocity	Deposit Hardness	Porosity	DE	FR
Particle diameter	1	0	0	0	−0.431	−0.187	−0.213	0.104	0.097
Particle hardness	0	1	0	0	0	0.935	0.109	0	−0.324
Gas pressure	0	0	1	0	0.594	0.417	−0.682	0.768	0.804
Gas temperature	0	0	0	1	0.498	0.297	−0.471	0.592	0.897
Particle velocity	−0.431	0	0.594	0.498	1	0.682	−0.734	0.803	0.817
Deposit hardness	−0.187	0.935	0.417	0.297	0.682	1	0	0	−0.352
Porosity	−0.213	0.109	−0.682	−0.471	−0.734	0	1	0	−0.819
DE	0.104	0	0.768	0.592	0.803	0	0	1	0
FR	0.097	−0.324	0.804	0.897	0.817	−0.352	−0.819	0	1

CS powders must be characterized before spraying, measuring their particle size distribution by the ASTM B214 standard [149], a sieving separation of the larger and smaller particles, or the laser scattering, classifying the particle size distribution by measuring the laser-illuminated flowing particles. The powder flowability is measured by the time elapsed to flow a certain powder mass through a certified Hall flowmeter, following the ASTM B213-20 standard [150], which is used to measure the powder's apparent density, as indicated by the ASTM B212-21 standard [151]. A previous characterization of the powder is imperative since powders with a flowrate higher than 1 g·s^{-1} tend to build up and block the gas flow in the nozzles for LPCS [146]. For HPCS, Vaz et al. [132] presented the flowability for different 316L, resulting in 9 and 17 g·s^{-1} for the irregular and spherical shapes, respectively. This powder characteristic impacted the CS powder feeding, which was 0.43 and 0.55 g·s^{-1} for the irregular and spherical shapes, respectively. By machine learning, Valente et al. [152] show how to predict a novel powder flowability on a per-particle basis, which can help scholars develop their alloys and powders for CSAM.

An irregular shape of the particles does not necessarily result in a coating or CSAM-ed part with worse properties [153–155]. The high deformation of the CS-ed particles at the impact can act as compensation for their shape irregularity and even for the particle size distribution, which enables using coarse particles, as presented by Singh et al. [153], who obtained similar material strength by coarse and fine Cu particles. CS-ed 316L coatings using water-atomized powders, which had an irregular shape, presented corrosion behavior and a wear-resistance performance very similar to, or even better than, the coatings obtained with spherical gas-atomized powders [132], indicating the viability of using a lower-cost raw material for CS, since the 316L gas-atomized powders are more expensive than the water-atomized ones. Wong et al. [155] obtained very similar porosity values (3.0 ± 0.5%), DE (100%), and hardness (200 ± 10 HV) for CS-ed Ti coatings employing irregular and spherical shape powders, but considering coating quality, the authors suggested the medium-sized spherical powder the best CS option. For Ti6Al4V, spherical particles presented a higher hardness and cohesive strength than a very irregular powder obtained by the Armstrong process, as shown by Munagala et al. [156]. In addition, the powder size distribution influences the CS-ed particles' velocity; smaller particles reach higher velocities than bigger and weightier ones, as presented in a simulation performed for 5, 25, and 50 µm Al particles. The first one resulted in a velocity higher than 600 m·s^{-1}, but the last one was lower than 500 m·s^{-1} [157]. For CS-ed Cu particles, small particles, 5 µm, reached a velocity of 700 m·s^{-1}, while big particles, 90 μm, accelerated up to 300 m·s^{-1}. Bagherifard et al. [119] presented a 316L fine powder, −29 + 12 μm, with a higher spraying velocity than coarse particles, −45 + 19 μm, which resulted in a material with higher particle deformation, mechanical properties, and electrical conductivity. Meanwhile, the V$_{cr}$ is dependent on the particle size, and smaller particles have a much higher V$_{cr}$ than the bigger ones, resulting in an even higher velocity, meaning small particles may not bond, and an optimum size range is achieved for each material, which is generally between 10 and 60 μm. When improving the temperature of particles, V$_{cr}$ is reduced, revealing the need to improve the CS working gas temperature to increase the temperature of smaller particles and the velocity of bigger particles [133,158,159]; however, higher gas temperatures put the equipment in an undesired condition, overloading it and promoting nozzle clogging.

The literature explains how the CS nozzle wall at a high temperature induces clogging because low-melting-point hot particles flow through the nozzle and collide against the nozzle's inner hot wall, inducing the bonding between the particles and nozzle wall, resulting in nozzle clogging [157,160]. Different solutions have been evaluated by researchers aiming to reduce the clogging and improve the nozzles' service life: the assembly of cooling systems surrounding the nozzle to reduce its temperature [157]; redesigning the nozzle for a bi-material component, using glass and WC [161]; aligning the sprayed particles by an electric field and avoiding them to touch the nozzle's hot wall [162]; and others. Clogging can be solved by cleaning methods, such as spraying hard particles at high temperatures or a chemical cleaning with acids. However, besides the monetary loss of clogging, it reduces the DE, can overload the gun chamber dangerously, and imposes maintenance stops during the deposition, generating undesired temperature transitions for large CSAM-ed parts. Sun et al. [10] comment that clogging has been one of the limitations of a more industrial CSAM application.

2.2. Challenges for CSAM

CSAM is a technique with great benefits compared to other AM methods. Therefore, it has excellent potential to be implemented in the solid-state AM industry to produce free-standing parts or repair worn components [2,163]. Yet, CSAM is still an emerging technology facing several challenges that need to be studied, such as low as-sprayed geometric tolerances, inferior mechanical properties compared to wrought materials, residual stresses, and low DE-depositing hard materials. In this section, these challenges are discussed, along with the strategies studied to overcome them. Figure 6 presents a scheme of the pros and cons of CSAM over other metal AM processes. It also indicates the advances studied and investigated to overcome the drawbacks.

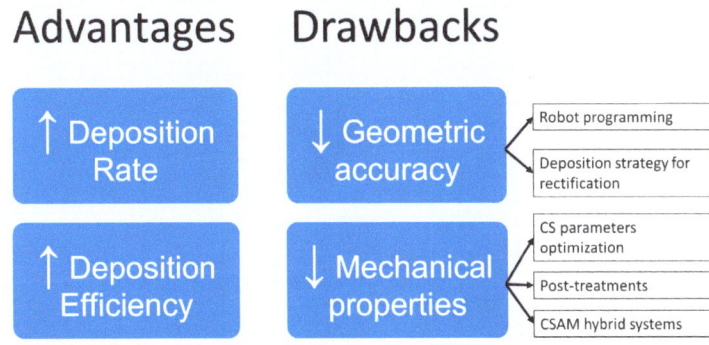

Figure 6. CSAM advantages and drawbacks, and the alternatives presented in the literature to overcome them.

2.2.1. Possible CSAM Geometries

The AM technologies rely on building an object layer by layer. Thus, controlling geometric tolerances is imperative to produce complex shapes or near-net-shape parts. Still, CSAM has to be more precise, and the literature presents some reasons for this CSAM limitation. First, because the velocity profile of the particles in the jet spot that exits the nozzle is uneven, the center of the laden-jet particles has a higher density of particles and greater velocities, promoting a superior deposition on this region than on the jet periphery. Cai et al. [138] simulated the single-track deposit profile, concluding that a 2D distribution profile approximately fits a Gaussian curve. Ikeuchi et al. [164] evaluated different machine learning approaches to accurately preview the CSAM track profile, saving much experimental time and CSAM spraying costs. Furthermore, Kotoban et al. [165] investigated the relationship between the shape of a single-track coating and the DE, concluding that in the first layer deposition, the particles on the jet periphery have a slight decrease in DE compared to the jet core, producing a triangle shape deposit that sharpens layer by layer. Finally, Wu et al. [166] developed a model to compensate the layer thickness by optimizing the robot velocity at the different regions on the substrate surface, resulting in a smoother CS-ed material surface.

Knowing that CSAM-ed deposits tend to produce pyramid-shaped coatings, some robot path trajectories and strategies have been developed to obtain near-net-shape parts [144,167–173]. For instance, Wu et al. [167] established a new stable layer-by-layer building strategy that sprays at a deflected angle towards the inclined walls of the pyramid-shaped coating, which allows building components with straight walls. Another example is the work of Vaz et al. [144], where a new method was implemented that consists of spraying with a circular movement at an angle different than the normal and allows free-standing building with controlled shapes, as presented schematically in Figure 7, but well described in the literature by Vaz et al. [144]. Yet, further studies on deposition strategies and the production of free-standing components are encouraged since they can expand the application areas of CSAM.

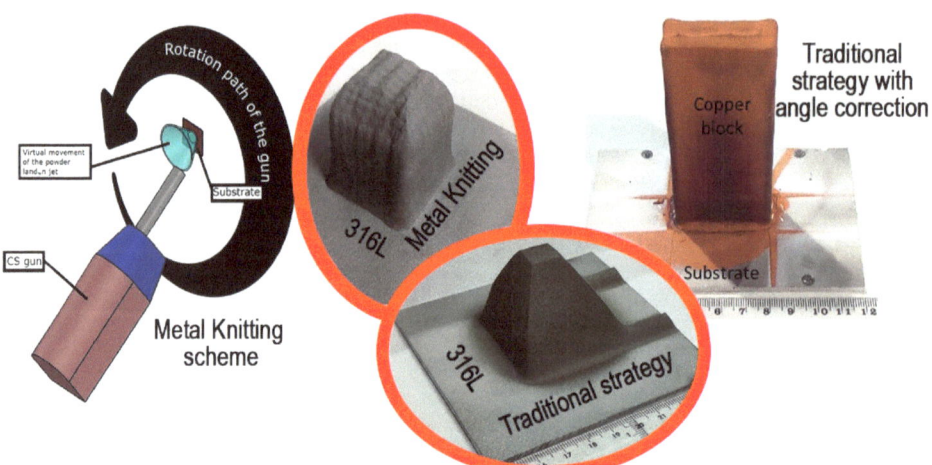

Figure 7. Metal Knitting, a CSAM alternative strategy to the traditional deposition. Reprinted with permission from Ref. [171], Springer Nature, 2020. Unit: mm.

CSAM can produce arrayed structural components, as presented in Figure 8 for CSAM-ed Ti on stainless steel. It was made by masking the substrate to shadow the areas where the sprayed material was not supposed to cover. Masking has been presented in the literature for other thermal spray processes, using tapes, pastes, shields, or other high-temperature resistant material removed after the coating deposition [174–177]. CFD has been developed to understand the influence of the masks on the CS gas flow, disturbances on the particle's trajectory, and the formation of bow shockwaves, which reduces the gas velocity [178]. It suggests using a higher particle velocity and setting the CS parameters to suppress this harmful effect of the masking strategy. Klinkov et al. [179] presented a model showing the impact of the mask on the particle behavior, velocity, and trajectory. The distance of the mask to the substrate cannot be excessive because it affects the deposition geometry, decreasing the width of the masked zone and diminishing the accuracy of the CSAM-ed geometry. An industrial application of the CSAM masking strategy is the fabrication of compact heat exchangers for electronic devices [180–183]. As well as array structures, diverse geometries are feasible, such as Braille impression for blind people or raised areas in molds for plastic injection, among others.

Figure 8. CSAM-ed Ti on stainless steel using masking strategy. Unit: mm.

Applying CSAM with other processes is a hot topic for industrial applications, e.g., for a unique component, some regions can be CSAM-produced, which is faster, and others can be DMLM-made, resulting in more details or complex geometries. Another CSAM use is joining dissimilar materials because CS has no metallurgic union with the substrate. It is helpful for composites, e.g., a carbon-fiber-reinforced polymer or a sintered SiC. This Al interlayer strategy was tested by Xie et al. [184] for joining TiN/Ti6Al4V, but using a hot dip to make a 25 µm-thick Al coating; however, it can promote an undesired atomic diffusion depending on the materials and process temperatures and time, which is prevented by using CSAM. These hard- or impossible-to-weld materials can have a surface coated by a thick CSAM weldable material, e.g., Al or 316L, which can be joined on other structures quickly. Champagne Jr and Champagne III [185] presented this method for using CS directly as the joining element and growing a CSAM-ed volume on the part to be arc-welded on another element. This joining was tested for a light-alloy magnesium ZE41A-T5, employing Al as the filler metal [186], joining Al 6061 to a ZE41A Mg alloy using CS sprayed Al as the transition material on the Mg alloy surface and welded to Al 6061 by FSW [187,188], and joining Al to Cu by FSW with a Ni interlayer [189]. Daroonpavar et al. [190] presented CS with the capability to make a corrosion-resistant coating on an AZ31B Mg alloy, employing the metallurgically incompatible Ta–Ti–Al layers, also reducing the wear rate from 10^{10} to 10^8 $\mu m^3 \cdot N^{-1} \cdot m^{-1}$.

2.2.2. Improving the Mechanical Properties

To date, CS processes have been used mainly in the aerospace, automotive, marine, and defense industries, where the performance requirements of the deposits are very demanding [99,163]. Therefore, one of the main issues with CSAM is the mechanical properties of the deposits. Apart from the hardness, which tends to be greater than the bulk due to the cold work hardening of particles during impact [132,144,191], the as-sprayed deposits present less favorable mechanical properties, such as lower strength, ductility, electrical and thermal conductivity, and wear resistance. It is attributed to the inherent microstructural defects of the CS process, such as micro-pores and interparticular boundaries [2,191]. Moreover, as the particles are arranged layer by layer, anisotropic responses have been reported in the literature for CSAM-ed deposits [141]. The literature presents anisotropy for other processes that deform the material in a preferential direction, e.g., cold rolling [192–194], extrusion [195], friction stir welding [196], or even laser AM processes [197,198]. For CSAM, high isotropy was observed in a plane parallel to the substrate surface [141,199–201], but in a vertical or Z-direction, the material had lower strength. This behavior is presented in the literature for CSAM-ed Cu [171], Al [202], and 316L [203].

Moreover, the use of CS is also limited by the intrinsic characteristics of the materials. For example, only soft and ductile materials, such as Cu and Al, are easily deposited, which is deducted from the number of papers linking "cold spraying" and "aluminum" or "copper" keywords. In contrast, hard materials (e.g., Maraging steels, Ti6Al4V, Inconel 718, etc.) with the poor capability to deform at a solid state can hinder the formation of a dense component [28]. Therefore, recent studies focus on optimizing the CS process parameters to obtain the ideal V_{cr} for each material so that quality coatings are produced [77,99,123,204–206]. For instance, Li et al. [205] did a literature review on the solid-state CS-ed Ti alloys, focusing on the process parameters, deposition characteristics, and limitations of these materials. Another example is the work of Pérez-Andrade et al. [123], which presents the optimization of parameters and post-treatment processes for obtaining high-quality thick deposits of Inconel 718 for AM applications.

CS-ed coatings also tend to be influenced by compressive residual stresses generated by the severe impact deformation of the particles. Such compressive stresses can be beneficial up to a certain point. However, if they are too high, the adhesion of the deposit is usually hindered, and a crack can nuclei and grow in the interface substrate/coating, or it can completely detach, de-coating from the substrate [207,208]. For CSAM, these residual stresses are accumulated layer by layer, and if the particles have poorly adhered to the substrate, the deposit separates from the substrate. Making freeform parts is not a problem because the substrate has to be eliminated and only acts as a base or support. Still, with the employment of CSAM as a repairing method, this detachment and low adhesion is highly prejudicial of the excellent performance of the repairing service.

These challenges represent a drawback for CSAM compared to other AM methods. Nevertheless, several process strategies have been successfully explored in the literature, such as post-processing methods (e.g., HT) or hybrid deposition technologies, such as Laser-Assisted Cold Spray (LACS) and Cold Spray Shot Peening (CS-SP).

Heat Treatments (HT) are one of the most effective ways to enhance the microstructure of CSAM-ed deposits [2]. The tailoring of the final properties of a broad range of materials with HT, such as Cu [136,208–210], Al alloys [208,211–213], Ti alloys [208,214–216], Ni alloys [81,123,124,204,217–219], 316L [119,208,220], among others, is reported in the literature. Furthermore, HT relieves residual stresses, reduces the microstructural defects (e.g., porosity and particle boundaries), and improves the cohesion between particles which significantly influences the material performance, since the failure mechanism during the stress loading changes from an interparticular mode to a cleavage-like and ductile mode. In the first one, the crack grows surrounding the particles and detaches one to the other. HT promoted a metallurgical bonding of particles, increasing cohesion, material strength, and plasticity or ductility. Dimples evidenced it in SEM images of the fracture surface [202,210,221–223]. For Inconel 718, Sun et al. [124] applied induction for heating the material, which represents a possibility to select the CSAM-ed part region to be HT-ed, instead of the whole material, e.g., treat only the component areas that are exposed to wear or friction. The induction HT promoted the cohesion of the particles by the eddy current, as well as the atomic diffusion, which resulted in higher mechanical properties due to higher dislocations and twin densities in the neck formed between the particles than in the particles' center [224]. Due to the hardness reduction, Zhang et al. [225] presented the HT-positive effects on the post-machining process of Al7075. Another heating process is Electric Pulse Processing (EPP), in which applying high-density electron charges through the material promotes changes in the microstructure and mechanical behavior of alloys, such as precipitates distribution, yield strength, elongation, and hardness [226]. For example, for CSAM-ed Cu, Li et al. [210] show an expressive improvement in its mechanical properties, reaching a UTS of 200 MPa over 100 MPa in the as-sprayed condition and elongation of 20% over 2%.

Particularly, annealing is considered a simple post-processing method that positively impacts the as-sprayed CSAM microstructure. It promotes diffusion and recrystallization processes that mitigate the undesired microstructural defects and change the mechanical properties; the work-hardened deposits are softened, increasing their ductility, but reducing their hardness compared to their as-sprayed counterparts.

Spark Plasma Sintering (SPS) is a technique developed for ceramics and powder metallurgy that has improved CSAM-ed density and mechanical properties. SPS is pressing compacted powder and applying a pulsed current discharge that can reach thousands of Amperes but low voltage under pressure. It generates plasma between the intimate close particles, which results in micro welding, forming necks at contact points, atomic diffusion, and plastic flow [224,227–229]. In addition, Joule heating and plastic deformation enhance the sinter's densification, improving the particles' cohesion and material strength [230]. For the CS-ed TiC–Cu composite, SPS eliminated the interparticular region [231]. The SPS temperature was directly related to improving the mechanical properties, ductility, and decreasing the hardness of the CSAM-ed Cu, as presented by Ito and Ogawa [230], who selected 50% of the Cu melting point as the maximum SPS temperature for 5 min. This short time is an advantage of SPS over annealing, which typically keeps the material in the furnace for hours. Figure 9 shows the microstructures of CSAM-ed Ti6Al4V as-sprayed and after SPS post-treatment.

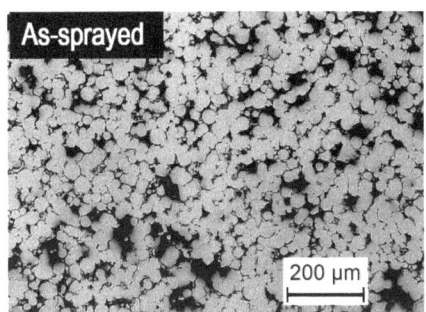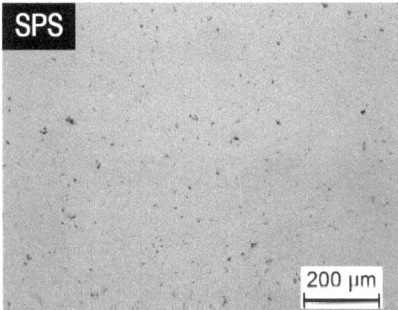

Figure 9. SPS effect on CSAM-ed Ti6Al4V microstructure.

Another HT that has recently drawn attention in the field of CSAM is Hot Isostatic Pressing (HIP), which is presented in detail by Bocanegra-Bernal [232] and by Atkinson and Davies [233]. The HIP technique can be used directly to consolidate a powder or supplementary to densify a cold-pressed, sintered, or cast part. This method can eliminate the pores and micro-cracks of the material by compressing the samples with high temperatures, e.g., 1000 °C for Ti alloys, to an isostatic pressure in the order of hundreds of MPa at the same time, resulting in fully isotropic material properties [234,235]. It has been successfully applied to metals, composites, and ceramics obtained by different processes. However, few studies are available in the literature for CSAM, and they are focused on hard materials that are difficult to deposit by CS, such as Ti [236,237], Ti6Al4V [237–239], and Inconel 718 [123].

Figure 10 presents the densification and phase changes, precipitating β in an α matrix, in CSAM-ed Ti6Al4V employing N_2 and He as the CS working gas. However, this post-treatment cannot close exposed porosity because the HIP gas fills these pores. A solution is a pre-HIP process of encapsulating the sample and converting those into internal pores to be removed by the HIP. The HIP also cannot remove large internal pores since diffusion bonding does not occur when metal/metal contact is not intimate. It happens when the

CS-ed material has low plasticity even in high temperatures, if the surfaces of the internal defect are oxidized, or if there is a gas inside the pore that does not diffuse, e.g., air, He, or N_2 [232]. It represents a limitation for CSAM HIP use if the CS-ed deposition process cannot produce parts with very low porosity and a very thin interparticular region, which occurs when spraying low-ductility powders.

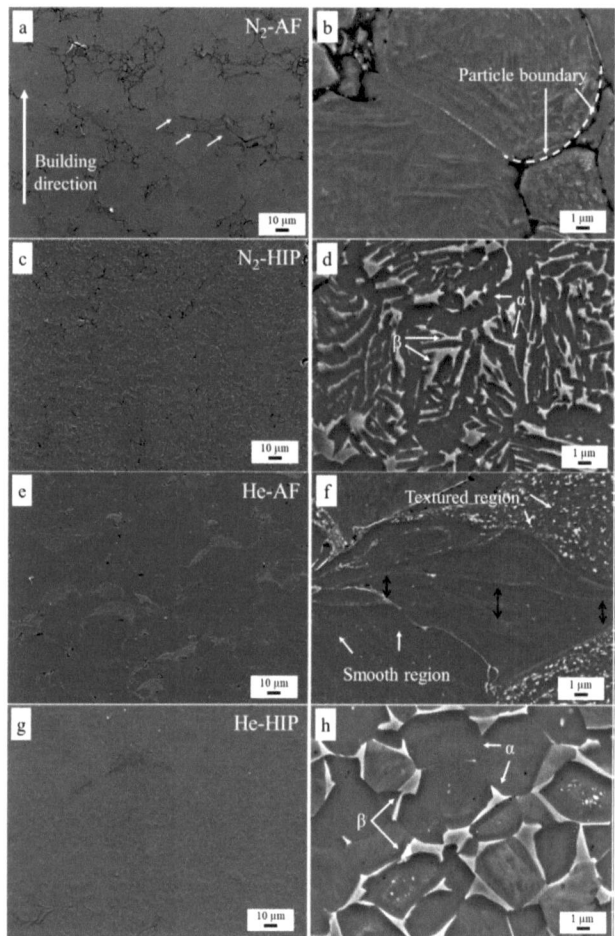

Figure 10. Effect of HIP on the microstructure of CSAM-ed Ti6Al4V, densification, and phase change. SEM images with (**a,c,e,g**) low magnification and (**b,d,f,h**) high magnification. Reprinted with permission from Ref. [238], Elsevier, 2019.

The melting or sintering AM processes drastically change the microstructure of the feedstock material during the processing, and CS represents an advantage over SLM or DMLM in this point. However, a hybrid process of coating a CSAM-ed part can significantly improve its wear and corrosion performance, as presented by Vaz et al. [240] coating CSAM-ed Maraging with HVOF-sprayed WC. In addition, Feng et al. [241] used induction heating to remelt AlCoCrCuFeNi HEA, improving the wear resistance by phase transformations. Laser remelting or glazing has been investigated as a post-treatment on CSAM, eliminating micropores within the deposit and enhancing the cohesion of particles. Remelting changes the ASI and other CS bonding mechanisms for metallurgical bonding. Laser remelting of CS-ed Al onto steel substrate presented an FeAl intermetallic formation, improving its wear resistance [242]. For Ti, Astarita et al. [243] and Marrocco et al. [244] obtained a thin and dense remelted layer, which improved the corrosion behavior in a 3.5% NaCl solution, reaching the same performance as a wrought Ti bulk. Kumar et al. [245] showed an improvement in the wear resistance for the Ti-based MMC.

The laser glazing applied on CS-ed Inconel 625 eliminated the cold-worked microstructure, generating a columnar dendritic one. It reduced the hardness but eliminated the interconnections between the pores, increasing the material corrosion performance [246]. Zybala et al. [247] improved the surface hydrophobicity of CSAM-ed irregular powders Ti6Al4V and Ti after laser surface post-treatment. This condition is attractive for CSAM-ed parts developed for the Oil and Gas sector, where Ni-based alloys have been employed as CO_2 and H_2S corrosion-resistant materials. Other cheaper heat sources can be studied, and it is a lack in the literature, e.g., PAW, used by Pukasiewicz et al. [248,249] for HVOF-sprayed FeMnCrSi coatings, or GTAW, applied by Zabihi and Soltani [250] for FS Al-based MMC coatings.

Plasma Electrolytic Oxidation (PEO) or micro-arc oxidation has been used to produce hard ceramic coatings on Al and other alloys [251]. It improves their wear and corrosion resistance due to the formation of the protective ceramic coating on the material enabled by the plasma discharges, supported by an aqueous electrolyte [252]. For CS-ed Al, Rao et al. [253] presented a stable, well-adhered, and harder PEO layer formed on the CS-ed Al7075 coating, 1353 and 144 HV, respectively, which resulted in a higher dry sliding wear resistance. It also improved the corrosion resistance, with a three-order lowered corrosion current density. Using PEO on CS-ed Al + Al_2O_3 on a Mg alloy substrate, Rao et al. [254] reduced the sliding and abrasion wear rates ten times, mainly due to the increase in hardness from 700 to 1300 HV.

The infrared irradiation as a heating source for the HT of CSAM-ed Cu alloys was tested by Chavan et al. [255]. This heat source is cheaper than laser equipment and has a wavelength range similar to lasers. Still, a significant advantage of infrared irradiation is the absence of a furnace, a chamber, or a controlled atmosphere, as occurs for annealing. It enables this system for CSAM in situ repairs or repairing large components that do not fit in conventional furnaces, e.g., oversized axles or injection molds. Infrared irradiation has been previously used for arc-sprayed Zn alloys coatings, improving wear and corrosion resistance [256].

Shot Peening (SP) is a post-treatment technique of cold working by propelling glass, ceramic, or steel balls against the material, reducing the material's surface roughness, inducing the surface compressive residual stress, and, consequently, increasing the part's fatigue resistance by retarding crack initiation [257]. Moridi et al. [258] showed the SP applied for CSAM-ed Al6082, reduced roughness from R_a 12.4 to 4.7 µm and improved the depth of the compressive stress layer from 350 to 400 µm, but without a significant compressive stress value improvement. Furthermore, due to porosity and plastic deformation reduction, the hardness and corrosion resistance were improved in CS-ed Zn by SP post-process [259]. Similar mechanisms and results were obtained by Ball-Burnishing (BB), a process in a ceramic or hard ball, with a diameter of <10 mm, which compresses and deforms the CS-ed material, as occurs with SP, but without impact, more similar to a rolling process. BB was applied to CS-ed 17-4PH stainless steel, improving the depth from 130 to 190 µm of 200 MPa compressive residual stress [260,261]. In general, any technique that improves the part's fatigue life is attractive for CSAM; however, SP has presented low effectiveness and does not indicate more research interest or industrial promisor use.

Waterjet Cavitation Impact (WCI) is another technique presented in the literature that should be tested for CSAM, since there is a lack of this in the literature, aiming to improve the material surface properties, especially the compressive residual stress. Cavitation is a phenomenon in which the static pressure of a liquid reduces to below the liquid's vapour pressure, forming microbubbles that collapse when subjected to a higher pressure. It generates shock waves that impact the material surface [262]. For WCI, the material is exposed to a water jet under controlled conditions, promoting its plastic deformation and densification, as occurs for SP. WCI has been applied in the industry since the 1990s for parts produced by different techniques and exposed to fatigue degradation, such as gears, shafts, and other [263]. Good results have been exposed in the literature, Zhang et al. [264] reached 175 MPa compressive residual stress in a 2A12 Al alloy by a WCI with a water jet under 75 MPa and 20 degrees off-normal inclined; Soyama and Okura [265] presented how WCI resulted in a significant improvement in the fatigue life of Ti6Al4V.

Cold Rolling (CR) was experimented with for CS-ed Cu on steel by Bobzin et al. [266], resulting, after a 14% thickness reduction, in cracks and delamination on the coating/substrate interface, which was resolved by annealing at 500 °C before CR, a Thermo-Mechanical Treatment (TMT) or Hot Rolling (HR). It resulted in good adhesion to the substrate without cracks in the Cu coating but lower hardness due to the annealing process that dwindled the cold working in the particles from the CS-ed deposition. Tariq et al. [267] employed TMT for CSAM-ed Al-B$_4$C MMC, reducing by 60% in thickness, improving the mechanical properties, increasing UTS from 35 to 131 MPa, and elongation from 0.5% to 5% due to the interfaces of the particles dramatically enhanced by the diffusion activity. TMT has been applied for materials that have poor formability at room temperatures, such as Mg alloys [268] or TiAl-based alloys [269,270], as well as the CSAM-ed Ni-Al [271], A380 alloy [272], and Al-B$_4$C [267]. Depending on the CSAM part's geometry designed, such as plate-like, TMT is adequate. However, TMT promotes a considerable anisotropy for a bulk-like shape due to the unidirectional plastic deformation induced by the post-treatment.

Friction Stir Processing (FSP) imposes a friction force on the CSAM-ed material that softens the surface, increasing the amount of shear straining in the processed region and promoting dynamic recrystallization. Microstructural changes and grain refinement showed this, altering the CSAM-ed material's mechanical properties, porosity, and cohesion of particles [273]. The literature presents the FSP applied for Al alloys [274], Mg alloys [275], MMC [276], and 316L [220], focusing on the improvement of their tribological and corrosion performance. Ralls et al. [220] studied CS-ed 316L + HT + FSP, showing that the post-treatments eliminated the δ-ferrite contained in the powder by atomic diffusion. In addition to that, the authors observed a reduction in porosity to values close to zero and hardness from 330 to 190 HV, as HT-ed; however, FSP increased it from 190 to 280 HV. On the other hand, FSP harmed the CSAM-ed 316L wear resistance, from 2.27×10^{-9} to 1.02×10^{-9} mm$^3 \cdot$N$^{-1} \cdot$mm^{-1}. Table 3 summarizes the CSAM post-treatments presented in the literature, considering their main effects and results studied by scholars.

Table 3. Post-treatments applied for CSAM.

Material	Post-Treatment	Post-Treatment Effects Obtained	Ref.
Cu	HT	Improved conductivity, mechanical properties, isotropy, and ductility; reduced hardness.	[208,210,230,277,278]
Cu	SPS	Improved mechanical properties and ductility, reduced hardness.	[230]
Cu	FSP	Microstructure changed, refining grain size, improved mechanical properties and ductility, reduced hardness	[210]
Cu	EPP	Microstructure changed, refining grain size, improved mechanical properties and ductility, reduced hardness	[210]
Cu-Al	Infrared irradiation HT	Improved electrical conductivity, maintained the elastic moduli, improved cohesion of particles, reduced hardness.	[255]
TiC-Cu	SPS	Promoted phase change and sintering Ti-C-Cu, eliminated interparticular region, increased hardness.	[231]
Al6082	SP	Improved compressive stress layer depth, changed the fatigue fracture mechanism from intercrystalline to transcrystalline.	[258]
Al-Mg-Sc-Zr	HIP	Maintained a very low porosity, improved mechanical properties, improved the compression resistance.	[237]

Table 3. *Cont.*

Material	Post-Treatment	Post-Treatment Effects Obtained	Ref.
Al-Al$_2$O$_3$	HT	Promoted phase change, reduced porosity and hardness, improved mechanical properties and ductility.	[213]
Al-B$_4$C	HT	Improved mechanical properties and ductility, reduced hardness.	[267]
Al-B$_4$C-	TMT	Improved adhesion, mechanical properties, and ductility, reduced hardness.	[267]
316L	HT	Reduced porosity and hardness, maintained phase composition, improved ductility and fatigue performance.	[119,208,220,279]
316L	HIP	Reduced porosity and hardness, maintained phase composition, improved ductility and fatigue performance.	[119,237]
316L	HT + FSP	Improved mechanical properties, reduced porosity, reduced hardness negligibly, reduced the wear resistance.	[220]
Ti	HT	Maintained the same porosity, increased the mechanical properties and ductility.	[208,216,280]
Ti	HIP	Reduced porosity from 4.3 to 2.2%, improved mechanical properties, changed pores morphology.	[236,237]
Ti	Remelting	Reduced hardness, transformed microstructure, eliminated interparticular region, improved corrosion behavior.	[243,244,281]
Ti6Al4V	HT	Reduced porosity, promoted phase changes, improved mechanical properties and ductility, reduced hardness,	[154,214,282]
Ti6Al4V	HIP	Reduced porosity, promoted phase changes, grain refine, improve mechanical, improve the ductility.	[237–239,282]
Ti6Al4V	Remelting	Improved hardness, increased surface roughness, coefficient of friction in wear testing, and tensile residual stress.	[247,283]
Invar 36	HT	Improved mechanical properties, ductility, reduced the compressive residual stress.	[284]
Inconel 625	HT	Increased hardness and the fatigue performance.	[217]
Inconel 625	Remelting	Reduced hardness, transformed cold worked microstructure in the particles to columnar dendritic, improved corrosion behavior.	[246]
Inconel 718	HT	Reduced porosity, improved mechanical properties and ductility, reduced the compressive residual stress.	[81,124,204,218,219, 285]
Inconel 718	HIP + solution HT + aging HT	Reduced porosity and compressive residual stress, improved conductivity.	[123]

2.2.3. Avoiding Post-Treatments

The first post-treatment needed for CSAM-ed parts is the machine processes because CSAM cannot produce parts with the final geometry or roughness, which is a challenge for CSAM, as stated by Kumar and Pandey [126]. However, with the development of more complex robot manipulations, the machining has been planned for specific and essential areas of the component, such as bearing houses, screws, or axles journals, among others.

Regarding the materials' properties and characteristics, Laser-assisted Cold Spray (LACS), also called Supersonic Laser Deposition (SLD), is a relatively recent manufacturing process that combines the CS process with a complementary laser that heats the deposition zone while spraying. This method combines the benefits of both technologies, the CS solid-state deposition of metals at short times with little material waste and the bonding strength by heating the deposition zone with a laser without increasing the oxygen levels within the deposit [108,286,287], even applied for the LPCS process [288]. Lupoi et al. [289] and Bray et al. [290] presented LACS as an option to suppress the disadvantage of N$_2$

as a working gas (with a low particle velocity) by the implementation of a laser source to illuminate the spraying location. It softens the substrate during the CS deposition, promoting particle plasticity, even by phase transformation. Barton et al. [291] showed an Fe-based alloy transforming the ferritic into an austenitic phase, which is more ductile; and Birt et al. [287] concluded that LCAS is capable of depositing Ti6Al4V using N_2 as the working gas with densities as high as or higher than those deposited using He without the laser assistance. Furthermore, the adhesion of CoNiCrAlY onto Inconel 625 and Cu onto Al were improved by particles/substrate local melting at a micro level and intermetallic formation [292], heating the particles to 80% of the powder melting point [293].

Another LACS option presented in the literature is using the laser to heat, clean, and ablate the substrate milliseconds before the CS deposition. It intends to soften the substrate, allowing the particles to deform and consolidate CS-ed material at an impact velocity lower than its V_{cr} [294]. The use of CSAM for hard materials has been one of the biggest challenges for the industry and researchers.

Regarding the selection of laser parameters, the laser (CO_2, Nd-YAG, or Yb-fiber), wavelength, pulse duration, and energy affect the penetration depth of the thermal energy transferred by phonons to the metal [295]. Therefore, their optimization and selection depend on the materials' characteristics and properties. For example, some authors presented experiments for a LACS employing power between 1 and 5 kW, with expressive benefits to the CSAM microstructure and DE using high-power laser assisting [293,296–298]. Still, an excessive heat input can result in grain growth and hardness reduction, i.e., the LACS process results in annealing effects on the cold-worked particles during deposition [299].

Overall, LACS increases the temperature of the particles at the impact, improving the DE and reducing porosity in the deposit microstructure, as presented by Olakanmi et al. [296], reaching pore- and crack-free Al-12Si CS-ed on stainless steel. LACS also broadens the range of CS-ed materials [290]. As a result, LACS has successfully deposited dense parts of hard materials with high DE, such as oxide-free Ti [290], Ti6Al4V [287], MMC [300], Stellite-6 [289], CrMnCoFeNi high entropy alloy (HEA) [301], $Fe_{91}Ni_8Zr_1$ [291,299], 15-5 PH stainless steel [302], and AISI 4340 [297]. Another methodology employed for substrate pre-heating and adhesion improvement was induction, presented by Ortiz-Fernandez and Jodoin [303], spraying Al onto Ti6Al4V, resulting in higher DE and lower porosity.

During CS, particles are accelerated and sprayed at high velocities. At the moment of impact of the first layer, these particles are deformed and remain attached to the substrate. In the subsequent layers, the particles now impact the deposited material, causing the so-called tampering or tamping effect: new particles crush the previous layers of deposited material, causing compaction of the coating, thereby reducing porosity, a peening effect [89,304]. This effect can also be activated by mixing larger particles with the CS feedstock particles to deform the deposited material [305], as presented by Ghelichi et al. [306], mixing $-30 + 5$ with $-90 + 45$ µm Al particles, and Lett et al. [140], mixing $-45 + 15$ with $-250 + 90$ µm Ti6Al4V particles. Luo et al. [219,307] studied the effects of in situ CS-SP on the microstructure of Ti, Ti6Al4V, and Inconel 718. They presented it as an effective way to increase the DE while reducing porosity and improving inter-particle bonding and cohesion.

However, the $-24 + 7$ µm Inconel 718 were mixed with $-187 + 127$ µm stainless steel particles in different concentrations, and this last did not participate in the final sprayed material. Still, it acted for the peening effect, resulting in a drastic porosity reduction from 5.5 to 0.2%, improving the DE from 22 to 33%, and improving the hardness from 420 to 510 $HV_{0.3}$ [219]. For CS-SP-ed Ti6Al4V, increasing the mass of larger particles from 50 to 90% vol. in the feedstock powder, the porosity reduced from 6 to 0.2% [140] and induced compressive residual stress of 444 MPa·m^{-1}, instead of the tensile residual stress of 126 MPa·m^{-1} obtained without SP. Hybrid use of CS was shown by Li et al. [308], spraying on the AA2219 alloy GTAW welded joint, altering the residual stress drastically and even promoting the compressive residual stress in some areas near the welding bead. Daroonparvar et al. [305] listed the use of CS-SP for different coatings on Mg alloy substrates, using other materials for the coating and SP; e.g., Ni with 410 stainless steel 150–200 µm and Al6061 with 1Cr18Ni stainless steel 200–300 µm, among others.

2.2.4. Measuring of Properties

Regarding the characterization, CSAM-ed samples have been evaluated as other thermal sprayed coatings. Conventional characterization techniques, such as optical or light microscopy, SEM, and microhardness have been seen in the literature [309]. These are potent tools for experts in the CSAM theme because the researchers can infer important materials' properties from the material microstructure. By image analysis, the Flattening Ratio (FR) is obtained, a measure of the compression of a sphere along a diameter to form an ellipsoid-like splat; the higher the FR, the higher the material plasticity. Electron Back Scattering Diffraction (EBSD) is a technique capable of identifying material phases at each analysis point and presenting the 3D orientation of the crystal lattice at each point. It has been used in CSAM to interpret the orientation in crystallographic planes in the different particles, which is even more important for the characterization of CSAM post-treatments [200]. Figure 11 shows CSAM-ed Ni/FeSiAl in as-sprayed and HT-ed conditions, Figure 11a,b, respectively. HT promoted a recrystallization, grain coarsening, and phase transformation in the material, as interpreted from the size of each colored area, which is more significant in Figure 11b. This image has fewer areas without a defined atomic lattice plane of the crystalline structure, as seen in the as-sprayed condition as dark green. These dark green areas represent patterns uninterpreted by the detector, which is related to the severe particle deformation during the CS deposition. The improvement in this indexing rate from 78 to 90% represents the recrystallization phenomena from HT post-treatment [310].

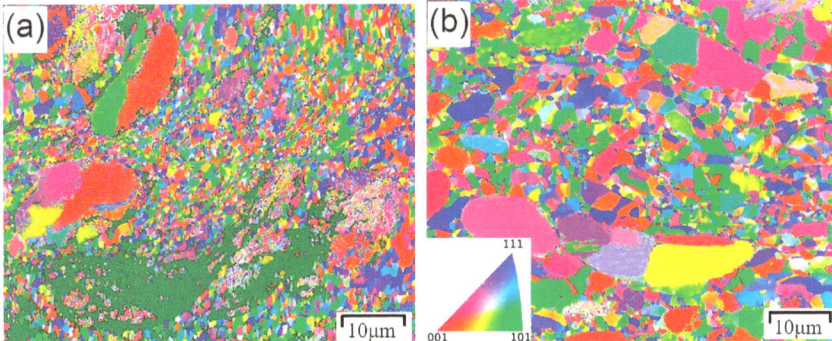

Figure 11. EBSD image of CSAM-ed Ni/FeSiAl before and after HT. (**a**) As-sprayed condition and (**b**) annealed condition. Reprinted with permission from Ref. [310], Elsevier, 2019.

Porosity has been calculated by CSAM cross-section image analysis [130,132,311], but other techniques have been presented in the literature as alternatives for higher accuracy in porosity measurements or a non-destructive approach, e.g., gas or He pycnometry [312,313], X-ray microtomography [314,315], laser-ultrasonic inspection [316], water absorption or the Archimedes method [317], and electrochemical impedance spectroscopy [318]. In addition, microhardness (Vickers and Knoop) employing low loading and nano-hardness techniques have been used to determine the hardness gradient in single particles. At the same time, microhardness utilizing higher loadings results in a macro evaluation of the material and fracture toughness by interpreting cracks grown due to the indenter loading [309,319]. Furthermore, the same Berkovich indenter used for the nano-hardness test provides the material elastic moduli, as described in the ISO 14577-1:2015 standard [320–322], an important property to preview the deformation of the material under the service loading.

Using CSAM for repairing processes or as a hybrid stage above a substrate made by other processes infers the need for good adhesion, which is the bonding strength between the CS-ed material and substrate. For thermally sprayed coatings, the ASTM C633-13 standard [323] is the most used technique to measure its adhesion, known as Tensile Adhesion Testing (TAT), which is basically comprised of a thermal-spray-coated disk, dia. 1 in., that is attached with epoxy to a complimentary uncoated plug and detached by a uniaxial tensile loading, the relation loading-area results in the bonding strength, in MPa or ksi [309,324]. However, for bulks, ASTM C633-13 [323] is inadequate. A technique within the sample machined from the CSAM-ed freeform part has been presented in the literature as a more effective method, modified tensile testing, based on the ASTM E8-22 standard [325]. Ichikawa et al. [326] machined adhesion samples of CSAM-ed Cu onto an Al substrate, eliminating the interference of a bonding agent, epoxy adhesive, and guaranteeing the rupture in the Cu/Al interface. Boruah et al. [327] used a similar technique, but with CS Ti6Al4V on a washer surrounding an exposed pin-like substrate, which are tensile together, rupturing in the coating/substrate interface. Figure 12 shows the schemes for TAT and ASTM E8-22 modified adhesive testing.

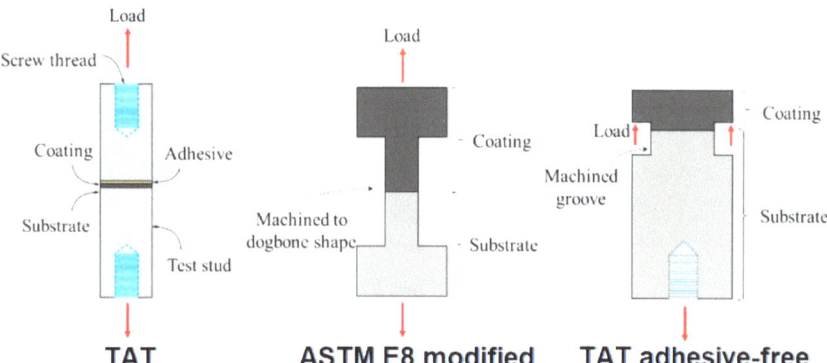

Figure 12. Schemes for CSAM adhesive testing. Reprinted with permission from Ref. [327], Elsivier, 2020.

The surface properties and the whole component quality are important for AM parts. Tensile testing has been done for different materials and fabrication strategies to measure the materials' strength and ductility [186,211,282]. Machining samples in different CSAM-ed part directions make the interpretation of mechanical isotropy possible, as performed by Yang et al. [200], Ren et al. [222], and Wu et al. [328]. The literature has presented the CSAM-ed mechanical resistance and ductility as a consequence of a good cohesion of particles, which TCT can easily measure. The TCT principle coats the circumference of two cylinders together head-to-head that are pulled in a universal testing machine. The stress or cohesion of particles is calculated as a relation between the loading collected and the coating thickness value, following the instructions of the EN 17,393:2020 standard [135].

Residual stress is crucial information for developing the CSAM as an industrial process, and the realization of the CSAM limitations is perhaps the main motivation behind the scholars' efforts to provide a reliable framework to study residual stresses in CS-ed deposits, initially by means of experimental and theoretical analyses, and later by finite element modeling. The residual stresses are divided into three types: the first order is macro-stresses homogeneous over multiple grains; the second order is micro-stresses over single grains; and the third order is micro-stresses in single grains, but with being inhomogeneous over the smallest areas such as unit cells [329]. Non-destructive diffraction measurement techniques for micro-stress have been used for CS-ed material, X-ray, and neutron diffraction. The first has a shallow penetration in the material, in order of micrometers, accrediting it just for superficial evaluation [258]. However, FEA was applied by Wang et al. [330] to simulate the residual stress along the CSAM-ed Cu part from X-ray diffractometry superficial residual stress results.

On the other hand, neutron diffraction penetrates the material in order of centimeters, but needs a long time exposition to achieve good results, in order of tens of minutes per measurement point [309]. Both methods are restricted to crystalline materials, and neutron diffraction has been studied for CSAM, as presented by Luzin et al. [331] and Vargas-Uscategui et al. [332] for Ti, Sinclair-Adamson et al. [333] for Cu, Loke et al. [334] for Al6061, and Boruah et al. [335] for Ti6Al4V. Despite being restricted to a few facilities worldwide and being an expensive technique, neutron diffraction has presented valid results in understanding the evolution of residual stress in CSAM deposition. In addition, it helps researchers to find new deposition strategies to reduce the regions with deleterious tensile fields.

Incremental Hole Drilling (IHD) is semi-destructive testing presented in the literature for the first-order residual stress measurement, which has been used for different materials and processes of fabrication for decades, including thermally sprayed coatings [336–338]. The IHD principle is based on drilling a small hole, <1 mm, into the material and collecting data about the deformations around the drilled hole using optical instruments or strain gauges. The material deformation or relaxation is related to the residual stress in the volume of the removed material through drilling [339], and the testing procedure is ruled by the ASTM E837-20 standard [340]. IHD is a technique routinely used for cast or rolled materials, and its use for CSAM promises high accuracy, easy sample preparation, and fast results. However, the literature still needs documents discussing the results and limitations and comparing IHD to other residual stress techniques, focusing on CSAM-ed bulks, which is a need to be filled by scholars.

In situ Coating Properties (ICP) measure the substrate curvature during and after deposition. The evolution of the sample curvature can be linked to the evolution of stresses in the thermally sprayed material using a variety of models [341]. Figure 13 shows an example of typical curves obtained by the ICP sensor, where there is evidence of the spraying time or deposition stress and the cooling time until room temperature, culminating in the residual stress. For HVOF sprayed coatings, normally, tensile residual stress is obtained, as indicated in Figure 13, with positive curvature values; however, for CS-ed coatings, the residual stress has negative curvature values, which is compressive. ICP has the advantages of being fast and not demanding the machining of samples, but it is limited to coatings, as shown by Sigh et al. [342], comparing ICP to X-ray for Inconel 718 coatings thinner than 1 mm, resulting in similar compressive residual stress values for both techniques. Furthermore, ICP does not apply to larger CSAM-ed parts, even though ICP results help the researchers optimize the CS parameters used for CSAM, mainly regarding improving adhesion.

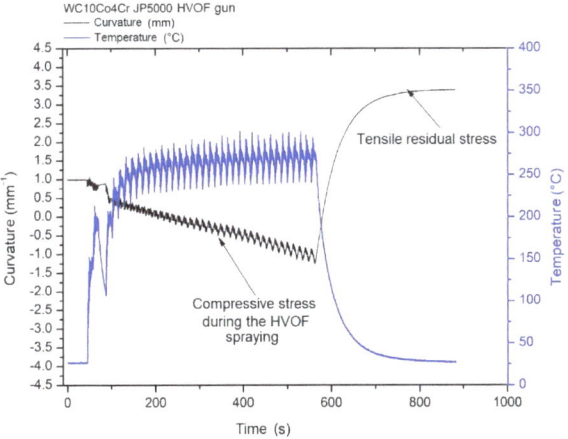

Figure 13. Typical curves obtained with ICP sensor.

Mechanical elements under cyclic loadings are subject to fatigue, reducing their life cycle. For CSAM-ed parts, Sample et al. [343] presented the influence of different CS properties, e.g., hardness, tensile properties, residual stress, etc., on the material fatigue performance. For CSAM-ed parts and CS-coated materials, different fatigue tests have been presented in the literature, which are designed and classified by the force or loading type: direct (axial) stress, plane bending, rotating beam, alternating torsion, or combined stress. Rotating beam and bending fatigue testing evaluate the parts exposed to revolutions under loading, such as axles, shafts, or wheels. Rotation bending exploits a rotating bending moment obtained through a rotating unbalanced mass, while the rotation beam places the load in the center of a supported sample at the ends. Applying CS as a coating improved the sample fatigue life by inserting compressive residual stress in the surface [258,344–348].

Using axial cyclic loading, three- and four-point bending fatigue testing have been presented in the literature. Xiong and Zhang [349] showed the improvement of mechanical resistance and fatigue life for an AZ91D Mg alloy after an LPCS-ed Al coating; Yamazaki, Fukuma, and Ohno [350] presented a low level of improvement in the fatigue life of CSAM repaired 316L samples, accrediting the repairing services for this material. Ševeček et al. [351] studied the benefits of CS-ed coatings on the Zircaloy-4 high-temperature fatigue life and displacement under cyclic loading. Considering the CSAM-ed bulk, Julien et al. [202] used compact tension specimens, following the ASTM E399-22 standard [352], to evaluate the fracture toughness (K_{IC}) of CSAM-ed Al6061. Wrought reference samples had much higher values than the CSAM-ed ones, 26.5 over 13.0 MPa·m$^{0.5}$, resulting from the CS-ed typical microstructure and the interparticular crack growth. A higher K_{IC} reduction was presented by Kovarik et al. [221] for CSAM repairing Al, Ti, Ni, and Cu compared to rolled materials. Making the CSAM-ed parts have similar properties to bulks produced by traditional processes represents a challenge for CSAM's industrial application. Scholars have employed efforts to find solutions and possibilities to achieve solutions for this, such as post-treatments. Regarding the material fatigue life, Li et al. [353] proposed a probabilistic fatigue modeling for a GH4169 Ni alloy, using the weakest link theory applied to calculate the number of cycles to crack initiation. Similar modeling should be performed for CSAM-ed materials to compare how their microstructure defects and characteristics influence the material performance, deviating the experimental results from the mathematical and statistical model formulated.

3. Bibliometric Analysis

This section presents CSAM from an academic viewpoint, considering how the literature, scholars, and institutions cover the theme of CSAM. Bibliometric analysis has gained immense popularity in many research areas in the last decade due to being a powerful tool for interpreting the massive amount of data available nowadays, which, depending on the theme studied, may reach hundreds or even thousands of relevant documents [354]. Scholars use bibliometric analysis for different reasons, such as uncovering emerging publishing and journal performance trends, looking for investigation collaborators, or exploring the intellectual structure of a specific domain in the study [355]. The exciting use of bibliometric research is to identify knowledge gaps in the literature, helping the researchers to generate a novelty character in their future works filling these gaps.

It is not a new technique, the term bibliometrics was presented in the 1960s [356], and the evaluation of metrics regarding an area of interest in scientific publishing has been developed for more than a century [357]. Nowadays, in the big data era, this tool has been even more helpful in filtering and interpreting a large amount of information and data available for scholars. For AM, it is not different, and the bibliometric analysis has been related to the AM impact on business [358], on the supply chain [26], on industry 4.0 [359], AM-specific applications in orthopedics [360], or the general AM overview and trends [361], among others. Regarding CS, the literature presents the use of bibliometric analysis for a general overview comparing CS to other thermal spray processes [362–364];

however, there is a gap in the literature presenting the evolution of publishing focused on CSAM, or who the researchers and the institutions involved in this important theme are.

This work aims to understand the research status and development trends in the CSAM field, and identify the most relevant themes of study in the CSAM field, as there are some remarkable challenges. Therefore, it is important to carry out a bibliometric analysis that maps the current guidelines in this domain, which can inspire scholars in their future research lines and works. Furthermore, it gives them insights into the most active authors and journals that publish this theme and the countries that invest more in AM-related research. It provides a scientific cartography that reveals the dynamics and structure of scientific fields. For this purpose, a bibliometric analysis is conducted to map CSAM R&D trends.

3.1. Data Mining Strategy

The bibliometric data was extracted from the Scopus database using a query string containing keywords to search in the title, abstract, and keyword fields. Since this work has aimed to see the trend in publishing on CSAM over the last decade, the query string was refined to exclude articles published before 2012 and those in other languages. The following string retrieved more than 450 items as of 27 December 2022: TITLE-ABS-KEY (cold AND spray* AND additive AND manufactur*) OR TITLE-ABS-KEY (cold AND spray* AND 3d AND print*) AND PUBYEAR > 2011. These documents were subjected to further text cleaning and bibliometric analyses.

Due to their irrelevance to the studied theme, some articles were eliminated after a manual screen or database cleaning. The articles were limited to the subject area "materials science" OR "engineering" OR "physics and astronomy" OR "chemistry" OR "chemical engineering" OR "energy" OR "mathematics". The articles listed were carefully reviewed by reading their abstracts or full paper. The documents with an unclear relationship with the theme studied were eliminated, refining the results, resulting in the number of works for the statistical analysis being 439. Finally, the bibliometric analysis software VOSviewer was used to analyze the publications. VOSviewer is a software that graphically presents the bibliometric network mapping, which facilitates the interpretation of maps and data. The main networks are co-citation, bibliographic coupling, co-author, and/or co-word analysis. The authors and index keywords were selected for the co-occurrence analysis. VOSviewer identified many similar keywords, and to make the data more coherent, they were classified manually, such as "cold spray", "cold spraying", "cold gas dynamic spray", and "cold gas spray", which were merged to "cold spray".

3.2. Results and Discussions

Figure 14 presents the scientific productivity regarding the CSAM theme, limited to the last decade (2012–2022). The number of published documents each year indicates this technology's academic impact or interest by researchers, funding institutions, and journals. The number of documents per year significantly rose from 4 documents in 2012 to a maximum of 84 papers in 2022. This trend remained steady from 2020 and 2021, keeping around 81 publications per year. It is reasonable because the number of research groups researching CSAM and their productivity has not maintained the growth rate, despite the increasing number of researchers, groups, and equipment observed in the last decade [68].

Furthermore, implementing the LPCS process demands less investment because the equipment is less expensive, and the noise level during the operation is low [2,99]. Additionally, an LPCS gun is light and typically uses compressed air as the working gas and can be manipulated manually or using a small robot. However, to operate with HPCS equipment, a reasonable noise-insulated booth is demanded, as a facility for dozens of N_2 or He bottles [10], as well as the fact that the equipment costs of hundreds of thousands of dollars and a large size robot to support the gun, following the robot classification proposed by Dobra [365].

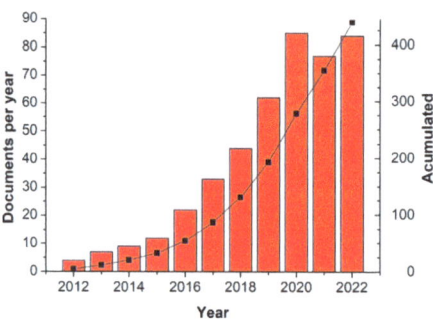

Figure 14. Year-wise publication of documents in CSAM field.

As seen in Figure 15, Halin Liao, who has an h-index of 61 in the Scopus database, is the researcher with more publications on the CSAM topic, with 35 documents. Liao has been a researcher at the Laboratoire Interdisciplinaire Carnot de Bourgogne/Université Bourgogne Franche-Comté (France) since 1994, and is co-author of 500 articles in diverse themes, such as materials characterization and performance, surface engineering, coatings, tribology, and corrosion, among others. Due to the relation of Liao with many other authors, his affiliation figures in the first position among the most important research groups, as seen in Figure 16, followed by Trinity College Dublin, due to the strong and numerous collaborations between Lupoi, Yin, and Chinese co-authors.

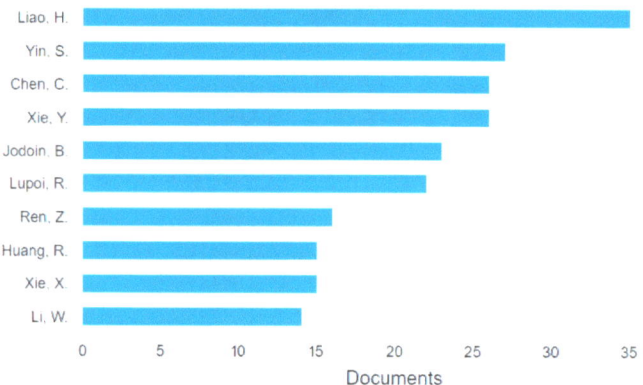

Figure 15. Number of documents per author.

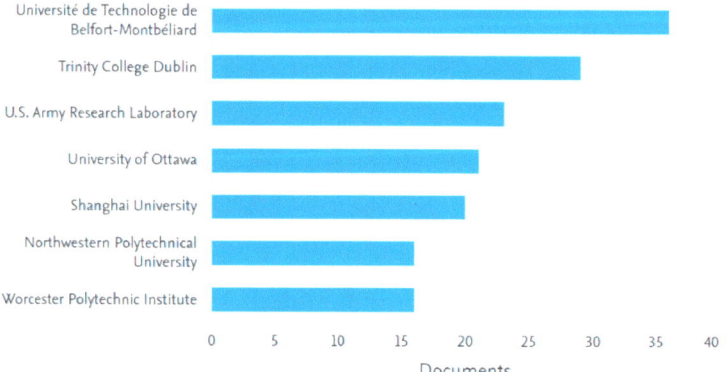

Figure 16. Number of documents by affiliation.

Most of the works published are in collaboration with researchers from Chinese institutions. For example, Shuo Yin, who has an h-index 36 in the Scopus database, is the second author in the number of published articles and worked with Halin Liao and Chaoyu Chen in France. Since 2015, he has been a researcher at Trinity College Dublin, where Rocco Lupoi, who has an h-index 31 in the Scopus database and is sixth on the publishers' list, develops his research too. Bertrand Jodoin, who has an h-index 33 in the Scopus database, is the fifth influential author in Figure 15 and works at the University of Ottawa (Canada).

China has the highest volume of documents published, mainly for collaborative works, as presented in Figure 17. China also leads this ranking, as it occurs in many other areas, due to the vast number of PhD students and active researchers at the Universities and research centers, as well as due to the massive amount of investments and R & D by governmental programs [366–369]. Even with an expressive high number of articles published by Chinese authors, the most cited articles in CSAM have only 6 Chinese figuring among the 25 co-authors enrolled in Table 4. However, the situation in the United States is worse because none of the important works presented in Table 4 has American co-authorship, for which the majority are Europeans. It reflects the importance given by the scientific community for the Chinese and American works, which could be by a lack of novelty seen in most of the hundreds of published works. Another consideration is that most works did not present new concepts but did an application and some important discussion on the concepts previously proposed by other original documents. Original documents or review articles have been cited more, as seen in Table 4. That article type is essential to consolidate the concepts but does not typically promote many citations, such as original or review articles [370,371]. It has caused a preoccupation by the Chinese institutions, which have looked at methodologies to make their work more recognized by the scientific community [372,373].

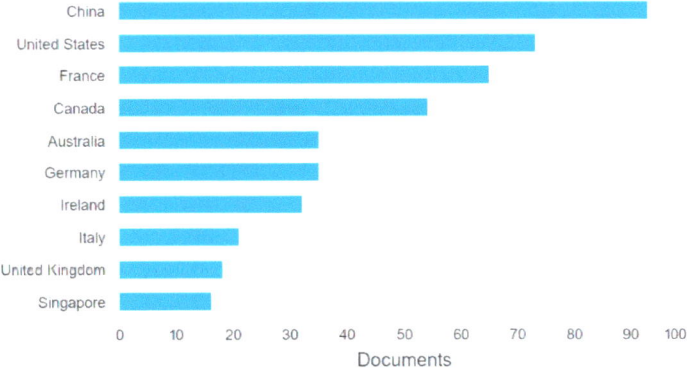

Figure 17. Number of documents by country.

Table 4. The articles most cited in CSAM theme.

Title	Citations	Contributions and Goals	Ref.
Cold spraying—A material's perspective	592	An overview regarding the CS principles, ASI bonding mechanism, materials characteristics, and applications.	[77]
Cold spray additive manufacturing and repair: Fundamentals and applications	372	Summarizing and reviewing the CSAM-related work, comparing CSAM to fusion-based AM techniques, presenting the effects of HT on a CSAM-ed material's properties, and CSAM real applications.	[2]

Table 4. *Cont.*

Title	Citations	Contributions and Goals	Ref.
Solid-state additive manufacturing and repairing by cold spraying: A review	235	Summarizing and reviewing the CSAM-related work, different possibilities of CSAM application, alloys, process parameters, post-treatments, and their effects on CSAM-ed material mechanical properties.	[374]
Cold gas dynamic manufacturing: A non-thermal approach to freeform fabrication	217	Introducing the application of CS as an AM technique to produce freeform parts, comparing CSAM to other AM processes and CSAM strategies.	[169]
Cold gas dynamic spray additive manufacturing today: Deposit possibilities, technological solutions and viable applications	212	Presenting the evolution in investments in CSAM, adhesion and cohesion mechanisms for CSAM-ed material, listing materials and applications, characteristics, and industrial applications.	[68]
Potential of cold gas dynamic spray as additive manufacturing technology	212	Presenting the CSAM principles, geometric characteristics, and materials' properties, as well as the potential in using CSAM and its compatibility with other metal AM techniques.	[97]

Collaborative works have characterized the articles and publishing in CSAM because of the mutual interests and the synergy in sharing equipment to develop the experiments and applying a kind of knowledge synergism to interpret the experimental results obtained. Regarding the authors' collaboration, the co-authorship relations were obtained by VOSviewer software, limiting the results to authors with more than ten articles published, reducing the total of 972 authors to the 16 presented in Figure 18. The circle size around the authors' names represents the number of articles in co-authorship, the color indicates a cluster of authors where the authors have more connections, and the line or link between the circles means the strength of their association; a thicker line means more collaborations. Chen and Xie are the leading authors in a cluster of Chinese cooperation, Yin and Lupoi are the most important authors in a cluster formed at Trinity College Dublin, and Liao is ahead of the French group. An interpretation of the map presented in Figure 18 is that its central persons are Liao, Xie, Chen, and Yin, indicating they act as bridges between the Chinese, Irish, and French institutions.

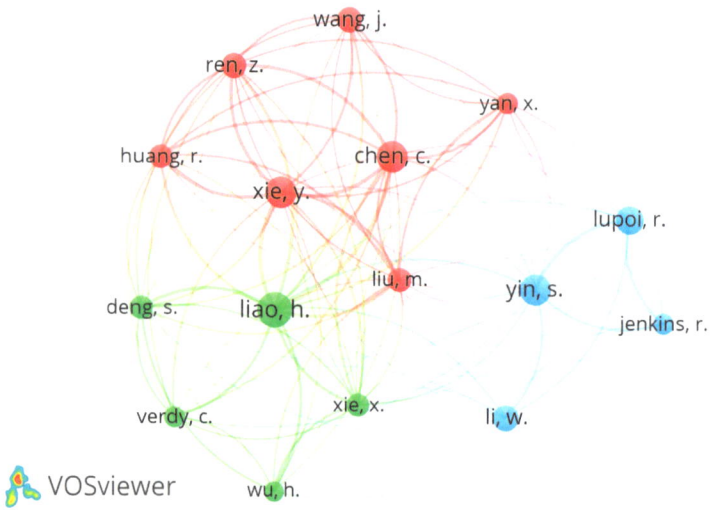

Figure 18. Authors' collaboration. Minimum of 10 articles per author.

The journal with more documents published, 48 articles, was the Journal of Thermal Spray Technology (JTST), a journal focused on surfaces, coatings, and films, justifying why the CS authors chose it to submit and publish their works. However, JTST places in the 70th percentile, or Q2, and has a cite score of 4.6 in the Scopus database. The second influential journal, with 29 papers, was Surface and Coatings Technology (SCT), older than JTST, in the 88th percentile, Q1, and with a cite score of 7.6 in the Scopus database. Open access journals have increased their contribution to CSAM publishing, attracting authors due to the faster publishing process and free access to the readers. Between the ten more relevant publishers, MDPI's journals Coatings, Materials, and Metals have 7, 10, and 10 documents published in the CSAM theme, respectively, from 2019 to 2022. MDPI's Metals has increased its relevance in the scientific community, publishing 6 documents only in 2021, reaching the 76th percentile, Q1, and 3.8 in the Scopus database.

Keywords represent the synthesis of the essential content of the documents, and their analysis aims to study the structure of the research related to the discipline. The analysis principle is based on the co-occurrence of keywords in the selected documents, revealing how closely they are connected in terms of the concepts they deal with, making it possible to understand the main themes of interest for the scholars. VOSviewer software identified more than 3000 keywords, and after a manual and critical evaluation, many of them were merged due to the similarity of their meaning. In addition, only keywords with at least 15 occurrences were considered, resulting in 70 keywords for the study, which are graphically presented in Figure 19b by their density, i.e., a darker and bigger circle represents more times the Keywords are listed: it results that "3d printing", "additive manufacturing", and "cold spraying" are the main terms, followed by "manufacturing processes", "additives", "coatings", and "microstructure" keywords.

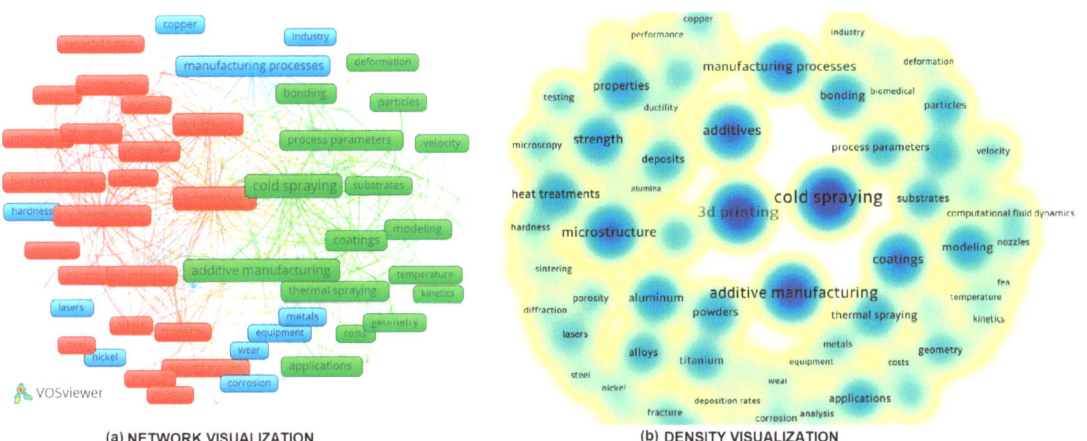

(a) NETWORK VISUALIZATION **(b) DENSITY VISUALIZATION**

Figure 19. Co-occurrences keywords. Minimum 15 occurrences.

By analyzing the mapping network of co-occurrence, three clusters were formed, identified by the colors, Figure 19a. The clusters result from the link strength between the keywords, i.e., a stronger link means the same keywords group is used in more documents. VOSviewer resolution was set to 1.00 to avoid too small clusters. The smallest cluster, the blue one, has 18 items. The primary term is "manufacturing processes", which is a more generalist approach to making parts by CS, laser, and hybrid processes and is also linked to AM-processed materials, such as Cu, Ni, and composites. The other two clusters have the same number of items, 27; the green one has the main terms "cold spraying" and "additive manufacturing" that are strongly linked to "coatings", which makes perfect sense since CS has been developed prior for coating. The authors usually present these keywords together.

From the query string used for searching in the Scopus database, it was predictable that CS and AM would figure as the primary keywords for the papers.

A relation is understood from the materials linked to CSAM, which are Cu, Ti, and Al, as observed reading the articles, but these are presented in Figure 19a with stronger links in the blue cluster. Additionally, material properties and process parameters are highlighted because most works are experimental and present the materials' evaluation and testing. The red cluster presents the keyword "3d printing" linked to material properties, such as the microstructure, porosity, and strength. The keyword "3d printing" could be merged with "additive manufacturing", mixing the red and green clusters, following the AM nomenclatures in the literature. It makes perfect sense, considering the content of the papers that present these keywords. However, the "3d printing" term is a legacy from the AM of polymers and has been presented as a friendlier expression to AM non-experts. On the other hand, "rapid prototyping" had been a keyword widely used for AM [78,89], but in the last decade, it has been substituted for "3d printing" and "additive manufacturing", as indicated by Jemghili, Ait Taleb, and Khalifa [361]. It was confirmed by the VOSviewer keywords list andFigure 19, where "rapid prototyping" was not mentioned, indicating that this keyword has not been linked to CSAM.

4. Summary and CSAM Future Trends

This article briefly introduces CSAM, its characteristics, advantages over other AM processes, limitations, and some answers or alternatives to overtake them based on the literature. In addition, the paper presents challenges that still have to be overcome. Nevertheless, the innovation potential of this research field is outcoming, and new applications have emerged in different industrial fields, supporting the crescent number of publications dedicated to CSAM industrialization. Based on the state-of-art and interpretation of the most recent literature contents, some trends are listed:

- CSAM for repairing services, with its application on expensive components or damages that do not need extensive restoration [2]. Improving the CSAM-ed geometries control generates a hot topic for research, including geometry construction simulation, robot programming, and robot self-learning for an adaptive path, spraying angle, or gun displacement velocity. Research on this theme has been done by the Italian group of Politecnico di Milano [375], the Spanish group of Thermal Spray Center [144], and the Australian company Speed3D, among others;
- CSAM for hard materials, improving the CSAM-ed deposit adherence on materials such as Inconel, Ti6Al4V, HEA, or martensitic steels. For this, studies on the optimization of pre- and process-heating or CS parameters must be exploited. Some examples are using the expensive He as a working gas only for the first layers and N_2 for the others, the CS-SP process, or introducing HT between the layers to reduce the tensile residual stress on the CSAM/substrate interface and improve the adhesion and repairing quality;
- Improve CSAM-ed properties, reaching close or better than the wrought reference materials. As well as the well-established HT and HIP, new post-treatments have to be investigated in this theme. SPS presented good properties, but strict limitations in the geometries are feasible, requiring more flexibility for more complex geometries;
- CS hybrid systems consolidation, such as CS-SP or LACS, to avoid post-treatments and eliminate steps in the AM production chain [286]. Most studies are related to CS-ed coatings, promoting a better adhesion to the substrate and cohesion of particles, besides a low porosity. Therefore, CSAM hybrid systems' use is a hot topic to provide a good performance CSAM-ed parts;
- CSAM applied with other AM processes, optimizing the manufacturing chain to make the low complexity part areas by the fast CSAM process and dedicate the slower but more accurate laser process to the areas that demand more geometrical control. It is feasible because other AM techniques have increased their maturity as industrial

processes; however, this mixing of methods is a lack in R & D, which is a hot topic for scholars.

Regarding the bibliometric analysis, the literature characteristics and metrics were studied, collecting data from more than 420 documents published in the last decade for CSAM and related themes. The analysis covered several dimensions, including subject areas through keyword analysis, productive journals, the most influential authors, most cited documents, and referent affiliations and countries. The main results of the bibliometric analysis can be summarized as follows:

- A total of 56% of the total publications in the CSAM theme were registered during the last three years, indicating the increase of academic interest in this research field, considering that in 2010 the number of documents published was zero. The main topics actively explored in the papers were related to the processing parameters' optimization and other experiments focused on improving the CSAM-ed material's performance to make this process more industrially mature;
- China is the country with more documents published, followed by the United States and France, where the most relevant research group in CSAM is from, the Université de Technologie de Belfort-Montbéliard, which is the affiliation of Liao, the author with the most documents published. The publishing mapping presents a collaboration between Chinese and European institutions, signing for a fast CSAM industry maturity since the Chinese founding objectives are scientific development and even more advances in mass production;
- The current scenario of publication in CSAM points to a future consolidation of CSAM as an industrial technique, first for specific applications in high-cost components, such as multi-alloy nozzles for rockets in the aerospace industry or repairing expensive components, such as turbine blades or vanes. However, in the medium-term and long-term, CSAM applications tend to expand their use;
- "Costs" is the keyword that indicates a crucial point for CSAM advances. For the feedstocks, scholars have studied less expensive materials and improved DE, reaching more than 95% for some materials. A considerable challenge and trend for reducing processing costs and improving CSAM reliability is making the processing more independent of experts but easier for industrialization.

Author Contributions: Conceptualization, R.F.V.; funding acquisition, J.S. and I.G.C.; investigation, R.F.V. and A.G.; methodology, R.F.V. and A.G.; project administration, J.S. and I.G.C.; writing—original draft preparation, R.F.V., A.G., J.S. and V.A.; writing—review and editing, R.F.V., A.G., J.S. and V.A.; supervision, I.C.C. All authors have read and agreed to the published version of the manuscript.

Funding: The grant PID2020-115508RB-C21 and EIN2020-112379 was funded by MCIN/AEI/10.13039/501100011033 and, as appropriate, by "ERDF A way of making Europe", by "European Union NextGenerationEU/PRTR". A.G. and R.F.V. have AGAUR Ph.D. grants 2021 FISDU 00300, and 2020 FISDU 00305, respectively.

Institutional Review Board Statement: Not applicable.

Informed Consent Statement: Not applicable.

Data Availability Statement: The data presented in this study are available on request from the corresponding author.

Conflicts of Interest: The authors declare no conflict of interest. The funders had no role in the design of the study; in the collection, analyses, or interpretation of data; in the writing of the manuscript, or in the decision to publish the results.

Abbreviations

The following abbreviations are used in this manuscript:

AM	Additive Manufacturing
APS	Air Plasma Spray
ASI	Adiabatic Shear Instability
BB	Ball-Burnishing
BJ	Binder Jetting
CFD	Computational Fluid Dynamics
CR	Cold Rolling
CS	Cold Spray
CSAM	Cold Spray Additive Manufacturing
CS-SP	Cold Spray Shot Peening
DE	Deposition Efficiency
DMLM	Direct Metal Laser Melting
DMLS	Direct Metal Laser Sintering
EBM	Electron Beam Melting
EBSD	Electron Back Scattering Diffraction
EPP	Electric Pulsing Processing
FR	Flattening Ratio
FS	Flame Spraying
FSP	Friction Stir Processing
FSAM	Friction Stir Additive Manufacturing
FSW	Friction Stir Welding
GMAW	Gas Metal Arc Welding
GTAW	Gas Tungsten Arc Welding
HEA	High Entropy Alloy
HIP	Hot Isostatic Pressing
HPCS	High-Pressure Cold Spray
HR	Hot Rolling
HT	Heat Treatment
HVOF	High-Velocity Oxy-Fuel
ICP	In situ Coating Properties
IHD	Incremental Hole Drilling
JTST	Journal of Thermal Spray Technology
K_{IC}	Fracture Toughness
LACS	Laser-Assisted Cold Spray
LMF	Laser Metal Fusion
LOM	Laminated Object Manufacturing
LPCS	Low-Pressure Cold Spray
MMC	Metal Matrix Composite
ME	Material Extrusion
MJ	Material Jetting
MMC	Metal Matrix Composite
MPCS	Medium-Pressure Cold Spray
PAW	Plasma Arc Welding
PEO	Plasma Electrolytic Oxidation
R&D	Research and Development
SCT	Surface and Coatings Technology
SD	Standoff Distance
SEM	Scanning Electron Microscopy
SL	Stereolithography
SLD	Supersonic Laser Deposition
SLM	Selective Laser Melting
SLS	Selective Laser Sintering
SP	Shot Peening

SPS	Spark Plasma Sintering
TAT	Tensile Adhesion Testing
TCT	Tubular Coating Tensile
TMT	Thermo-Mechanical Treatment
UAM	Ultrasonic Additive Manufacturing
UTS	Ultimate Tensile Strength
V_{cr}	Critical Velocity
V_{er}	Erosion Velocity
WAAM	Wire Arc Additive Manufacturing
WCI	Waterjet Cavitation Impact

References

1. *ISO/ASTM 52900-21*; Standard Terminology for Additive Manufacturing—General Principles—Terminology. ASTM International: West Conshohocken, PA, USA, 2022.
2. Yin, S.; Cavaliere, P.; Aldwell, B.; Jenkins, R.; Liao, H.; Li, W.; Lupoi, R. Cold Spray Additive Manufacturing and Repair: Fundamentals and Applications. *Addit. Manuf.* **2018**, *21*, 628–650. [CrossRef]
3. Saboori, A.; Aversa, A.; Marchese, G.; Biamino, S.; Lombardi, M.; Fino, P. Application of Directed Energy Deposition-Based Additive Manufacturing in Repair. *Appl. Sci.* **2019**, *9*, 3316. [CrossRef]
4. Campbell, I.; Bourell, D.; Gibson, I. Additive Manufacturing: Rapid Prototyping Comes of Age. *Rapid Prototyp. J.* **2012**, *18*, 255–258. [CrossRef]
5. Dilberoglu, U.M.; Gharehpapagh, B.; Yaman, U.; Dolen, M. The Role of Additive Manufacturing in the Era of Industry 4.0. *Procedia Manuf.* **2017**, *11*, 545–554. [CrossRef]
6. Horst, D.J.; Duvoisin, C.A.; Vieira, R.D.A.; Horst, D.J.; Vieira, R.D.A. Additive Manufacturing at Industry 4.0: A Review. *Int. J. Eng. Tech. Res.* **2018**, *8*, 3–8.
7. Parupelli, S.K.; Desai, S. A Comprehensive Review of Additive Manufacturing (3D Printing): Processes, Applications and Future Potential. *Am. J. Appl. Sci.* **2019**, *16*, 244–272. [CrossRef]
8. Wong, K.V.; Hernandez, A. A Review of Additive Manufacturing. *ISRN Mech. Eng.* **2012**, *2012*, 208760. [CrossRef]
9. Huang, S.H.; Liu, P.; Mokasdar, A.; Hou, L. Additive Manufacturing and Its Societal Impact: A Literature Review. *Int. J. Adv. Manuf. Technol.* **2013**, *67*, 1191–1203. [CrossRef]
10. Sun, W.; Chu, X.; Lan, H.; Huang, R.; Huang, J.; Xie, Y.; Huang, J.; Huang, G. Current Implementation Status of Cold Spray Technology: A Short Review. *J. Therm. Spray Technol.* **2022**, *31*, 848–865. [CrossRef]
11. Kruth, J.; Vandenbroucke, B.; Van Vaerenbergh, J.; Naert, I. Rapid Manufacturing of Dental Prostheses by Means of Selective Laser Sintering/Melting. In Proceedings of the Les 11emes Assises Europeennes du Prototypage Rapide, Paris, France, 4–5 October 2005.
12. Jamieson, R.; Holmer, B.; Ashby, A. How Rapid Prototyping Can Assist in the Development of New Orthopaedic Products—A Case Study. *Rapid Prototyp. J.* **1995**, *1*, 38–41. [CrossRef]
13. Hieu, L.C.; Bohez, E.; Vander Sloten, J.; Phien, H.N.; Vatcharaporn, E.; Binh, P.H.; An, P.V.; Oris, P. Design for Medical Rapid Prototyping of Cranioplasty Implants. *Rapid Prototyp. J.* **2003**, *9*, 175–186. [CrossRef]
14. Esses, S.J.; Berman, P.; Bloom, A.I.; Sosna, J. Clinical Applications of Physical 3D Models Derived from MDCT Data and Created by Rapid Prototyping. *Am. J. Roentgenol.* **2011**, *196*, 683–688. [CrossRef]
15. Sun, J.; Zhang, F.Q. The Application of Rapid Prototyping in Prosthodontics. *J. Prosthodont.* **2012**, *21*, 641–644. [CrossRef]
16. Popov, V.V.; Muller-Kamskii, G.; Kovalevsky, A.; Dzhenzhera, G.; Strokin, E.; Kolomiets, A.; Ramon, J. Design and 3D-Printing of Titanium Bone Implants: Brief Review of Approach and Clinical Cases. *Biomed. Eng. Lett.* **2018**, *8*, 337–344. [CrossRef]
17. Nickels, L. AM and Aerospace: An Ideal Combination. *Met. Powder Rep.* **2015**, *70*, 300–303. [CrossRef]
18. Liou, F.; Liu, R.; Wang, Z.; Sparks, T.; Newkink, J. Aerospace Applications of Laser Additive Manufacturing. In *Laser Additive Manufacturing: Materials, Design, Technologies, and Applications*; Brandt, M., Ed.; Woodhead Publishing: Duxford, UK, 2017; pp. 351–371.
19. Uriondo, A.; Esperon-Miguez, M.; Perinpanayagam, S. The Present and Future of Additive Manufacturing in the Aerospace Sector: A Review of Important Aspects. *J. Aerosp. Eng.* **2015**, *229*, 2132–2147. [CrossRef]
20. Singamneni, S.; Yifan, L.V.; Hewitt, A.; Chalk, R.; Thomas, W.; Jordison, D. Additive Manufacturing for the Aircraft Industry: A Review. *J. Aeronaut. Aerosp. Eng.* **2019**, *8*, 351–371. [CrossRef]
21. Mantovani, S.; Barbieri, S.G.; Giacopini, M.; Croce, A.; Sola, A.; Bassoli, E. Synergy between Topology Optimization and Additive Manufacturing in the Automotive Field. *J. Eng. Manuf.* **2021**, *235*, 555–567. [CrossRef]
22. Juechter, V.; Franke, M.M.; Merenda, T.; Stich, A.; Körner, C.; Singer, R.F. Additive Manufacturing of Ti-45Al-4Nb-C by Selective Electron Beam Melting for Automotive Applications. *Addit. Manuf.* **2018**, *22*, 118–126. [CrossRef]
23. Böckin, D.; Tillman, A. Environmental Assessment of Additive Manufacturing in the Automotive Industry. *J. Clean. Prod.* **2019**, *226*, 977–987. [CrossRef]
24. Thomas, D. Costs, Benefits, and Adoption of Additive Manufacturing: A Supply Chain Perspective. *Int. J. Adv. Manuf. Technol.* **2016**, *85*, 1857–1876. [CrossRef] [PubMed]

25. Delic, M.; Eyers, D.R. The Effect of Additive Manufacturing Adoption on Supply Chain Flexibility and Performance: An Empirical Analysis from the Automotive Industry. *Int. J. Prod. Econ.* **2020**, *228*, 107689. [CrossRef]
26. Nuñez, J.; Ortiz, Á.; Ramírez, M.A.J.; González Bueno, J.A.; Briceño, M.L. Additive Manufacturing and Supply Chain: A Review and Bibliometric Analysis. In *Engineering Digital Transformation. Lecture Notes in Management and Industrial Engineering*; Ortiz, Á., Romano, C.A., Poler, R., García-Sabater, J.-P., Eds.; Springer: Berlin/Heidelberg, Germany, 2019; pp. 323–331.
27. Kozior, T. The Influence of Selected Selective Laser Sintering Technology Process Parameters on Stress Relaxation, Mass of Models, and Their Surface Texture Quality. *3D Print. Addit. Manuf.* **2020**, *7*, 126–138. [CrossRef] [PubMed]
28. Prashar, G.; Vasudev, H. A Comprehensive Review on Sustainable Cold Spray Additive Manufacturing: State of the Art, Challenges and Future Challenges. *J. Clean. Prod.* **2021**, *310*, 127606. [CrossRef]
29. Gutowski, T.G.; Branham, M.S.; Dahmus, J.B.; Jones, A.J.; Thiriez, A.; Sekulic, D.P. Thermodynamic Analysis of Resources Used in Manufacturing Processes. *Environ. Sci. Technol.* **2009**, *43*, 1584–1590. [CrossRef]
30. Peng, T.; Kellens, K.; Tang, R.; Chen, C.; Chen, G. Sustainability of Additive Manufacturing: An Overview on Its Energy Demand and Environmental Impact. *Addit. Manuf.* **2018**, *21*, 694–704. [CrossRef]
31. Agnusdei, L.; Del Prete, A. Additive Manufacturing for Sustainability: A Systematic Literature Review. *Sustain. Future* **2022**, *4*, 100098. [CrossRef]
32. Bártalo, P.J.; Gibson, I. History of Stereolithographic Processes. In *Stereolithography: Materials, Processes and Applications*; Bártalo, P.J., Ed.; Springer: Boston, MA, USA, 2011; pp. 37–56.
33. Mohd Yusuf, S.; Cutler, S.; Gao, N. Review: The Impact of Metal Additive Manufacturing on the Aerospace Industry. *Metals* **2019**, *9*, 1286. [CrossRef]
34. Gokuldoss, P.K.; Kolla, S.; Eckert, J. Additive Manufacturing Processes: Selective Laser Melting, Electron Beam Melting and Binder Jetting—Selection Guidelines. *Materials* **2017**, *10*, 672. [CrossRef]
35. Gülcan, O.; Günaydın, K.; Tamer, A. The State of the Art of Material Jetting—A Critical Review. *Polymers* **2021**, *13*, 2829. [CrossRef]
36. Dickens, P.M. Research Developments in Rapid Prototyping. *Proc. Inst. Mech. Eng. Part B J. Eng. Manuf.* **1995**, *209*, 261–266. [CrossRef]
37. Ziaee, M.; Crane, N.B. Binder Jetting: A Review of Process, Materials, and Methods. *Addit. Manuf.* **2019**, *28*, 781–801. [CrossRef]
38. Sachs, E.; Cima, M.; Williams, P.; Brancazio, D.; Cornie, J. Three Dimensional Printing: Rapid Tooling and Prototypes Directly from a CAD Model. *J. Eng. Ind.* **1992**, *114*, 481–488. [CrossRef]
39. Turner, B.N.; Strong, R.; Gold, S.A. A Review of Melt Extrusion Additive Manufacturing Processes: I. Process Design and Modeling. *Rapid Prototyp. J.* **2014**, *20*, 192–204. [CrossRef]
40. Bochnia, J.; Blasiak, M.; Kozior, T. A Comparative Study of the Mechanical Properties of FDM 3D Prints Made of PLA and Carbon Fiber-Reinforced PLA for Thin-Walled Applications. *Materials* **2021**, *14*, 7062. [CrossRef]
41. Molitch-Hou, M. Overview of Additive Manufacturing Process. In *Additive Manufacturing*; Elsevier: Amsterdam, The Netherlands, 2018; pp. 1–38.
42. Gonzalez, J.A.; Mireles, J.; Lin, Y.; Wicker, R.B. Characterization of Ceramic Components Fabricated Using Binder Jetting Additive Manufacturing Technology. *Ceram. Int.* **2016**, *42*, 10559–10564. [CrossRef]
43. Allahverdi, M.; Danforth, S.; Jafari, M.; Safari, A. Processing of Advanced Electroceramic Components by Fused Deposition Technique. *J. Eur. Ceram. Soc.* **2001**, *21*, 1485–1490. [CrossRef]
44. Yi, S.; Liu, F.; Zhang, J.; Xiong, S. Study of the Key Technologies of LOM for Functional Metal Parts. *J. Mater. Process. Technol.* **2004**, *150*, 175–181. [CrossRef]
45. Prechtl, M.; Otto, A.; Geiger, M. Rapid Tooling by Laminated Object Manufacturing of Metal Foil. *Adv. Mater. Res.* **2005**, *6–8*, 303–312. [CrossRef]
46. Ferrández-Montero, A.; Lieblich, M.; Benavente, R.; González-Carrasco, J.L.; Ferrari, B. Study of the Matrix-Filler Interface in PLA/Mg Composites Manufactured by Material Extrusion Using a Colloidal Feedstock. *Addit. Manuf.* **2020**, *33*, 101142. [CrossRef]
47. Ngo, T.D.; Kashani, A.; Imbalzano, G.; Nguyen, K.T.Q.; Hui, D. Additive Manufacturing (3D Printing): A Review of Materials, Methods, Applications and Challenges. *Compos. Part B* **2018**, *143*, 172–196. [CrossRef]
48. Lewandowski, J.J.; Seifi, M. Metal Additive Manufacturing: A Review of Mechanical Properties. *Annu. Rev. Mater. Res.* **2016**, *46*, 151–186. [CrossRef]
49. Murr, L.E.; Gaytan, S.M.; Ramirez, D.A.; Martinez, E.; Hernandez, J.; Amato, K.N.; Shindo, P.W.; Medina, F.R.; Wicker, R.B. Metal Fabrication by Additive Manufacturing Using Laser and Electron Beam Melting Technologies. *J. Mater. Sci. Technol.* **2012**, *28*, 1–14. [CrossRef]
50. Negi, S.; Nambolan, A.A.; Kapil, S.; Joshi, P.S.; Manivannan, R.; Karunakaran, K.P.; Bhargava, P. Review on Electron Beam Based Additive Manufacturing. *Rapid Prototyp. J.* **2020**, *26*, 485–498. [CrossRef]
51. Parthasarathy, J.; Starly, B.; Raman, S.; Christensen, A. Mechanical Evaluation of Porous Titanium (Ti6Al4V) Structures with Electron Beam Melting (EBM). *J. Mech. Behav. Biomed. Mater.* **2010**, *3*, 249–259. [CrossRef]
52. Ma, Y.; Cuiuri, D.; Shen, C.; Li, H.; Pan, Z. Effect of Interpass Temperature on In-Situ Alloying and Additive Manufacturing of Titanium Aluminides Using Gas Tungsten Arc Welding. *Addit. Manuf.* **2015**, *8*, 71–77. [CrossRef]
53. Spaniol, E.; Ungethüm, T.; Trautmann, M.; Andrusch, K.; Hertel, M.; Füssel, U. Development of a Novel TIG Hot-Wire Process for Wire and Arc Additive Manufacturing. *Weld. World* **2020**, *64*, 1329–1340. [CrossRef]

54. Moreira, A.F.; Ribeiro, K.S.B.; Mariani, F.E.; Coelho, R.T. An Initial Investigation of Tungsten Inert Gas (TIG) Torch as Heat Source for Additive Manufacturing (AM) Process. *Procedia Manuf.* **2020**, *48*, 671–677. [CrossRef]
55. Tabernero, I.; Paskual, A.; Álvarez, P.; Suárez, A. Study on Arc Welding Processes for High Deposition Rate Additive Manufacturing. *Procedia CIRP* **2018**, *68*, 358–362. [CrossRef]
56. Jin, W.; Zhang, C.; Jin, S.; Tian, Y.; Wellmann, D.; Liu, W. Wire Arc Additive Manufacturing of Stainless Steels: A Review. *Appl. Sci.* **2020**, *10*, 1563. [CrossRef]
57. Artaza, T.; Suárez, A.; Murua, M.; García, J.C.; Tabernero, I.; Lamikiz, A. Wire Arc Additive Manufacturing of Mn4Ni2CrMo Steel: Comparison of Mechanical and Metallographic Properties of PAW and GMAW. *Procedia Manuf.* **2019**, *41*, 1071–1078. [CrossRef]
58. Pattanayak, S.; Sahoo, S.K. Gas Metal Arc Welding Based Additive Manufacturing—A Review. *CIRP J. Manuf. Sci. Technol.* **2021**, *33*, 398–442. [CrossRef]
59. Veiga, F.; Del Val, A.G.; Suárez, A.; Alonso, U. Analysis of the Machining Process of Titanium Ti6Al-4V Parts Manufactured by Wire Arc Additive Manufacturing (WAAM). *Materials* **2020**, *13*, 766. [CrossRef] [PubMed]
60. Poolperm, P.; Nakkiew, W.; Naksuk, N. Experimental Investigation of Additive Manufacturing Using a Hot-Wire Plasma Welding Process on Titanium Parts. *Materials* **2021**, *14*, 1270. [CrossRef]
61. Mercado Rojas, J.G.; Wolfe, T.; Fleck, B.A.; Qureshi, A.J. Plasma Transferred Arc Additive Manufacturing of Nickel Metal Matrix Composites. *Manuf. Lett.* **2018**, *18*, 31–34. [CrossRef]
62. Kumar Srivastava, A.; Kumar, N.; Rai Dixit, A. Friction Stir Additive Manufacturing—An Innovative Tool to Enhance Mechanical and Microstructural Properties. *Mater. Sci. Eng. B* **2021**, *263*, 114832. [CrossRef]
63. Mishra, R.S.; Palanivel, S. Building without Melting: A Short Review of Friction-Based Additive Manufacturing Techniques. *Int. J. Addit. Subtractive Mater. Manuf.* **2017**, *1*, 82. [CrossRef]
64. Hehr, A.; Norfolk, M. A Comprehensive Review of Ultrasonic Additive Manufacturing. *Rapid Prototyp. J.* **2019**, *26*, 445–458. [CrossRef]
65. Friel, R.J.; Harris, R.A. Ultrasonic Additive Manufacturing—A Hybrid Production Process for Novel Functional Products. *Procedia CIRP* **2013**, *6*, 35–40. [CrossRef]
66. Váz, R.F.; Tristante, R.; Pukasiewicz, A.G.M.; Capra, A.R.; Chicoski, A.; Filippin, C.G.; Paredes, R.S.C.; Henke, S.L. Welding and Thermal Spray Processes for Maintenance of Hydraulic Turbine Runners: Case Studies. *Soldag. Insp.* **2021**, *26*, 1–13. [CrossRef]
67. Keshavarz, M.K.; Gontcharov, A.; Lowden, P.; Chan, A.; Kulkarni, D.; Brochu, M. Turbine Blade Tip Repair by Laser Directed Energy Deposition Additive Manufacturing Using a Rene 142–MERL 72 Powder Blend. *J. Manuf. Mater. Process.* **2021**, *5*, 21. [CrossRef]
68. Raoelison, R.N.; Verdy, C.; Liao, H. Cold Gas Dynamic Spray Additive Manufacturing Today: Deposit Possibilities, Technological Solutions and Viable Applications. *Mater. Des.* **2017**, *133*, 266–287. [CrossRef]
69. Srikanth, A.; Mohammed Thalib Basha, G.; Venkateshwarlu, B. A Brief Review on Cold Spray Coating Process. *Mater. Today Proc.* **2019**, *22*, 1390–1397. [CrossRef]
70. Bagherifard, S.; Roscioli, G.; Zuccoli, M.V.; Hadi, M.; D'Elia, G.; Demir, A.G.; Previtali, B.; Kondás, J.; Guagliano, M. Cold Spray Deposition of Freestanding Inconel Samples and Comparative Analysis with Selective Laser Melting. *J. Therm. Spray Technol.* **2017**, *26*, 1517–1526. [CrossRef]
71. Gamon, A.; Arrieta, E.; Gradl, P.R.; Katsarelis, C.; Murr, L.E.; Wicker, R.B.; Medina, F. Microstructure and Hardness Comparison of As-Built Inconel 625 Alloy Following Various Additive Manufacturing Processes. *Results Mater.* **2021**, *12*, 100239. [CrossRef]
72. Additive Manufacturing Metal Technology Available online: https://additive-manufacturing-report.com/additive-manufacturing-metal-technology (accessed on 18 November 2022).
73. Guerges, M. NASA 3D-Printed Manufactured Rocket Engine Hardware Passes Cold Spray, Hot Fire Tests. Available online: https://www.nasa.gov/centers/marshall/news/releases/2021/nasa-additively-manufactured-rocket-engine-hardware-passes-cold-spray-hot-fire-tests.html (accessed on 29 November 2022).
74. Buchanan, C.; Gardner, L. Metal 3D Printing in Construction: A Review of Methods, Research, Applications, Opportunities and Challenges. *Eng. Struct.* **2019**, *180*, 332–348. [CrossRef]
75. Pathak, S.; Saha, G.C. Development of Sustainable Cold Spray Coatings and 3D Additive Manufacturing Components for Repair/Manufacturing Applications: A Critical Review. *Coatings* **2017**, *7*, 122. [CrossRef]
76. Ashokkumar, M.; Thirumalaikumarasamy, D.; Sonar, T.; Deepak, S.; Vignesh, P.; Anbarasu, M. An Overview of Cold Spray Coating in Additive Manufacturing, Component Repairing and Other Engineering Applications. *J. Mech. Behav. Mater.* **2022**, *31*, 514–534. [CrossRef]
77. Assadi, H.; Kreye, H.; Gärtner, F.; Klassen, T. Cold Spraying—A Materials Perspective. *Acta Mater.* **2016**, *116*, 382–407. [CrossRef]
78. Crawmer, D.E. Cold Spray Process. In *Thermal Spray Technology*; Davis, J.R., Ed.; ASM International: Materials Park, OH, USA, 2013; pp. 77–84.
79. Schmidt, T.; Gärtner, F.; Assadi, H.; Kreye, H. Development of a Generalized Parameter Window for Cold Spray Deposition. *Acta Mater.* **2006**, *54*, 729–742. [CrossRef]
80. Canales, H.; Cano, I.G.; Dosta, S. Window of Deposition Description and Prediction of Deposition Efficiency via Machine Learning Techniques in Cold Spraying. *Surf. Coat. Technol.* **2020**, *401*, 126143. [CrossRef]
81. Bagherifard, S.; Monti, S.; Zuccoli, M.V.; Riccio, M.; Kondás, J.; Guagliano, M. Cold Spray Deposition for Additive Manufacturing of Freeform Structural Components Compared to Selective Laser Melting. *Mater. Sci. Eng. A* **2018**, *721*, 339–350. [CrossRef]

82. Li, W.; Liu, P.; Liaw, P.K. Microstructures and Properties of High-Entropy Alloy Films and Coatings: A Review. *Mater. Res. Lett.* **2018**, *6*, 199–229. [CrossRef]
83. Jeandin, M.; Rolland, G.; Descurninges, L.L.; Berger, M.H. Which Powders for Cold Spray? *Surf. Eng.* **2014**, *30*, 291–298. [CrossRef]
84. Silvello, A.; Cavaliere, P.D.; Albaladejo, V.; Martos, A.; Dosta, S.; Cano, I.G. Powder Properties and Processing Conditions Affecting Cold Spray Deposition. *Coatings* **2020**, *10*, 91. [CrossRef]
85. Assadi, H.; Schmidt, T.; Richter, H.; Kliemann, J.O.; Binder, K.; Gärtner, F.; Klassen, T.; Kreye, H. On Parameter Selection in Cold Spraying. *J. Therm. Spray Technol.* **2011**, *20*, 1161–1176. [CrossRef]
86. Kosarev, V.F.; Klinkov, S.V.; Alkhimov, A.P.; Papyrin, A.N. On Some Aspects of Gas Dynamics of the Cold Spray Process. *J. Therm. Spray Technol.* **2003**, *12*, 265–281. [CrossRef]
87. Karthikeyan, J. The Advantages and Disadvantages of the Cold Spray Coating Process. In *The Cold Spray Materials Deposition Process*; Champagne, V.K., Ed.; Elsevier: Cambridge, UK, 2007; pp. 62–71.
88. Wu, H.; Xie, X.; Liu, S.; Xie, S.; Huang, R.; Verdy, C.; Liu, M.; Liao, H.; Deng, S.; Xie, Y. Bonding Behavior of Bi-Metal-Deposits Produced by Hybrid Cold Spray Additive Manufacturing. *J. Mater. Process. Technol.* **2022**, *299*, 117375. [CrossRef]
89. Guo, D.; Kazasidis, M.; Hawkins, A.; Fan, N.; Leclerc, Z.; MacDonald, D.; Nastic, A.; Nikbakht, R.; Ortiz-Fernandez, R.; Rahmati, S.; et al. Cold Spray: Over 30 Years of Development Toward a Hot Future. *J. Therm. Spray Technol.* **2022**, *31*, 866–907. [CrossRef]
90. Singh, H.; Sidhu, T.S.; Kalsi, S.B.S.; Karthikeyan, J. Development of Cold Spray from Innovation to Emerging Future Coating Technology. *J. Braz. Soc. Mech. Sci. Eng.* **2013**, *35*, 231–245. [CrossRef]
91. Munsch, M.; Schmidt-Lehr, M.; Wycisk, E. *Additive Manufacturing New Metal Technologies*; AMPOWER GmbH & Co. KG: Hamburg, Germany, 2020; Volume 6.
92. Lynch, M.E.; Gu, W.; El-Wardany, T.; Hsu, A.; Viens, D.; Nardi, A.; Klecka, M. Design and Topology/Shape Structural Optimisation for Additively Manufactured Cold Sprayed Components. *Virtual Phys. Prototyp.* **2013**, *8*, 213–231. [CrossRef]
93. Wang, X.; Feng, F.; Klecka, M.A.; Mordasky, M.D.; Garofano, J.K.; El-Wardany, T.; Nardi, A.; Champagne, V.K. Characterization and Modeling of the Bonding Process in Cold Spray Additive Manufacturing. *Addit. Manuf.* **2015**, *8*, 149–162. [CrossRef]
94. Benenati, G.; Lupoi, R. Development of a Deposition Strategy in Cold Spray for Additive Manufacturing to Minimize Residual Stresses. *Procedia CIRP* **2016**, *55*, 101–108. [CrossRef]
95. Villafuerte, J. Considering Cold Spray for Additive Manufacturing. *Adv. Mater. Process.* **2014**, *172*, 50–52.
96. Řehořek, L.; Dlouhý, I.; Jan, V. Cold Gas Dynamic Spray Deposition as Additive Manufacturing of Architectured Materials. *Mater. Eng. Mater. Inž.* **2017**, *24*, 115–123.
97. Sova, A.; Grigoriev, S.; Okunkova, A.; Smurov, I. Potential of Cold Gas Dynamic Spray as Additive Manufacturing Technology. *Int. J. Adv. Manuf. Technol.* **2013**, *69*, 2269–2278. [CrossRef]
98. Al-Mangour, B. Fundamentals of Cold Spray Processing: Evolution and Future Perspectives. In *Cold-Spray Coatings: Recent Trends and Future Perspectives*; Cavaliere, P., Ed.; Elsevier: Cham, Switzerland, 2018; pp. 3–24.
99. Rokni, M.R.; Nutt, S.R.; Widener, C.A.; Champagne, V.K.; Hrabe, R.H. Review of Relationship Between Particle Deformation, Coating Microstructure, and Properties in High-Pressure Cold Spray. *J. Therm. Spray Technol.* **2017**, *26*, 1308–1355. [CrossRef]
100. Moridi, A.; Hassani-Gangaraj, S.M.; Guagliano, M.; Dao, M. Cold Spray Coating: Review of Material Systems and Future Perspectives. *Surf. Eng.* **2014**, *30*, 369–395. [CrossRef]
101. Suo, X.; Yin, S.; Planche, M.P.; Liu, T.; Liao, H. Strong Effect of Carrier Gas Species on Particle Velocity during Cold Spray Processes. *Surf. Coat. Technol.* **2015**, *268*, 90–93. [CrossRef]
102. Gilmore, D.L.; Dykhuizen, R.C.; Neiser, R.A.; Roemer, T.J.; Smith, M.F. Particle Velocity and Deposition Efficiency in the Cold Spray Process. *J. Therm. Spray Technol.* **1999**, *8*, 576–582. [CrossRef]
103. Schmidt, T.; Assadi, H.; Gärtner, F.; Richter, H.; Stoltenhoff, T.; Kreye, H.; Klassen, T. From Particle Acceleration to Impact and Bonding in Cold Spraying. *J. Therm. Spray Technol.* **2009**, *18*, 794–808. [CrossRef]
104. Maev, R.G.; Leshchynsky, V. Air Gas Dynamic Spraying of Powder Mixtures: Theory and Application. *J. Therm. Spray Technol.* **2006**, *15*, 198–205. [CrossRef]
105. Alonso, L.; Garrido, M.A.; Poza, P. An Optimisation Method for the Cold-Spray Process: On the Nozzle Geometry. *Mater. Des.* **2022**, *214*, 110387. [CrossRef]
106. Da Silva, F.S.; Cinca, N.; Dosta, S.; Cano, I.G.; Benedetti, A.V.; Guilemany, J.M. Cold Gas Spray Coatings: Basic Principles Corrosion Protection and Applications. *Eclét. Quím. J.* **2017**, *42*, 9–32. [CrossRef]
107. Grujicic, M.; Zhao, C.L.; DeRosset, W.S.; Helfritch, D. Adiabatic Shear Instability Based Mechanism for Particles/Substrate Bonding in the Cold-Gas Dynamic-Spray Process. *Mater. Des.* **2004**, *25*, 681–688. [CrossRef]
108. Assadi, H.; Gärtner, F.; Stoltenhoff, T.; Kreye, H. Bonding Mechanism in Cold Gas Spraying. *Acta Mater.* **2003**, *51*, 4379–4394. [CrossRef]
109. Hassani-Gangaraj, S.M.; Moridi, A.; Guagliano, M. Critical Review of Corrosion Protection by Cold Spray Coatings. *Surf. Eng.* **2015**, *31*, 803–815. [CrossRef]
110. Champagne, V.K.; Helfritch, D.; Leyman, P.; Grendahl, S.; Klotz, B. Interface Material Mixing Formed by the Deposition of Copper on Aluminum by Means of the Cold Spray Process. *J. Therm. Spray Technol.* **2005**, *14*, 330–334. [CrossRef]
111. Hussain, T.; McCartney, D.G.; Shipway, P.H.; Zhang, D. Bonding Mechanisms in Cold Spraying: The Contributions of Metallurgical and Mechanical Components. *J. Therm. Spray Technol.* **2009**, *18*, 364–379. [CrossRef]

112. Hassani-Gangaraj, M.; Veysset, D.; Champagne Jr, V.K.; Nelson, K.A. Adiabatic Shear Instability Is Not Necessary for Adhesion in Cold Spray. *Acta Mater.* **2018**, *158*, 430–439. [CrossRef]
113. Assadi, H.; Gärtner, F.; Klassen, T.; Kreye, H. Comment on 'Adiabatic Shear Instability Is Not Necessary for Adhesion in Cold Spray. *Scr. Mater.* **2019**, *162*, 512–514. [CrossRef]
114. Chen, C.; Su, H.; Wang, X.; Liu, Y.; Zhao, L.; Wei, X.; Zhao, Y.; Pan, J.; Qiu, X. Impact-Induced Bonding Process of Copper at Low Velocity and Room Temperature. *Mater. Des.* **2023**, *226*, 111603. [CrossRef]
115. Macdonald, D.; Nastic, A.; Jodoin, B. Understanding Adhesion. In *Cold-Spray Coatings: Recent Trends and Future Perspectives*; Cavaliere, P.D., Ed.; Springer: Cham, Switzerland, 2016; pp. 421–450.
116. Li, C.J.; Wang, H.T.; Zhang, Q.; Yang, G.J.; Li, W.Y.; Liao, H.L. Influence of Spray Materials and Their Surface Oxidation on the Critical Velocity in Cold Spraying. *J. Therm. Spray Technol.* **2010**, *19*, 95–101. [CrossRef]
117. Wang, Z.; Cai, S.; Chen, W.; Ali, R.A.; Jin, K. Analysis of Critical Velocity of Cold Spray Based on Machine Learning Method with Feature Selection. *J. Therm. Spray Technol.* **2021**, *30*, 1213–1225. [CrossRef]
118. Gärtner, F.; Stoltenhoff, T.; Schmidt, T.; Kreye, H. The Cold Spray Process and Its Potential for Industrial Applications. *J. Therm. Spray Technol.* **2006**, *15*, 223–232. [CrossRef]
119. Bagherifard, S.; Kondas, J.; Monti, S.; Cizek, J.; Perego, F.; Kovarik, O.; Lukac, F.; Gaertner, F.; Guagliano, M. Tailoring Cold Spray Additive Manufacturing of Steel 316 L for Static and Cyclic Load-Bearing Applications. *Mater. Des.* **2021**, *203*, 109575. [CrossRef]
120. Raletz, F.; Vardelle, M.; Ezo'o, G. Critical Particle Velocity under Cold Spray Conditions. *Surf. Coat. Technol.* **2006**, *201*, 1942–1947. [CrossRef]
121. Binder, K.; Gottschalk, J.; Kollenda, M.; Gärtner, F.; Klassen, T. Influence of Impact Angle and Gas Temperature on Mechanical Properties of Titanium Cold Spray Deposits. *J. Therm. Spray Technol.* **2011**, *20*, 234–242. [CrossRef]
122. Vidaller, M.V.; List, A.; Gaertner, F.; Klassen, T.; Dosta, S.; Guilemany, J.M. Single Impact Bonding of Cold Sprayed Ti-6Al-4V Powders on Different Substrates. *J. Therm. Spray Technol.* **2015**, *24*, 644–658. [CrossRef]
123. Pérez-Andrade, L.I.; Gärtner, F.; Villa-Vidaller, M.; Klassen, T.; Muñoz-Saldaña, J.; Alvarado-Orozco, J.M. Optimization of Inconel 718 Thick Deposits by Cold Spray Processing and Annealing. *Surf. Coat. Technol.* **2019**, *378*, 124997. [CrossRef]
124. Sun, W.; Bhowmik, A.; Tan, A.W.; Li, R.; Xue, F.; Marinescu, I.; Liu, E. Improving Microstructural and Mechanical Characteristics of Cold-Sprayed Inconel 718 Deposits via Local Induction Heat Treatment. *J. Alloy. Compd.* **2019**, *797*, 1268–1279. [CrossRef]
125. Oyinbo, S.T.; Jen, T.-C. A Comparative Review on Cold Gas Dynamic Spraying Processes and Technologies. *Manuf. Rev.* **2019**, *6*, 25. [CrossRef]
126. Kumar, S.; Pandey, S.M. The Study of Assessment Parameters and Performance Measurement of Cold Spray Technique: A Futuristic Approach Towards Additive Manufacturing. *MAPAN* **2022**, *37*, 859–879. [CrossRef]
127. Wong, W.; Irissou, E.; Ryabinin, A.N.; Legoux, J.G.; Yue, S. Influence of Helium and Nitrogen Gases on the Properties of Cold Gas Dynamic Sprayed Pure Titanium Coatings. *J. Therm. Spray Technol.* **2011**, *20*, 213–226. [CrossRef]
128. Al-Mangour, B.; Vo, P.; Mongrain, R.; Irissou, E.; Yue, S. Effect of Heat Treatment on the Microstructure and Mechanical Properties of Stainless Steel 316L Coatings Produced by Cold Spray for Biomedical Applications. *J. Therm. Spray Technol.* **2014**, *23*, 641–652. [CrossRef]
129. Borchers, C.; Schmidt, T.; Gärtner, F.; Kreye, H. High Strain Rate Deformation Microstructures of Stainless Steel 316L by Cold Spraying and Explosive Powder Compaction. *Appl. Phys. A* **2008**, *90*, 517–526. [CrossRef]
130. Pukasiewicz, A.G.M.; de Oliveira, W.R.; Váz, R.F.; de Souza, G.B.; Serbena, F.C.; Dosta, S.; Cano, I.G. Influence of the Deposition Parameters on the Tribological Behavior of Cold Gas Sprayed FeMnCrSi Alloy Coatings. *Surf. Coat. Technol.* **2021**, *428*, 127888. [CrossRef]
131. Belgroune, A.; Alhussein, A.; Aissani, L.; Zaabat, M.; Obrosov, A.; Verdy, C.; Langlade, C. Effect of He and N2 Gas on the Mechanical and Tribological Assessment of SS316L Coating Deposited by Cold Spraying Process. *J. Mater. Sci.* **2022**, *57*, 5258–5274. [CrossRef]
132. Vaz, R.F.; Silvello, A.; Sanchez, J.; Albaladejo, V.; Cano, I.G.G. The Influence of the Powder Characteristics on 316L Stainless Steel Coatings Sprayed by Cold Gas Spray. *Coatings* **2021**, *11*, 168. [CrossRef]
133. Yin, S.; Meyer, M.; Li, W.; Liao, H.; Lupoi, R. Gas Flow, Particle Acceleration, and Heat Transfer in Cold Spray: A Review. *J. Therm. Spray Technol.* **2016**, *25*, 874–896. [CrossRef]
134. Lee, M.-W.; Park, J.-J.; Kim, D.-Y.; Yoon, S.S.; Kim, H.-Y.; James, S.C.; Chandra, S.; Coyle, T. Numerical Studies on the Effects of Stagnation Pressure and Temperature on Supersonic Flow Characteristics in Cold Spray Applications. *J. Therm. Spray Technol.* **2011**, *20*, 1085–1097. [CrossRef]
135. UNE EN 17393; Thermal Spraying. Tubular Coating Testing. Asociación Española de Normalización: Madrid, Spain, 2020.
136. Gärtner, F.; Stoltenhoff, T.; Voyer, J.; Kreye, H.; Riekehr, S.; Koçak, M. Mechanical Properties of Cold-Sprayed and Thermally Sprayed Copper Coatings. *Surf. Coat. Technol.* **2006**, *200*, 6770–6782. [CrossRef]
137. Meng, X.; Zhang, J.; Zhao, J.; Liang, Y.; Zhang, Y. Influence of Gas Temperature on Microstructure and Properties of Cold Spray 304SS Coating. *J. Mater. Sci. Technol.* **2011**, *27*, 809–815. [CrossRef]
138. Cai, Z.; Deng, S.; Liao, H.; Zeng, C.; Montavon, G. The Effect of Spray Distance and Scanning Step on the Coating Thickness Uniformity in Cold Spray Process. *J. Therm. Spray Technol.* **2014**, *23*, 354–362. [CrossRef]
139. Li, W.-Y.; Zhang, C.; Guo, X.P.; Zhang, G.; Liao, H.L.; Li, C.-J.; Coddet, C. Effect of Standoff Distance on Coating Deposition Characteristics in Cold Spraying. *Mater. Des.* **2008**, *29*, 297–304. [CrossRef]

140. Lett, S.; Quet, A.; Hémery, S.; Cormier, J.; Meillot, E.; Villechaise, P. Residual Stresses Development during Cold Spraying of Ti-6Al-4V Combined with In Situ Shot Peening. *J. Therm. Spray Technol.* **2022**, 1–15. [CrossRef]
141. Yin, S.; Jenkins, R.; Yan, X.; Lupoi, R. Microstructure and Mechanical Anisotropy of Additively Manufactured Cold Spray Copper Deposits. *Mater. Sci. Eng. A* **2018**, *734*, 67–76. [CrossRef]
142. Baek, M.-S.; Kim, H.-J.; Lee, K.-A. Anisotropy of Compressive Deformation Behavior in Cold Sprayed Cu Bulk Material. *J. Nanosci. Nanotechnol.* **2019**, *19*, 3935–3942. [CrossRef]
143. Vargas-Uscategui, A.; King, P.C.; Yang, S.; Chu, C.; Li, J. Toolpath Planning for Cold Spray Additively Manufactured Titanium Walls and Corners: Effect on Geometry and Porosity. *J. Mater. Process. Technol.* **2021**, *298*, 117272. [CrossRef]
144. Vaz, R.F.; Albaladejo-Fuentes, V.; Sanchez, J.; Ocaña, U.; Corral, Z.G.Z.G.; Canales, H.; Cano, I.G. Metal Knitting: A New Strategy for Cold Gas Spray. *Materials* **2022**, *15*, 6785. [CrossRef]
145. Nardi, A.T.; El-Wardany, T.I.; Viens, D.V.; Lynch, M.E.; Hsu, A.; Klecka, M.A.; Gu, W. Additive Topology Optimized Manufacturing for Multi-Functional Components. U.S. Patent 2014/0277669 A1, 2014.
146. Brodmann, F.J. Cold Spray Process Parameters: Powders. In *The Cold Spray Materials Deposition Process*; Champagne, V.K., Ed.; Elsevier: Cambridge, UK, 2007; pp. 105–116.
147. Poirier, D.; Thomas, Y.; Guerreiro, B.; Martin, M.; Aghasibeig, M.; Irissou, E. Novel Powder Modification Method for the Cold Spray of Hard Steels. In Proceedings of the International Thermal Spray Conference, Quebec City, QC, Canada, 24–28 May 2021; ASM Thermal Spray Society: Materials Park, OH, USA, 2021; pp. 603–610.
148. Story, W.A.; Brewer, L.N. Heat Treatment of Gas-Atomized Powders for Cold Spray Deposition. *Metall. Mater. Trans. A* **2018**, *49*, 446–449. [CrossRef]
149. *ASTM B214-22*; Standard Test Method for Sieve Analysis of Metal Powders. ASTM International: West Conshohocken, PA, USA, 2022.
150. *ASTM B213-20*; Standard Test Method for Flow Rate of Metal Powders. ASTM International: West Conshohocken, PA, USA, 2020.
151. *ASTM B212-21*; Standard Test Method for Apparent Density of Powders Using the Hall Flowmeter Funnel. ASTM International: West Conshohocken, PA, USA, 2021.
152. Valente, R.; Ostapenko, A.; Sousa, B.C.; Grubbs, J.; Massar, C.J.; Cote, D.L.; Neamtu, R. Classifying Powder Flowability for Cold Spray Additive Manufacturing Using Machine Learning. In Proceedings of the 2020 IEEE International Conference on Big Data (Big Data), Atlanta, GA, USA, 10–13 December 2020; IEEE: Atlanta, GA, USA, 2020; pp. 2919–2928.
153. Singh, R.; Kondás, J.; Bauer, C.; Cizek, J.; Medricky, J.; Csaki, S.; Čupera, J.; Procházka, R.; Melzer, D.; Konopík, P. Bulk-like Ductility of Cold Spray Additively Manufactured Copper in the as-Sprayed State. *Addit. Manuf. Lett.* **2022**, *3*, 100052. [CrossRef]
154. Wong, W.; Irissou, E.; Legoux, J.G.; Vo, P.; Yue, S. Powder Processing and Coating Heat Treatment on Cold Sprayed Ti-6Al-4V Alloy. *Mater. Sci. Forum* **2012**, *706–709*, 258–263. [CrossRef]
155. Wong, W.; Vo, P.; Irissou, E.; Ryabinin, A.N.; Legoux, J.-G.; Yue, S. Effect of Particle Morphology and Size Distribution on Cold-Sprayed Pure Titanium Coatings. *J. Therm. Spray Technol.* **2013**, *22*, 1140–1153. [CrossRef]
156. Munagala, V.N.V.; Akinyi, V.; Vo, P.; Chromik, R.R. Influence of Powder Morphology and Microstructure on the Cold Spray and Mechanical Properties of Ti6Al4V Coatings. *J. Therm. Spray Technol.* **2018**, *27*, 827–842. [CrossRef]
157. Wang, X.; Zhang, B.; Lv, J.; Yin, S. Investigation on the Clogging Behavior and Additional Wall Cooling for the Axial-Injection Cold Spray Nozzle. *J. Therm. Spray Technol.* **2015**, *24*, 696–701. [CrossRef]
158. Li, C.J.; Li, W.Y.; Wang, Y.Y.; Yang, G.J.; Fukanuma, H. A Theoretical Model for Prediction of Deposition Efficiency in Cold Spraying. *Thin Solid Films* **2005**, *489*, 79–85. [CrossRef]
159. Schmidt, T.; Gaertner, F.; Kreye, H. New Developments in Cold Spray Based on Higher Gas and Particle Temperatures. *J. Therm. Spray Technol.* **2006**, *15*, 488–494. [CrossRef]
160. Foelsche, A.F. Nozzle Clogging Prevention and Analysis in Cold Spray. Master's Thesis, University of Massachusetts, Amherst, MA, USA, September 2020.
161. Fukanuma, H. Cold-Spray Nozzle and Cold-Spray Device Using Cold-Spray Nozzle. U.S. Patent 9,095,858B2, 4 August 2015.
162. Tronsson, H.C. Feasibility of Electric Field Assisted Clogging Reduction in Cold Gas Spraying Nozzle. Bachelor's Thesis, Dartmouth College, Hanover, NH, USA, June 2020.
163. Saboori, A.; Biamino, S.; Valente, A.; Gitardi, D.; Basile, G.; Lombardi, M.; Fino, P. The Capacity of Cold Spray Additive Manufacturing Technology for Metallic Part Repairing. In Proceedings of the Euro PM 2018 Congress and Exhibition, Bilbao, Spain, 14–18 October 2018.
164. Ikeuchi, D.; Vargas-Uscategui, A.; Wu, X.; King, P. Data-Efficient Neural Network for Track Profile Modelling in Cold Spray Additive Manufacturing. *Appl. Sci.* **2021**, *11*, 1654. [CrossRef]
165. Kotoban, D.; Grigoriev, S.; Okunkova, A.; Sova, A. Influence of a Shape of Single Track on Deposition Efficiency of 316L Stainless Steel Powder in Cold Spray. *Surf. Coat. Technol.* **2017**, *309*, 951–958. [CrossRef]
166. Wu, H.; Liu, S.; Lewke, M.; Li, W.; Raoelison, R.-N.; List, A.; Gärtner, F.; Liao, H.; Klassen, T.; Deng, S. Strategies and Analyses for Robot Trajectory Optimization in Thermal and Kinetic Spraying. In Proceedings of the International Thermal Spray Conference, Vienna, Austria, 4–6 May 2022; ASM Thermal Spray Society: Vienna, Austria, 2022; pp. 299–305.
167. Wu, H.; Xie, X.; Liu, M.; Verdy, C.; Zhang, Y.; Liao, H.; Deng, S. Stable Layer-Building Strategy to Enhance Cold-Spray-Based Additive Manufacturing. *Addit. Manuf.* **2020**, *35*, 101356. [CrossRef]

168. Chen, C.; Gojon, S.; Xie, Y.; Yin, S.; Verdy, C.; Ren, Z.; Liao, H.; Deng, S. A Novel Spiral Trajectory for Damage Component Recovery with Cold Spray. *Surf. Coat. Technol.* **2017**, *309*, 719–728. [CrossRef]
169. Pattison, J.; Celotto, S.; Morgan, R.; Bray, M.; O'Neill, W. Cold Gas Dynamic Manufacturing: A Non-Thermal Approach to Freeform Fabrication. *Int. J. Mach. Tools Manuf.* **2007**, *47*, 627–634. [CrossRef]
170. Sokore, M.; Wu, H.; Li, W.; Raoelison, R.-N.; Deng, S.; Liao, H. Perspective of 3D Near-Net-Shape Additive Manufacturing by Cold Spraying: An Empirical Study Using Pure Al Powders. In Proceedings of the International Thermal Spray Conference, Vienna, Austria, 4–6 May 2022; ASM Thermal Spray Society: Vienna, Austria, 2022; pp. 306–313.
171. Hutasoit, N.; Rashid, R.A.R.; Palanisamy, S.; Duguid, A. Effect of Build Orientation and Post-Build Heat Treatment on the Mechanical Properties of Cold Spray Additively Manufactured Copper Parts. *Int. J. Adv. Manuf. Technol.* **2020**, *110*, 2359. [CrossRef]
172. Wu, H.; Raoelison, R.N.; Zhang, Y.; Deng, S.; Liao, H. Cold Spraying of 3D Parts—Challenges. In *Thermal Spray Coatings*; Thakur, L., Vasudev, H., Eds.; CRC Press: Boca Raton, FL, USA, 2021; pp. 37–58.
173. Nguyen, X.A.; Vargas-Uscategui, A.; Lohr, H.; Chu, C. A Continuous Toolpath Strategy from Offset Contours for Robotic Additive Manufacturing. *Research Square* **2022**, 1–17. [CrossRef]
174. Prudhomme, C.A.; Holtzinger, J.; Goldstein, G.H.; Tzivanis, M.J.; Noonan, W.E.; Austin, R.J. Thermal Spray Masking Tape. U.S. Patent 9.434,137B2, 6 September 2016.
175. Pergande, P.E.; Kinane, J.A.; Pank, D.R.; Collins, D.R. Making and Using Thermal Spray Masks Carrying Thermoset Epoxy Coating 2000. U.S. Patent 6,060,117, 9 May 2000.
176. Beck, J.E.; Prinz, F.B.; Siewiorek, D.P.; Weiss, L. Manufacturing Mechatronics Using Thermal Spray Shape Deposition. In Proceedings of the Solid Freeform Fabrication Symposium, Austin, TX, USA, 3–5 August 1992; pp. 272–279.
177. Tejero-Martin, D.; Rezvani Rad, M.; McDonald, A.; Hussain, T. Beyond Traditional Coatings: A Review on Thermal-Sprayed Functional and Smart Coatings. *J. Therm. Spray Technol.* **2019**, *28*, 598–644. [CrossRef]
178. Kosarev, V.F.; Klinkov, S.V.; Melamed, B.M.; Trubacheev, G.V.; Usynin, S.Y.; Shikalov, V.S. Investigation of Gas Flow through a Mask at Cold Spraying. *AIP Conf. Proc.* **2020**, *2288*, 030013. [CrossRef]
179. Klinkov, S.V.; Kosarev, V.F.; Ryashin, N.S. Comparison of Experiments and Computations for Cold Gas Spraying Through a Mask. Part 2. *Thermophys. Aeromech.* **2017**, *24*, 213–224. [CrossRef]
180. Cormier, Y.; Dupuis, P.; Farjam, A.; Corbeil, A.; Jodoin, B. Additive Manufacturing of Pyramidal Pin Fins: Height and Fin Density Effects Under Forced Convection. *Int. J. Heat Mass Transf.* **2014**, *75*, 235–244. [CrossRef]
181. Cormier, Y.; Dupuis, P.; Jodoin, B.; Corbeil, A. Mechanical Properties of Cold Gas Dynamic-Sprayed Near-Net-Shaped Fin Arrays. *J. Therm. Spray Technol.* **2015**, *24*, 476–488. [CrossRef]
182. Cormier, Y.; Dupuis, P.; Jodoin, B.; Corbeil, A. Pyramidal Fin Arrays Performance Using Streamwise Anisotropic Materials by Cold Spray Additive Manufacturing. *J. Therm. Spray Technol.* **2016**, *25*, 170–182. [CrossRef]
183. Dupuis, P.; Cormier, Y.; Fenech, M.; Jodoin, B. Heat Transfer and Flow Structure Characterization for Pin Fins Produced by Cold Spray Additive Manufacturing. *Int. J. Heat Mass Transf.* **2016**, *98*, 650–661. [CrossRef]
184. Xie, J.; Chen, Y.; Yin, L.; Zhang, T.; Wang, S.; Wang, L. Microstructure and Mechanical Properties of Ultrasonic Spot Welding TiNi/Ti6Al4V Dissimilar Materials Using Pure Al Coating. *J. Manuf. Process.* **2021**, *64*, 473–480. [CrossRef]
185. Champagne, V.K., Jr.; Champagne, V.K., III. Method to Join Dissimilar Materials by the Cold Spray Process. U.S. Patent 2016/0089750A1, 31 March 2016.
186. Champagne, V.; Kaplowitz, D.; Champagne, V.K.; Howe, C.; West, M.K.; McNally, B.; Rokni, M. Dissimilar Metal Joining and Structural Repair of ZE41A-T5 Cast Magnesium by the Cold Spray (CS) Process. *Mater. Manuf. Process.* **2018**, *33*, 130–139. [CrossRef]
187. Curtis, T.R.; Champagne, V.K.; West, M.K.; Rokni, R.; Widener, C.A. Joining Al 6061 to ZE41A Mg Alloy by Friction Stir Welding Using a Cold Spray Transition Joint. In *Friction Stir Welding and Processing IX*; Hovanski, Y., Mishra, R., Sato, Y., Upadhyay, P., Yan, D., Eds.; Springer: Cham, Switzerland, 2017; pp. 221–236.
188. Champagne, V.K.; West, M.K.; Reza Rokni, M.; Curtis, T.; Champagne, V.; McNally, B. Joining of Cast ZE41A Mg to Wrought 6061 Al by the Cold Spray Process and Friction Stir Welding. *J. Therm. Spray Technol.* **2016**, *25*, 143–159. [CrossRef]
189. Hou, W.; Shen, Z.; Huda, N.; Oheil, M.; Shen, Y.; Jahed, H.; Gerlich, A.P. Enhancing Metallurgical and Mechanical Properties of Friction Stir Butt Welded Joints of Al–Cu Via Cold Sprayed Ni Interlayer. *Mater. Sci. Eng. A* **2021**, *809*, 140992. [CrossRef]
190. Daroonparvar, M.; Khan, M.U.F.; Saadeh, Y.; Kay, C.M.; Kasar, A.K.; Kumar, P.; Esteves, L.; Misra, M.; Menezes, P.; Kalvala, P.R.; et al. Modification of Surface Hardness, Wear Resistance and Corrosion Resistance of Cold Spray Al Coated AZ31B Mg Alloy Using Cold Spray Double Layered Ta/Ti Coating in 3.5 Wt% NaCl Solution. *Corros. Sci.* **2020**, *176*, 109029. [CrossRef]
191. Poza, P.; Garrido-Maneiro, M.Á. Cold-Sprayed Coatings: Microstructure, Mechanical Properties, and Wear Behaviour. *Prog. Mater. Sci.* **2022**, *123*, 100839. [CrossRef]
192. Montheillet, F.; Jonas, J.J.; Benferrah, M. Development of Anisotropy During the Cold Rolling of Aluminium Sheet. *Int. J. Mech. Sci.* **1991**, *33*, 197–209. [CrossRef]
193. Park, M.; Kang, M.S.; Park, G.-W.; Kim, H.C.; Moon, H.-S.; Kim, B.; Jeon, J.B.; Kim, H.; Park, H.-S.; Kwon, S.-H.; et al. Effects of Annealing Treatment on the Anisotropy Behavior of Cold-Rolled High-Manganese Austenite Stainless Steels. *Met. Mater. Int.* **2021**, *27*, 3839–3855. [CrossRef]

194. You, Z.; Fu, H.; Qu, S.; Bao, W.; Lu, L. Revisiting Anisotropy in the Tensile and Fracture Behavior of Cold-Rolled 316L Stainless Steel with Heterogeneous Nano-Lamellar Structures. *Nano Mater. Sci.* **2020**, *2*, 72–79. [CrossRef]
195. Park, C.M.; Jung, J.; Yu, B.C.; Park, Y.H. Anisotropy of the Wear and Mechanical Properties of Extruded Aluminum Alloy Rods (AA2024-T4). *Met. Mater. Int.* **2019**, *25*, 71–82. [CrossRef]
196. Zhang, Z.H.; Li, W.Y.; Feng, Y.; Li, J.L.; Chao, Y.J. Global Anisotropic Response of Friction Stir Welded 2024 Aluminum Sheets. *Acta Mater.* **2015**, *92*, 117–125. [CrossRef]
197. Kok, Y.; Tan, X.P.; Wang, P.; Nai, M.L.S.; Loh, N.H.; Liu, E.; Tor, S.B. Anisotropy and Heterogeneity of Microstructure and Mechanical Properties in Metal Additive Manufacturing: A Critical Review. *Mater. Des.* **2018**, *139*, 565–586. [CrossRef]
198. Deev, A.A.; Kuznetcov, P.A.; Petrov, S.N. Anisotropy of Mechanical Properties and Its Correlation with the Structure of the Stainless Steel 316L Produced by the SLM Method. *Phys. Procedia* **2016**, *83*, 789–796. [CrossRef]
199. Seiner, H.; Cizek, J.; Sedlák, P.; Huang, R.; Cupera, J.; Dlouhy, I.; Landa, M. Elastic Moduli and Elastic Anisotropy of Cold Sprayed Metallic Coatings. *Surf. Coat. Technol.* **2016**, *291*, 342–347. [CrossRef]
200. Yang, K.; Li, W.; Yang, X.; Xu, Y. Anisotropic Response of Cold Sprayed Copper Deposits. *Surf. Coat. Technol.* **2018**, *335*, 219–227. [CrossRef]
201. Yang, K.; Li, W.; Guo, X.; Yang, X.; Xu, Y. Characterizations and Anisotropy of Cold-Spraying Additive-Manufactured Copper Bulk. *J. Mater. Sci. Technol.* **2018**, *34*, 1570–1579. [CrossRef]
202. Julien, S.E.; Nourian-Avval, A.; Liang, W.; Schwartz, T.; Ozdemir, O.C.; Müftü, S. Bulk Fracture Anisotropy in Cold-Sprayed Al 6061 Deposits. *Eng. Fract. Mech.* **2022**, *263*, 108301. [CrossRef]
203. Groarke, R.; Danilenkoff, C.; Karam, S.; McCarthy, E.; Michel, B.; Mussatto, A.; Sloane, J.; O'Neill, A.; Raghavendra, R.; Brabazon, D. 316L Stainless Steel Powders for Additive Manufacturing: Relationships of Powder Rheology, Size, Size Distribution to Part Properties. *Materials* **2020**, *13*, 5537. [CrossRef]
204. Wong, W.; Irissou, E.; Vo, P.; Sone, M.; Bernier, F.; Legoux, J.-G.; Fukanuma, H.; Yue, S. Cold Spray Forming of Inconel 718. *J. Therm. Spray Technol.* **2013**, *22*, 413–421. [CrossRef]
205. Li, W.; Cao, C.; Yin, S. Solid-State Cold Spraying of Ti and Its Alloys: A Literature Review. *Prog. Mater. Sci.* **2020**, *110*, 100633. [CrossRef]
206. Chen, C.; Yan, X.; Xie, Y.; Huang, R.; Kuang, M.; Ma, W.; Zhao, R.; Wang, J.; Liu, M.; Ren, Z.; et al. Microstructure Evolution and Mechanical Properties of Maraging Steel 300 Fabricated by Cold Spraying. *Mater. Sci. Eng. A* **2019**, *743*, 482–493. [CrossRef]
207. Luzin, V.; Spencer, K.; Zhang, M.X. Residual Stress and Thermo-Mechanical Properties of Cold Spray Metal Coatings. *Acta Mater.* **2011**, *59*, 1259–1270. [CrossRef]
208. Huang, R.; Sone, M.; Ma, W.; Fukanuma, H. The Effects of Heat Treatment on the Mechanical Properties of Cold-Sprayed Coatings. *Surf. Coat. Technol.* **2015**, *261*, 278–288. [CrossRef]
209. Li, W.Y.; Li, C.J.; Liao, H. Effect of Annealing Treatment on the Microstructure and Properties of Cold-Sprayed Cu Coating. *J. Therm. Spray Technol.* **2006**, *15*, 206–211. [CrossRef]
210. Li, W.; Wu, D.; Hu, K.; Xu, Y.; Yang, X.; Zhang, Y. A Comparative Study on the Employment of Heat Treatment, Electric Pulse Processing and Friction Stir Processing to Enhance Mechanical Properties of Cold-Spray-Additive-Manufactured Copper. *Surf. Coat. Technol.* **2021**, *409*, 126887. [CrossRef]
211. Hall, A.C.; Cook, D.J.; Neiser, R.A.; Roemer, T.J.; Hirschfeld, D.A. The Effect of a Simple Annealing Heat Treatment on the Mechanical Properties of Cold-Sprayed Aluminum. *J. Therm. Spray Technol.* **2006**, *15*, 233–238. [CrossRef]
212. Al-Hamdani, K.S.; Murray, J.W.; Hussain, T.; Clare, A.T. Heat-Treatment and Mechanical Properties of Cold-Sprayed High Strength Al Alloys from Satellited Feedstocks. *Surf. Coat. Technol.* **2019**, *374*, 21–31. [CrossRef]
213. Wu, D.; Li, W.; Liu, K.; Yang, Y.; Hao, S. Optimization of Cold Spray Additive Manufactured AA2024/Al2O3 Metal Matrix Composite with Heat Treatment. *J. Mater. Sci. Technol.* **2022**, *106*, 211–224. [CrossRef]
214. Vo, P.; Irissou, E.; Legoux, J.G.; Yue, S. Mechanical and Microstructural Characterization of Cold-Sprayed Ti-6Al-4V after Heat Treatment. *J. Therm. Spray Technol.* **2013**, *22*, 954–964. [CrossRef]
215. Li, W.Y.; Zhang, C.; Guo, X.; Xu, J.; Li, C.J.; Liao, H.; Coddet, C.; Khor, K.A. Ti and Ti-6Al-4V Coatings by Cold Spraying and Microstructure Modification by Heat Treatment. *Adv. Eng. Mater.* **2007**, *9*, 418–423. [CrossRef]
216. Yu, J.S.; Kim, H.J.; Oh, I.H.; Lee, K.A. Densification and Purification of Cold Sprayed Ti Coating Layer by Using Annealing in Different Heat Treatment Environments. *Adv. Mater. Res.* **2013**, *602–604*, 1604–1608. [CrossRef]
217. Cavaliere, P.; Perrone, A.; Silvello, A. Fatigue Behaviour of Inconel 625 Cold Spray Coatings. *Surf. Eng.* **2018**, *34*, 380–391. [CrossRef]
218. Shrestha, D.; Azarmi, F.; Tangpong, X.W. Effect of Heat Treatment on Residual Stress of Cold Sprayed Nickel-Based Superalloys. *J. Therm. Spray Technol.* **2022**, *31*, 197–205. [CrossRef]
219. Luo, X.T.; Yao, M.L.; Ma, N.; Takahashi, M.; Li, C.J. Deposition Behavior, Microstructure and Mechanical Properties of an in-Situ Micro-Forging Assisted Cold Spray Enabled Additively Manufactured Inconel 718 Alloy. *Mater. Des.* **2018**, *155*, 384–395. [CrossRef]
220. Ralls, A.M.; Daroonparvar, M.; Kasar, A.K.; Misra, M.; Menezes, P.L. Influence of Friction Stir Processing on the Friction, Wear and Corrosion Mechanisms of Solid-State Additively Manufactured 316L Duplex Stainless Steel. *Tribol. Int.* **2023**, *178*, 108033. [CrossRef]

221. Kovarik, O.; Siegl, J.; Cizek, J.; Chraska, T.; Kondas, J. Fracture Toughness of Cold Sprayed Pure Metals. *J. Therm. Spray Technol.* **2020**, *29*, 147–157. [CrossRef]

222. Ren, Y.; Tariq, N.U.H.; Liu, H.; Zhao, L.; Cui, X.; Shen, Y.; Wang, J.; Xiong, T. Study of Microstructural and Mechanical Anisotropy of 7075 Al Deposits Fabricated by Cold Spray Additive Manufacturing. *Mater. Des.* **2021**, *212*, 110271. [CrossRef]

223. Cavaliere, P.; Perrone, A.; Silvello, A.; Laska, A.; Blasi, G.; Cano, I.G. Fatigue Bending of V-Notched Cold-Sprayed FeCoCrNiMn Coatings. *Metals* **2022**, *12*, 780. [CrossRef]

224. Trzaska, Z.; Couret, A.; Monchoux, J. Spark Plasma Sintering Mechanisms at the Necks between TiAl Powder Particles. *Acta Mater.* **2016**, *118*, 100–108. [CrossRef]

225. Zhang, P.; Liu, J.; Gao, Y.; Liu, Z.; Mai, Q. Effect of Heat Treatment Process on the Micro Machinability of 7075 Aluminum Alloy. *Vacuum* **2023**, *207*, 111574. [CrossRef]

226. Fard, R.A.; Kazeminezhad, M. Effect of Electropulsing on Microstructure and Hardness of Cold-Rolled Low Carbon Steel. *J. Mater. Res. Technol.* **2019**, *8*, 3114–3125. [CrossRef]

227. Shen, Z.; Johnsson, M.; Zhao, Z.; Nygren, M. Spark Plasma Sintering of Alumina. *J. Am. Ceram. Soc.* **2002**, *85*, 1921–1927. [CrossRef]

228. Miriyev, A.; Stern, A.; Tuval, E.; Kalabukhov, S.; Hooper, Z.; Frage, N. Titanium to Steel Joining by Spark Plasma Sintering (SPS) Technology. *J. Mater. Process. Technol.* **2013**, *213*, 161–166. [CrossRef]

229. Matizamhuka, W.R. Spark Plasma Sintering (SPS)—An Advanced Sintering Technique for Structural Nanocomposite Materials. *J. S. Afr. Inst. Min. Metall.* **2016**, *116*, 1171–1180. [CrossRef]

230. Ito, K.; Ogawa, K. Effects of Spark-Plasma Sintering Treatment on Cold-Sprayed Copper Coatings. *J. Therm. Spray Technol.* **2014**, *23*, 104–113. [CrossRef]

231. Vidyuk, T.M.; Dudina, D.V.; Korchagin, M.A.; Gavrilov, A.I.; Bokhonov, B.B.; Ukhina, A.V.; Esikov, M.A.; Shikalov, V.S.; Kosarev, V.F. Spark Plasma Sintering Treatment of Cold Sprayed Materials for Synthesis and Structural Modification: A Case Study Using TiC-Cu Composites. *Mater. Lett. X* **2022**, *14*, 100140. [CrossRef]

232. Bocanegra-Bernal, M.H. Hot Isostatic Pressing (HIP) Technology and Its Applications to Metals and Ceramics. *J. Mater. Sci.* **2004**, *39*, 6399–6420. [CrossRef]

233. Atkinson, H.V.; Davies, S. Fundamental Aspects of Hot Isostatic Pressing: An Overview. *Metall. Mater. Trans. A* **2000**, *31*, 2981–3000. [CrossRef]

234. Brian James, W. New Shaping Methods for Powder Metallurgy Components. *Mater. Des.* **1987**, *8*, 187–197. [CrossRef]

235. Benzing, J.; Hrabe, N.; Quinn, T.; White, R.; Rentz, R.; Ahlfors, M. Hot Isostatic Pressing (HIP) to Achieve Isotropic Microstructure and Retain as-Built Strength in an Additive Manufacturing Titanium Alloy (Ti-6Al-4V). *Mater. Lett.* **2019**, *257*, 126690. [CrossRef] [PubMed]

236. Petrovskiy, P.; Sova, A.; Doubenskaia, M.; Smurov, I. Influence of Hot Isostatic Pressing on Structure and Properties of Titanium Cold-Spray Deposits. *Int. J. Adv. Manuf. Technol.* **2019**, *102*, 819–827. [CrossRef]

237. Petrovskiy, P.; Khomutov, M.; Cheverikin, V.; Travyanov, A.; Sova, A.; Smurov, I. Influence of Hot Isostatic Pressing on the Properties of 316L Stainless Steel, Al-Mg-Sc-Zr Alloy, Titanium and Ti6Al4V Cold Spray Deposits. *Surf. Coat. Technol.* **2021**, *405*, 126736. [CrossRef]

238. Chen, C.; Xie, Y.; Yan, X.; Yin, S.; Fukanuma, H.; Huang, R.; Zhao, R.; Wang, J.; Ren, Z.; Liu, M.; et al. Effect of Hot Isostatic Pressing (HIP) on Microstructure and Mechanical Properties of Ti6Al4V Alloy Fabricated by Cold Spray Additive Manufacturing. *Addit. Manuf.* **2019**, *27*, 595–605. [CrossRef]

239. Petrovskiy, P.; Travyanov, A.; Cheverikin, V.V.; Chereshneva, A.A.; Sova, A.; Smurov, I. Effect of Encapsulated Hot Isostatic Pressing on Properties of Ti6Al4V Deposits Produced by Cold Spray. *Int. J. Adv. Manuf. Technol.* **2020**, *107*, 437–449. [CrossRef]

240. Vaz, R.F.; Silvello, A.; Albaladejo, V.; Sanchez, J.; Cano, I.G. Improving the Wear and Corrosion Resistance of Maraging Part Obtained by Cold Gas Spray Additive Manufacturing. *Metals* **2021**, *11*, 1092. [CrossRef]

241. Feng, L.; Yang, W.-J.; Ma, K.; Yuan, Y.-D.; An, G.-S.; Li, W.-S. Microstructure and Properties of Cold Spraying AlCoCrCuFeNix HEA Coatings Synthesized by Induction Remelting. *Mater. Technol.* **2022**, *37*, 2567–2579. [CrossRef]

242. Jing, Z.; Dejun, K. Effect of Laser Remelting on Friction-Wear Behaviors of Cold Sprayed Al Coatings in 3.5% NaCl Solution. *Materials* **2018**, *11*, 283. [CrossRef]

243. Astarita, A.; Genna, S.; Leone, C.; Minutolo, F.M.C.; Rubino, F.; Squillace, A. Study of the Laser Remelting of a Cold Sprayed Titanium Layer. *Procedia CIRP* **2015**, *33*, 452–457. [CrossRef]

244. Marrocco, T.; Hussain, T.; McCartney, D.G.; Shipway, P.H. Corrosion Performance of Laser Posttreated Cold Sprayed Titanium Coatings. *J. Therm. Spray Technol.* **2011**, *20*, 909–917. [CrossRef]

245. Kumar, A.; Kant, R.; Singh, H. Microstructural and Tribological Properties of Laser-Treated Cold-Sprayed Titanium/Baghdadite Deposits. *J. Mater. Res.* **2022**, *37*, 2698–2709. [CrossRef]

246. Poza, P.; Múnez, C.J.; Garrido-Maneiro, M.A.; Vezzù, S.; Rech, S.; Trentin, A. Mechanical Properties of Inconel 625 Cold-Sprayed Coatings after Laser Remelting. Depth Sensing Indentation Analysis. *Surf. Coat. Technol.* **2014**, *243*, 51–57. [CrossRef]

247. Zybała, R.; Bucholc, B.; Kaszyca, K.; Kowiorski, K.; Soboń, D.; Żórawski, W.; Moszczyńska, D.; Molak, R.; Pakieła, Z. Properties of Cold Sprayed Titanium and Titanium Alloy Coatings after Laser Surface Treatment. *Materials* **2022**, *15*, 9014. [CrossRef]

248. Vaz, R.F.; Pukasiewicz, A.G.M.; Siqueira, I.B.A.F.; Sucharski, G.B.; Chicoski, A.; Tristante, R. Thermal Spraying of FeMnCrSi Alloys: An Overview. In Proceedings of the International Thermal Spray Conference, Quebec City, QC, Canada, 24–28 May 2021; pp. 431–439.
249. Pukasiewicz, A.G.M.; Alcover, P.R.C.; Capra, A.R.; Paredes, R.S.C. Influence of Plasma Remelting on the Microstructure and Cavitation Resistance of Arc-Sprayed Fe-Mn-Cr-Si Alloy. *J. Therm. Spray Technol.* **2014**, *23*, 51–59. [CrossRef]
250. Zabihi, A.; Soltani, R. Tribological Properties of B4C Reinforced Aluminum Composite Coating Produced by TIG Re-Melting of Flame Sprayed Al-Mg-B4C Powder. *Surf. Coat. Technol.* **2018**, *349*, 707–718. [CrossRef]
251. Simchen, F.; Sieber, M.; Kopp, A.; Lampke, T. Introduction to Plasma Electrolytic Oxidation—An Overview of the Process and Applications. *Coatings* **2020**, *10*, 628. [CrossRef]
252. Egorkin, V.S.; Gnedenkov, S.V.; Sinebryukhov, S.L.; Vyaliy, I.E.; Gnedenkov, A.S.; Chizhikov, R.G. Increasing Thickness and Protective Properties of PEO-Coatings on Aluminum Alloy. *Surf. Coat. Technol.* **2018**, *334*, 29–42. [CrossRef]
253. Rao, Y.; Wang, Q.; Oka, D.; Ramachandran, C.S. On the PEO Treatment of Cold Sprayed 7075 Aluminum Alloy and Its Effects on Mechanical, Corrosion and Dry Sliding Wear Performances Thereof. *Surf. Coat. Technol.* **2020**, *383*, 125271. [CrossRef]
254. Rao, Y.; Wang, Q.; Chen, J.; Ramachandran, C.S. Abrasion, Sliding Wear, Corrosion, and Cavitation Erosion Characteristics of a Duplex Coating Formed on AZ31 Mg Alloy by Sequential Application of Cold Spray and Plasma Electrolytic Oxidation Techniques. *Mater. Today Commun.* **2021**, *26*, 101978. [CrossRef]
255. Chavan, N.M.; Pant, P.; Sundararajan, G.; Suresh Babu, P. Post Treatment of Cold Sprayed Coatings Using High-Energy Infrared Radiation: First Comprehensive Study on Structure-Property Correlation. *Surf. Coat. Technol.* **2022**, *448*, 128902. [CrossRef]
256. Pokhmurska, H.; Wielage, B.; Lampke, T.; Grund, T.; Student, M.; Chervinska, N. Post-Treatment of Thermal Spray Coatings on Magnesium. *Surf. Coat. Technol.* **2008**, *202*, 4515–4524. [CrossRef]
257. Kostilnik, T. Shot Peening. In *Surface Engineering*; Cotell, C.M., Sprague, J.A., Smidt, F.A., Jr., Eds.; ASM International: Materials Park, OH, USA, 1994; Volume 1, pp. 126–135.
258. Moridi, A.; Hassani-Gangaraj, S.M.; Vezzú, S.; Trško, L.; Guagliano, M. Fatigue Behavior of Cold Spray Coatings: The Effect of Conventional and Severe Shot Peening as Pre-/Post-Treatment. *Surf. Coat. Technol.* **2015**, *283*, 247–254. [CrossRef]
259. Yao, H.-L.; Hu, X.-Z.; Yi, Z.-H.; Xia, J.; Tu, X.-Y.; Li, S.-B.; Yu, B.; Zhang, M.-X.; Bai, X.-B.; Chen, Q.-Y.; et al. Microstructure and Improved Anti-Corrosion Properties of Cold-Sprayed Zn Coatings Fabricated by Post Shot-Peening Process. *Surf. Coat. Technol.* **2021**, *422*, 127557. [CrossRef]
260. Sova, A.; Courbon, C.; Valiorgue, F.; Rech, J.; Bertrand, P. Effect of Turning and Ball Burnishing on the Microstructure and Residual Stress Distribution in Stainless Steel Cold Spray Deposits. *J. Therm. Spray Technol.* **2017**, *26*, 1922–1934. [CrossRef]
261. Courbon, C.; Sova, A.; Valiorgue, F.; Pascal, H.; Sijobert, J.; Kermouche, G.; Bertrand, P.; Rech, J. Near Surface Transformations of Stainless Steel Cold Spray and Laser Cladding Deposits after Turning and Ball-Burnishing. *Surf. Coat. Technol.* **2019**, *371*, 235–244. [CrossRef]
262. Mayer, A.R.; Bertuol, K.; Siqueira, I.B.A.F.A.F.; Chicoski, A.; Váz, R.F.; de Sousa, M.J.; Pukasiewicz, A.G.M.M. Evaluation of Cavitation/Corrosion Synergy of the Cr3C2-25NiCr Coating Deposited by HVOF Process. *Ultrason. Sonochem.* **2020**, *69*, 105271. [CrossRef]
263. Soyama, H. Cavitation Peening: A Review. *Metals* **2020**, *10*, 270. [CrossRef]
264. Zhang, P.; Liu, Z.; Yue, X.; Wang, P.; Zhai, Y. Water Jet Impact Damage Mechanism and Dynamic Penetration Energy Absorption of 2A12 Aluminum Alloy. *Vacuum* **2022**, *206*, 111532. [CrossRef]
265. Soyama, H.; Okura, Y. The Use of Various Peening Methods to Improve the Fatigue Strength of Titanium Alloy Ti6Al4V Manufactured by Electron Beam Melting. *AIMS Mater. Sci.* **2018**, *5*, 1000–1015. [CrossRef]
266. Bobzin, K.; Öte, M.; Wiesner, S.; Gerdt, L.; Senge, S.; Hirt, G. Investigation on the Cold Rolling and Structuring of Cold Sprayed Copper-Coated Steel Sheets. *IOP Conf. Ser. Mater. Sci. Eng.* **2017**, *181*, 012028. [CrossRef]
267. Tariq, N.H.; Gyansah, L.; Qiu, X.; Du, H.; Wang, J.Q.; Feng, B.; Yan, D.S.; Xiong, T.Y. Thermo-Mechanical Post-Treatment: A Strategic Approach to Improve Microstructure and Mechanical Properties of Cold Spray Additively Manufactured Composites. *Mater. Des.* **2018**, *156*, 287–299. [CrossRef]
268. Wu, T.; Jin, L.; Wu, W.X.; Gao, L.; Wang, J.; Zhang, Z.Y.; Dong, J. Improved Ductility of Mg–Zn–Ce Alloy by Hot Pack-Rolling. *Mater. Sci. Eng. A* **2013**, *584*, 97–102. [CrossRef]
269. Huang, H.; Liao, M.; Yu, Q.; Liu, G.; Wang, Z. The Effects of Hot-Pack Coating Materials on the Pack Rolling Process and Microstructural Characteristics during Ti-46Al-8Nb Sheet Fabrication. *Materials* **2020**, *13*, 762. [CrossRef]
270. Shen, Z.Z.; Lin, J.P.; Liang, Y.F.; Zhang, L.Q.; Shang, S.L.; Liu, Z.K. A Novel Hot Pack Rolling of High Nb–TiAl Sheet from Cast Ingot. *Intermetallics* **2015**, *67*, 19–25. [CrossRef]
271. Zhao, H.; Tan, C.; Yu, X.; Ning, X.; Nie, Z.; Cai, H.; Wang, F.; Cui, Y. Enhanced Reactivity of Ni-Al Reactive Material Formed by Cold Spraying Combined with Cold-Pack Rolling. *J. Alloys Compd.* **2018**, *741*, 883–894. [CrossRef]
272. Qiu, X.; Tariq, N.U.H.; Qi, L.; Zan, Y.; Wang, Y.; Wang, J.; Du, H.; Xiong, T. In-Situ Sip/A380 Alloy Nano/Micro Composite Formation through Cold Spray Additive Manufacturing and Subsequent Hot Rolling Treatment: Microstructure and Mechanical Properties. *J. Alloys Compd.* **2019**, *780*, 597–606. [CrossRef]
273. Li, K.; Liu, X.; Zhao, Y. Research Status and Prospect of Friction Stir Processing Technology. *Coatings* **2019**, *9*, 129. [CrossRef]
274. Mondal, M.; Das, H.; Hong, S.-T.; Jeong, B.-S.; Han, H.N. Local Enhancement of the Material Properties of Aluminium Sheets by a Combination of Additive Manufacturing and Friction Stir Processing. *CIRP Ann.* **2019**, *68*, 289–292. [CrossRef]

275. Hasani, B.M.; Hedaiatmofidi, H.; Zarebidaki, A. Effect of Friction Stir Process on the Microstructure and Corrosion Behavior of AZ91 Mg Alloy. *Mater. Chem. Phys.* **2021**, *267*, 124672. [CrossRef]
276. Han, P.; Wang, W.; Liu, Z.; Zhang, T.; Liu, Q.; Guan, X.; Qiao, K.; Ye, D.; Cai, J.; Xie, Y.; et al. Modification of Cold-Sprayed High-Entropy Alloy Particles Reinforced Aluminum Matrix Composites Via Friction Stir Processing. *J. Alloys Compd.* **2022**, *907*, 164426. [CrossRef]
277. Sudharshan Phani, P.; Srinivasa Rao, D.; Joshi, S.V.; Sundararajan, G. Effect of Process Parameters and Heat Treatments on Properties of Cold Sprayed Copper Coatings. *J. Therm. Spray Technol.* **2007**, *16*, 425–434. [CrossRef]
278. Yang, K.; Li, W.; Yang, X.; Xu, Y.; Vairis, A. Effect of Heat Treatment on the Inherent Anisotropy of Cold Sprayed Copper Deposits. *Surf. Coat. Technol.* **2018**, *350*, 519–530. [CrossRef]
279. Yin, S.; Cizek, J.; Yan, X.; Lupoi, R. Annealing Strategies for Enhancing Mechanical Properties of Additively Manufactured 316L Stainless Steel Deposited by Cold Spray. *Surf. Coat. Technol.* **2019**, *370*, 353–361. [CrossRef]
280. Ren, Y.Q.; King, P.C.; Yang, Y.S.; Xiao, T.Q.; Chu, C.; Gulizia, S.; Murphy, A.B. Characterization of Heat Treatment-Induced Pore Structure Changes in Cold-Sprayed Titanium. *Mater. Charact.* **2017**, *132*, 69–75. [CrossRef]
281. Rubino, F.; Astarita, A.; Carlone, P.; Genna, S.; Leone, C.; Memola Capece Minutolo, F.; Squillace, A. Selective Laser Post-Treatment on Titanium Cold Spray Coatings. *Mater. Manuf. Process.* **2016**, *31*, 1500–1506. [CrossRef]
282. Boruah, D.; Zhang, X.; McNutt, P.; Khan, R.; Begg, H. Effect of Post-Deposition Thermal Treatments on Tensile Properties of Cold Sprayed Ti6Al4V. *Metals* **2022**, *12*, 1908. [CrossRef]
283. Khun, N.W.; Tan, A.W.Y.; Sun, W.; Liu, E. Effects of Nd:YAG Laser Surface Treatment on Tribological Properties of Cold-Sprayed Ti-6Al-4V Coatings Tested against 100Cr6 Steel under Dry Condition. *Tribol. Trans.* **2019**, *62*, 391–402. [CrossRef]
284. Chen, C.; Xie, Y.; Liu, L.; Zhao, R.; Jin, X.; Li, S.; Huang, R.; Wang, J.; Liao, H.; Ren, Z. Cold Spray Additive Manufacturing of Invar 36 Alloy: Microstructure, Thermal Expansion and Mechanical Properties. *J. Mater. Sci. Technol.* **2021**, *72*, 39–51. [CrossRef]
285. Ma, W.; Xie, Y.; Chen, C.; Fukanuma, H.; Wang, J.; Ren, Z.; Huang, R. Microstructural and Mechanical Properties of High-Performance Inconel 718 Alloy by Cold Spraying. *J. Alloys Compd.* **2019**, *792*, 456–467. [CrossRef]
286. Li, W.; Cao, C.; Wang, G.; Wang, F.; Xu, Y.; Yang, X. 'Cold Spray+' as a New Hybrid Additive Manufacturing Technology: A Literature Review. *Sci. Technol. Weld. Join.* **2019**, *24*, 420–445. [CrossRef]
287. Birt, A.M.; Champagne, V.K.; Sisson, R.D.; Apelian, D. Statistically Guided Development of Laser-Assisted Cold Spray for Microstructural Control of Ti-6Al-4V. *Metall. Mater. Trans. A* **2017**, *48*, 1931–1943. [CrossRef]
288. Kulmala, M.; Vuoristo, P. Influence of Process Conditions in Laser-Assisted Low-Pressure Cold Spraying. *Surf. Coat. Technol.* **2008**, *202*, 4503–4508. [CrossRef]
289. Lupoi, R.; Cockburn, A.; Bryan, C.; Sparkes, M.; Luo, F.; O'Neill, W. Hardfacing Steel with Nanostructured Coatings of Stellite-6 by Supersonic Laser Deposition. *Light Sci. Appl.* **2012**, *1*, e10. [CrossRef]
290. Bray, M.; Cockburn, A.; O'Neill, W. The Laser-Assisted Cold Spray Process and Deposit Characterisation. *Surf. Coat. Technol.* **2009**, *203*, 2851–2857. [CrossRef]
291. Barton, D.J.; Hornbuckle, B.C.; Darling, K.A.; Brewer, L.N.; Thompson, G.B. Influence of Surface Temperature in the Laser Assisted Cold Spray Deposition of Sequential Oxide Dispersion Strengthened Layers: Microstructure and Hardness. *Mater. Sci. Eng. A* **2021**, *811*, 141027. [CrossRef]
292. Christoulis, D.K.; Jeandin, M.; Irissou, É.; Legoux, J.-G.; Knapp, W. Laser-Assisted Cold Spray (LACS). In *Nd-YAG Laser*; Dumitras, D.C., Ed.; IntechOpen: Rijeka, Croatia, 2012; pp. 59–96.
293. Olakanmi, E.O.; Doyoyo, M. Laser-Assisted Cold-Sprayed Corrosion- and Wear-Resistant Coatings: A Review. *J. Therm. Spray Technol.* **2014**, *23*, 765–785. [CrossRef]
294. Christoulis, D.K.; Guetta, S.; Irissou, E.; Guipont, V.; Berger, M.H.; Jeandin, M.; Legoux, J.-G.; Moreau, C.; Costil, S.; Boustie, M.; et al. Cold-Spraying Coupled to Nano-Pulsed Nd-YaG Laser Surface Pre-Treatment. *J. Therm. Spray Technol.* **2010**, *19*, 1062–1073. [CrossRef]
295. Lee, H.; Lim, C.H.J.; Low, M.J.; Tham, N.; Murukeshan, V.M.; Kim, Y.J. Lasers in Additive Manufacturing: A Review. *Int. J. Precis. Eng. Manuf. Green Technol.* **2017**, *4*, 307–322. [CrossRef]
296. Olakanmi, E.O.; Tlotleng, M.; Meacock, C.; Pityana, S.; Doyoyo, M. Deposition Mechanism and Microstructure of Laser-Assisted Cold-Sprayed (LACS) Al-12 Wt.%Si Coatings: Effects of Laser Power. *JOM* **2013**, *65*, 776–783. [CrossRef]
297. Barton, D.J.; Bhattiprolu, V.S.; Thompson, G.B.; Brewer, L.N. Laser Assisted Cold Spray of AISI 4340 Steel. *Surf. Coat. Technol.* **2020**, *400*, 126218. [CrossRef]
298. Olakanmi, E.O. Optimization of the Quality Characteristics of Laser-Assisted Cold-Sprayed (LACS) Aluminum Coatings with Taguchi Design of Experiments (DOE). *Mater. Manuf. Process.* **2016**, *31*, 1490–1499. [CrossRef]
299. Story, W.A.; Barton, D.J.; Hornbuckle, B.C.; Darling, K.A.; Thompson, G.B.; Brewer, L.N. Laser Assisted Cold Spray of Fe–Ni–Zr Oxide Dispersion Strengthened Steel. *Materialia* **2018**, *3*, 239–242. [CrossRef]
300. Shi, J.; Wang, Y. Development of Metal Matrix Composites by Laser-Assisted Additive Manufacturing Technologies: A Review. *J. Mater. Sci.* **2020**, *55*, 9883–9917. [CrossRef]
301. Nikbakht, R.; Cojocaru, C.V.; Aghasibeig, M.; Irissou, É.; Kim, T.S.; Kim, H.S.; Jodoin, B. Cold Spray and Laser-Assisted Cold Spray of CrMnCoFeNi High Entropy Alloy Using Nitrogen as the Propelling Gas. *J. Therm. Spray Technol.* **2022**, *31*, 1129–1142. [CrossRef]

302. Dey, D.; Sarkar, S.; Mahata, A.; Roy Choudhury, A.; Nath, A.K. Laser Assisted Cold Spray of 15–5 PH Stainless Steel in a Designed and Developed Setup. *Opt. Laser Technol.* **2023**, *158*, 108902. [CrossRef]
303. Ortiz-Fernandez, R.; Jodoin, B. Hybrid Additive Manufacturing Technology: Induction Heating Cold Spray—Part I: Fundamentals of Deposition Process. *J. Therm. Spray Technol.* **2020**, *29*, 684–699. [CrossRef]
304. Li, C.J.; Li, W.Y. Deposition Characteristics of Titanium Coating in Cold Spraying. *Surf. Coat. Technol.* **2003**, *167*, 278–283. [CrossRef]
305. Daroonparvar, M.; Bakhsheshi-Rad, H.R.; Saberi, A.; Razzaghi, M.; Kasar, A.K.; Ramakrishna, S.; Menezes, P.L.; Misra, M.; Ismail, A.F.; Sharif, S.; et al. Surface Modification of Magnesium Alloys Using Thermal and Solid-State Cold Spray Processes: Challenges and Latest Progresses. *J. Magnes. Alloy.* **2022**, *10*, 2025–2061. [CrossRef]
306. Ghelichi, R.; Bagherifard, S.; Parienete, I.F.; Guagliano, M.; Vezzù, S. Experimental Study of Shot Peening Followed by Cold Spray Coating on Residual Stresses of the Treated Parts. *SDHM Struct. Durab. Health Monit.* **2010**, *6*, 17–29.
307. Luo, X.T.; Wei, Y.K.; Wang, Y.; Li, C.J. Microstructure and Mechanical Property of Ti and Ti6Al4V Prepared by an In-Situ Shot Peening Assisted Cold Spraying. *Mater. Des.* **2015**, *85*, 527–533. [CrossRef]
308. Li, W.Y.; Zou, Y.F.; Wang, F.F.; Yang, X.W.; Xu, Y.X.; Hu, K.W.; Yan, D.Y. Employing Cold Spray to Alter the Residual Stress Distribution of Workpieces: A Case Study on Fusion-Welded AA2219 Joints. *J. Therm. Spray Technol.* **2020**, *29*, 1538–1549. [CrossRef]
309. Ang, A.S.M.; Berndt, C.C. A Review of Testing Methods for Thermal Spray Coatings. *Int. Mater. Rev.* **2014**, *59*, 179–223. [CrossRef]
310. Xie, X.; Chen, C.; Ma, Y.; Xie, Y.; Wu, H.; Ji, G.; Aubry, E.; Ren, Z.; Liao, H. Influence of Annealing Treatment on Microstructure and Magnetic Properties of Cold Sprayed Ni-Coated FeSiAl Soft Magnetic Composite Coating. *Surf. Coat. Technol.* **2019**, *374*, 476–484. [CrossRef]
311. Deshpande, S.; Kulkarni, A.; Sampath, S.; Herman, H. Application of Image Analysis for Characterization of Porosity in Thermal Spray Coatings and Correlation with Small Angle Neutron Scattering. *Surf. Coat. Technol.* **2004**, *187*, 6–16. [CrossRef]
312. Baker, A.A.; Thuss, R.; Woollett, N.; Maich, A.; Stavrou, E.; McCall, S.K.; Radousky, H.B. Cold Spray Deposition of Thermoelectric Materials. *JOM* **2020**, *72*, 2853–2859. [CrossRef]
313. Van Steenkiste, T.; Smith, J.R. Evaluation of Coatings Produced via Kinetic and Cold Spray Processes. *J. Therm. Spray Technol.* **2004**, *13*, 274–282. [CrossRef]
314. Wang, Y.; Adrien, J.; Normand, B. Porosity Characterization of Cold Sprayed Stainless Steel Coating Using Three-Dimensional X-Ray Microtomography. *Coatings* **2018**, *8*, 326. [CrossRef]
315. Zahiri, S.H.; Mayo, S.C.; Jahedi, M. Characterization of Cold Spray Titanium Deposits by X-Ray Microscopy and Microtomography. *Microsc. Microanal.* **2008**, *14*, 260–266. [CrossRef]
316. Lévesque, D.; Bescond, C.; Cojocaru, C. Laser-Ultrasonic Inspection of Cold Spray Additive Manufacturing Components. *AIP Conf. Proc.* **2019**, *2102*, 020026. [CrossRef]
317. Li, Y.-J.; Luo, X.-T.; Li, C.-J. Dependency of Deposition Behavior, Microstructure and Properties of Cold Sprayed Cu on Morphology and Porosity of the Powder. *Surf. Coat. Technol.* **2017**, *328*, 304–312. [CrossRef]
318. Zhu, Q.J.; Wang, K.; Wang, X.H.; Hou, B.R. Electrochemical Impedance Spectroscopy Analysis of Cold Sprayed and Arc Sprayed Aluminium Coatings Serviced in Marine Environment. *Surf. Eng.* **2012**, *28*, 300–305. [CrossRef]
319. Dosta, S.; Bolelli, G.; Candeli, A.; Lusvarghi, L.; Cano, I.G.; Guilemany, J.M. Plastic Deformation Phenomena During Cold Spray Impact of WC-Co Particles onto Metal Substrates. *Acta Mater.* **2017**, *124*, 173–181. [CrossRef]
320. ISO 14577-1:2015; Metallic Materials–Instrumented Indentation Test for Hardness and Materials Parameters–Part 1: Test Method. ISO: Geneva, Switzerland, 2015.
321. Váz, R.F.; Silvello, A.; Cavalière, P.D.; Dosta, S.; Cano, I.G.; Capodieci, L.; Rizzo, A.; Valerini, D. Fretting Wear and Scratch Resistance of Cold-Sprayed Pure Cu and Ti. *Metallogr. Microstruct. Anal.* **2021**, *10*, 496–513. [CrossRef]
322. Ctibor, P.; Boháč, P.; Stranyánek, M.; Čtvrtlík, R. Structure and Mechanical Properties of Plasma Sprayed Coatings of Titania and Alumina. *J. Eur. Ceram. Soc.* **2006**, *26*, 3509–3514. [CrossRef]
323. ASTM C633-13; Standard Test Method for Adhesion or Cohesion Strength of Thermal Spray Coatings. ASTM International: West Conshohocken, PA, USA, 2021.
324. Shinde, S.; Sampath, S. A Critical Analysis of the Tensile Adhesion Test for Thermally Sprayed Coatings. *J. Therm. Spray Technol.* **2022**, *31*, 2247–2279. [CrossRef]
325. ASTM E8-22; Standard Test Methods for Tension Testing of Metallic Materials. ASTM International: West Conshohocken, PA, USA, 2022.
326. Ichikawa, Y.; Tokoro, R.; Tanno, M.; Ogawa, K. Elucidation of Cold-Spray Deposition Mechanism by Auger Electron Spectroscopic Evaluation of Bonding Interface Oxide Film. *Acta Mater.* **2019**, *164*, 39–49. [CrossRef]
327. Boruah, D.; Robinson, B.; London, T.; Wu, H.; de Villiers-Lovelock, H.; McNutt, P.; Doré, M.; Zhang, X. Experimental Evaluation of Interfacial Adhesion Strength of Cold Sprayed Ti-6Al-4V Thick Coatings Using an Adhesive-Free Test Method. *Surf. Coat. Technol.* **2020**, *381*, 125130. [CrossRef]
328. Wu, H.; Huang, C.; Xie, X.; Liu, S.; Wu, T.; Niendorf, T.; Xie, Y.; Deng, C.; Liu, M.; Liao, H.; et al. Influence of Spray Trajectories on Characteristics of Cold-Sprayed Copper Deposits. *Surf. Coat. Technol.* **2021**, *405*, 126703. [CrossRef]
329. Hauer, M.; Krebs, S.; Kroemmer, W.; Henkel, K.M. Correlation of Residual Stresses and Coating Properties in Arc-Sprayed Coatings on Different Substrates for Maritime Applications. *J. Therm. Spray Technol.* **2020**, *29*, 1289–1299. [CrossRef]

330. Wang, Q.; Luo, X.; Tsutsumi, S.; Sasaki, T.; Li, C.; Ma, N. Measurement and Analysis of Cold Spray Residual Stress Using Arbitrary Lagrangian–Eulerian Method. *Addit. Manuf.* **2020**, *35*, 101296. [CrossRef]
331. Luzin, V.; Kirstein, O.; Zahiri, S.H.; Fraser, D. Residual Stress Buildup in Ti Components Produced by Cold Spray Additive Manufacturing (CSAM). *J. Therm. Spray Technol.* **2020**, *29*, 1498–1507. [CrossRef]
332. Vargas-Uscategui, A.; King, P.C.; Styles, M.J.; Saleh, M.; Luzin, V.; Thorogood, K. Residual Stresses in Cold Spray Additively Manufactured Hollow Titanium Cylinders. *J. Therm. Spray Technol.* **2020**, *29*, 1508–1524. [CrossRef]
333. Sinclair-Adamson, R.; Luzin, V.; Duguid, A.; Kannoorpatti, K.; Murray, R. Residual Stress Distributions in Cold-Sprayed Copper 3D-Printed Parts. *J. Therm. Spray Technol.* **2020**, *29*, 1525–1537. [CrossRef]
334. Loke, K.; Zhang, Z.-Q.; Narayanaswamy, S.; Koh, P.K.; Luzin, V.; Gnaupel-Herold, T.; Ang, A.S.M. Residual Stress Analysis of Cold Spray Coatings Sprayed at Angles Using Through-Thickness Neutron Diffraction Measurement. *J. Therm. Spray Technol.* **2021**, *30*, 1810–1826. [CrossRef]
335. Boruah, D.; Ahmad, B.; Lee, T.L.; Kabra, S.; Syed, A.K.; McNutt, P.; Doré, M.; Zhang, X. Evaluation of Residual Stresses Induced by Cold Spraying of Ti-6Al-4V on Ti-6Al-4V Substrates. *Surf. Coat. Technol.* **2019**, *374*, 591–602. [CrossRef]
336. Ajovalasit, A. Review of Some Development of the Hole Drilling Method. In *Applied Stress Analysis*; Hyde, T.H., Ollerton, E., Eds.; Springer: Dordrecht, The Netherlands, 1990; pp. 60–71.
337. Niku-Lari, A.; Lu, J.; Flavenot, J.F. Measurement of Residual-Stress Distribution by the Incremental Hole-Drilling Method. *Exp. Mech.* **1985**, *25*, 175–185. [CrossRef]
338. Santana, Y.Y.; La Barbera-Sosa, J.G.; Staia, M.H.; Lesage, J.; Puchi-Cabrera, E.S.; Chicot, D.; Bemporad, E. Measurement of Residual Stress in Thermal Spray Coatings by the Incremental Hole Drilling Method. *Surf. Coat. Technol.* **2006**, *201*, 2092–2098. [CrossRef]
339. Schajer, G.S.; Whitehead, P.S. Hole-Drilling Method Concept and Development. In *Hole-Drilling Method for Measuring Residual Stresses*; Springer: Cham, Switzerland, 2018; pp. 47–68.
340. *ASTM E837-20*; Standard Test Method for Determining Residual Stresses by the Hole-Drilling Strain-Gages. ASTM International: West Conshohocken, PA, USA, 2020.
341. Clyne, T.W.; Gill, S.C. Residual Stresses in Thermal Spray Coatings and Their Effect on Interfacial Adhesion: A Review of Recent Work. *J. Therm. Spray Technol.* **1996**, *5*, 401–418. [CrossRef]
342. Singh, R.; Schruefer, S.; Wilson, S.; Gibmeier, J.; Vassen, R. Influence of Coating Thickness on Residual Stress and Adhesion-Strength of Cold-Sprayed Inconel 718 Coatings. *Surf. Coat. Technol.* **2018**, *350*, 64–73. [CrossRef]
343. Sample, C.M.; Champagne, V.K.; Nardi, A.T.; Lados, D.A. Factors Governing Static Properties and Fatigue, Fatigue Crack Growth, and Fracture Mechanisms in Cold Spray Alloys and Coatings/Repairs: A Review. *Addit. Manuf.* **2020**, *36*, 101371. [CrossRef]
344. Price, T.S.; Shipway, P.H.; McCartney, D.G. Effect of Cold Spray Deposition of a Titanium Coating on Fatigue Behavior of a Titanium Alloy. *J. Therm. Spray Technol.* **2006**, *15*, 507–512. [CrossRef]
345. Dayani, S.B.; Shaha, S.K.; Ghelichi, R.; Wang, J.F.; Jahed, H. The Impact of AA7075 Cold Spray Coating on the Fatigue Life of AZ31B Cast Alloy. *Surf. Coat. Technol.* **2018**, *337*, 150–158. [CrossRef]
346. Cizek, J.; Matejkova, M.; Dlouhy, I.; Siska, F.; Kay, C.M.; Karthikeyan, J.; Kuroda, S.; Kovarik, O.; Siegl, J.; Loke, K.; et al. Influence of Cold-Sprayed, Warm-Sprayed, and Plasma-Sprayed Layers Deposition on Fatigue Properties of Steel Specimens. *J. Therm. Spray Technol.* **2015**, *24*, 758–768. [CrossRef]
347. *ISO 1143:2021*; Metallic Materials—Rotating Bar Bending Fatigue Testing. ISO: Geneva, Switzerland, 2021.
348. Ziemian, C.W.; Sharma, M.M.; Bouffard, B.D.; Nissley, T.; Eden, T.J. Effect of Substrate Surface Roughening and Cold Spray Coating on the Fatigue Life of AA2024 Specimens. *Mater. Des.* **2014**, *54*, 212–221. [CrossRef]
349. Xiong, Y.; Zhang, M.-X. The Effect of Cold Sprayed Coatings on the Mechanical Properties of AZ91D Magnesium Alloys. *Surf. Coat. Technol.* **2014**, *253*, 89–95. [CrossRef]
350. Yamazaki, Y.; Fukanuma, H.; Ohno, N. Anisotropic Mechanical Properties of the Free-Standing Cold Sprayed SUS316 Coating and Effect of the Post-Spray Heat Treatment on It. *J. Japan Therm. Spray Soc.* **2016**, *53*, 91–95. [CrossRef]
351. Ševeček, M.; Krejčí, J.; Shahin, M.H.; Petrik, J.; Ballinger, R.G.; Shirvan, K. Fatigue Behavior of Cold Spray-Coated Accident Tolerant Cladding. In Proceedings of the TopFuel 2018, Prague, Czech Republic, 30 September–4 October 2018; pp. 1–15.
352. *ASTM E399-22*; Standard Test Method for Linear-Elastic Plane-Strain Fracture Toughness of Metallic Materials. ASTM International: West Conshohocken, PA, USA, 2022.
353. Li, X.-K.; Zhu, S.-P.; Liao, D.; Correia, J.A.F.O.; Berto, F.; Wang, Q. Probabilistic Fatigue Modelling of Metallic Materials Under Notch and Size Effect Using the Weakest Link Theory. *Int. J. Fatigue* **2022**, *159*, 106788. [CrossRef]
354. Batistič, S.; der Laken, P. History, Evolution and Future of Big Data and Analytics: A Bibliometric Analysis of Its Relationship to Performance in Organizations. *Br. J. Manag.* **2019**, *30*, 229–251. [CrossRef]
355. Donthu, N.; Kumar, S.; Mukherjee, D.; Pandey, N.; Lim, W.M. How to Conduct a Bibliometric Analysis: An Overview and Guidelines. *J. Bus. Res.* **2021**, *133*, 285–296. [CrossRef]
356. Broadus, R.N. Toward a Definition of "Bibliometrics". *Scientometrics* **1987**, *12*, 373–379. [CrossRef]
357. Hulme, E.W. Statistical Bibliography in Relation to the Growth of Modern Civilization: Two Lectures Delivered in the University of Cambridge in May 1922. *Nature* **1923**, *112*, 585–586. [CrossRef]
358. Caviggioli, F.; Ughetto, E. A Bibliometric Analysis of the Research Dealing with the Impact of Additive Manufacturing on Industry, Business and Society. *Int. J. Prod. Econ.* **2019**, *208*, 254–268. [CrossRef]

359. Hernandez Korner, M.E.; Lambán, M.P.; Albajez, J.A.; Santolaria, J.; Ng Corrales, L.D.C.; Royo, J. Systematic Literature Review: Integration of Additive Manufacturing and Industry 4.0. *Metals* **2020**, *10*, 1061. [CrossRef]
360. Javaid, M.; Haleem, A. Current Status and Challenges of Additive Manufacturing in Orthopaedics: An Overview. *J. Clin. Orthop. Trauma* **2019**, *10*, 380–386. [CrossRef]
361. Jemghili, R.; Ait Taleb, A.; Khalifa, M. A Bibliometric Indicators Analysis of Additive Manufacturing Research Trends From 2010 to 2020. *Rapid Prototyp. J.* **2021**, *27*, 1432–1454. [CrossRef]
362. Khor, K.A.; Yu, L.G. Global Research Trends in Thermal Sprayed Coatings Technology Analyzed with Bibliometrics Tools. *J. Therm. Spray Technol.* **2015**, *24*, 1346–1354. [CrossRef]
363. Li, R.-T.; Khor, K.A.; Yu, L.-G. Identifying Indicators of Progress in Thermal Spray Research Using Bibliometrics Analysis. *J. Therm. Spray Technol.* **2016**, *25*, 1526–1533. [CrossRef]
364. Milanez, D.H.; de Oliveira, B.S.; Noyons, E.C.M.; Faria, L.I.L.; Botta, W.J. Assessing Collaboration and Knowledge Flow on Coatings of Metallic Glasses Obtained from Thermal Spraying Processes Using Bibliometrics and Science Mapping. *Mater. Res.* **2017**, *20*, 71–80. [CrossRef]
365. Dobra, A. General Classification of Robots. Size Criteria. In Proceedings of the 23rd International Conference on Robotics in Alpe-Adria-Danube Region, Smolenice, Slovakia, 3–5 September 2014; IEEE: Smolenice Castle, Slovakia, 2014; pp. 1–6.
366. Xie, Y.; Zhang, C.; Lai, Q. China's Rise as a Major Contributor to Science and Technology. *Proc. Natl. Acad. Sci. USA* **2014**, *111*, 9437–9442. [CrossRef]
367. Tollefson, J. China Declared World's Largest Producer of Scientific Articles. *Nature* **2018**, *553*, 390. [CrossRef] [PubMed]
368. Zhou, P.; Leydesdorff, L. The Emergence of China as a Leading Nation in Science. *Res. Policy* **2006**, *35*, 83–104. [CrossRef]
369. Wang, J.; Halffman, W.; Zwart, H. The Chinese Scientific Publication System: Specific Features, Specific Challenges. *Learn. Publ.* **2021**, *34*, 105–115. [CrossRef]
370. Stevens, M.R.; Park, K.; Tian, G.; Kim, K.; Ewing, R. Why Do Some Articles in Planning Journals Get Cited More than Others? *J. Plan. Educ. Res.* **2022**, *42*, 442–463. [CrossRef]
371. Meho, L.I. The Rise and Rise of Citation Analysis. *Phys. World* **2007**, *20*, 32–36. [CrossRef]
372. Larivière, V.; Gong, K.; Sugimoto, C.R. Citations Strength Begins at Home. *Nature* **2018**, *564*, S70–S71. [CrossRef]
373. Zhou, P.; Leydesdorff, L. A Comparative Study of the Citation Impact of Chinese Journals with Government Priority Support. *Front. Res. Metrics Anal.* **2016**, *1*, 3. [CrossRef]
374. Li, W.; Yang, K.; Yin, S.; Yang, X.; Xu, Y.; Lupoi, R. Solid-State Additive Manufacturing and Repairing by Cold Spraying: A Review. *J. Mater. Sci. Technol.* **2018**, *34*, 440–457. [CrossRef]
375. Vanerio, D.; Kondas, J.; Guagliano, M.; Bagherifard, S. 3D Modelling of the Deposit Profile in Cold Spray Additive Manufacturing. *J. Manuf. Process.* **2021**, *67*, 521–534. [CrossRef]

Communication

High-Performance Pure Aluminum Coatings on Stainless Steels by Cold Spray

Jialin Fan [1,2,3,†], **Haitao Yun** [4,†], **Xiaoqiang Zhang** [1,2], **Dongxu Chang** [1], **Xin Chu** [2,*], **Yingchun Xie** [2,*] and **Guosheng Huang** [3,*]

1   School of Materials Science and Engineering, Shenyang University of Technology, Shenyang 110870, China
2   National Engineering Laboratory of Modern Materials Surface Engineering Technology, Guangdong Provincial Key Laboratory of Modern Surface Engineering Technology, Institute of New Materials, Guangdong Academy of Sciences, Guangzhou 510650, China
3   State Key Laboratory for Marine Corrosion and Protection, Luoyang Ship Material Research Institute (LSMRI), Qingdao 266237, China
4   Aecc South Industry Co., Ltd., Zhuzhou 412002, China
*   Correspondence: xin.chu@hotmail.com (X.C.); xieyingchun@gdinm.com (Y.X.); huanggs@sunrui.net (G.H.)
†   These authors contributed equally to this work.

Abstract: Aluminum target material is an important target material and is widely used in preparations of semiconductor films, integrated circuits, display circuits, protective films, decorative films, etc. In this study, pure aluminum coatings were deposited on stainless steel substrates by cold-spray technology as part of an overall project to produce large-size pure aluminum sputtering target materials. The results show that pure aluminum coatings exhibit high adhesive strength (~98 MPa), high deposition efficiency (~95%), and low porosity (~0.3%) on stainless steel substrates. The bonding mechanisms of pure aluminum coatings on stainless steel are a combination of metallurgical and mechanical interlocking. The evolutions of microstructure and mechanical properties of pure aluminum coatings under different heat treatments were also studied. With the increase of heat treatment temperature, it is found that cold-sprayed aluminum coatings become more homogenous in microstructure, the microhardness is reduced, and the adhesive strength seems to be slightly reduced. Overall, this study demonstrates significant advantages of cold-spray technology in depositing high-performance pure aluminum coatings on stainless steels.

Keywords: cold spraying; aluminum; stainless steel; bond strength; heat treatment

Citation: Fan, J.; Yun, H.; Zhang, X.; Chang, D.; Chu, X.; Xie, Y.; Huang, G. High-Performance Pure Aluminum Coatings on Stainless Steels by Cold Spray. *Coatings* **2023**, *13*, 738. https://doi.org/10.3390/coatings13040738

Academic Editor: Cecilia Bartuli

Received: 9 March 2023
Revised: 27 March 2023
Accepted: 3 April 2023
Published: 4 April 2023

1. Introduction

Aluminum target material is an important target material and has a wide range of applications in preparations of semiconductor films, integrated circuits, display circuits, protective films, decorative films, etc. [1–4]. With the development of the market industry, the requirements for the quality of aluminum targets are also increasing. In order to prevent fracture or rupture in the target preparation or transportation process, the target is usually connected to a stainless-steel backing tube, which improves the mechanical properties of the target and also ensures good electrical contact with the target. The traditional methods of joining aluminum with stainless steel include overlay welding and thermal spraying. The overlay-welding process is tedious and costly. The heat input is large and easily produces lots of welding slag and flying chips [5–7]. The thermal-spraying processes, such as HVOF, are mature technology and have benefits such as unrestricted workpiece size and high efficiency; however, due to the high process temperature, metallic elements are easy to oxidize, the coating porosity is relatively high, the microstructural uniformity is poor due to local corrosion, and thick deposits are difficult due to large residual stresses [8–12]. At present, it is still difficult to prepare high-quality, high-bond-strength pure aluminum coatings directly on stainless steel surfaces.

In recent years, cold-spray technology has developed rapidly and has wide application prospects in the fields of advanced functional coatings and additive manufacturing [13–19]. The main advantages of cold-spray technology are as follows: (1) the particle temperature is below melting point, and therefore it can effectively avoid oxidation and phase transformation and retain the high purity of feedstock materials; (2) low thermal effect, high adhesive strength, and high coating density of the deposits; (3) high powder deposition efficiency (DE), high deposition rate, and thick coatings can be obtained. In the literature, there are a few studies on cold spraying pure aluminum on different substrates (e.g., Mg alloys and Al alloys) and their properties. For instance, Bu et al. [20] reported dense and thick cold-sprayed aluminum coatings on AZ91D Mg substrates and found that the coating adhesion strength decreases after heat treatment due to the formation of a brittle Al_3Mg_2 intermetallic layer. Blochet et al. [21] investigated the effect of surface treatment on the bond strength of cold-sprayed pure aluminum coatings on aluminum alloys and showed that the bond strength can reach ~35 MPa with the use of coarse grit sandblasting, which is 40% higher than non-blasted or blasted with fine grits aluminum coatings, indicating that substrate surface roughness has a significant effect on particle adhesion. However, there are few works in the literature on the cold spraying of pure aluminum coating on stainless steel surfaces. Due to the high hardness and high strain-hardening rate of stainless steel, it is generally considered difficult to directly deposit aluminum particles onto it ("soft on hard" mode) by using cold spray due to the lack of substrate deformation [22]. In this study, cold-spray technology was used to assess the feasibility of depositing pure aluminum coatings directly onto stainless steels, and the microstructure, mechanical properties, and the effects of heat treatment on the above metrics were studied.

2. Materials and Methods

2.1. Experimental Materials

The substrate material was 304 stainless steel plates with the size of 100 mm × 100 mm × 1.8 mm. The substrate surface was slightly polished with #180 sandpaper, ultrasonicated to remove surface dirt, rinsed with acetone, and then blown dry. The feedstock powder was pure aluminum powder (XCLL401.1, Institute of New Materials, Guangdong Academy of Sciences). Figure 1a,b show the Scanning Electron Microscope (SEM) morphology of Al powders at 500× and 2000× magnifications. Most of the powder particles have high sphericity, with a small number of irregular satellite particles being present. The particle size distribution of the Al powder measured by a laser particle size analyzer is shown in Figure 1c. The particle size range is Dv (10) 15.4 μm, Dv (50) 25.0 μm, and Dv (90) 44.4 μm.

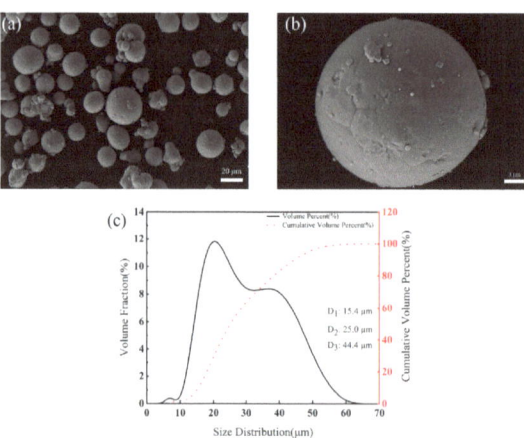

Figure 1. (**a**) 500× SEM, (**b**) 2000× SEM, and (**c**) particle size distribution of the Al powder.

2.2. Coating Preparation

The coating deposition was carried out using a high-pressure cold-spraying system PCS 800 (Plasma Giken Co., Ltd., Hiroshima, Japan), which is located at the Institute of New Materials, Guangdong Academy of Sciences. A plastic nozzle was used which was specifically designed to deposit pure aluminum coatings. Nitrogen was used as the carrier gas. The gas pressure and temperature are two key process parameters for cold spray. In previous trial tests, we studied a wide range of process parameters to cover the entire spraying window of aluminum on stainless steel. Therefore, in this study, we report only the four representative parameters and the deposit performance. For the ease of comparison, we only vary one parameter in each condition, and the 4# is the highest parameters possible to deposit aluminum without any nozzle-clogging issue. The specific process parameters are shown in Table 1.

Table 1. Process parameters for cold spray of Al powder.

No.	Carrier Gas	Gas Pressure (MPa)	Gas Temperature (°C)	Spray Distance (mm)
1#	N_2	3	400	20
2#	N_2	4	400	20
3#	N_2	5	400	20
4#	N_2	5	600	20

2.3. Sample Characterization

The DE was calculated by the weight of powder deposited on the substrate divided by the weight of powder over the substrate. The coating cross-sections were observed using an optical microscope and a field-emission SEM (Gemini SEM300, ZEISS, Jena, Germany). The coating porosity was determined according to the ASTM E2109 standard and was calculated using ImageJ software. The coatings were also etched by Keller's reagent for 15 s to reveal the particle boundaries and grain features. The coating microhardness was measured using a Vickers microhardness tester with a load of 50 g and loading time of 15 s.

In this study, the coating bond strength was measured using two different methods. The first is the usual ASTM C633-2013 method in which coating specimens were sectioned into 25.4 mm diameter cylinders. Then the specimen and the counterpart were glued together with E7 epoxy resin adhesive and then cured at 110 °C for 4 h in fixture; after that, tensile tests were carried out using a universal testing machine (GP-TS2000M, Gopoint, Shenzhen, China). However, this method can be limited by the strength of the epoxy itself (the specified maximum strength of E-7 epoxy adhesive is ~70–80 MPa, and such values also depend on coating materials being tested, surface conditions, curing process, etc.). The second method is a micro-tensile setup reported by our group previously, which was designed to measure the bonding strength of highly adhesive coatings, as shown in Figure 2. In the test, cylinders with the same material as the substrate are machined into the special geometry, as shown in Figure 2a. The coating is then deposited onto the substrate, and then the entire setup is subjected to tension pull-off tests. The failure at the interface is considered to be the coating bond strength and is then calculated as the force (F) divided by the area of the conical end surface.

After the cold-spraying treatment, a heat treatment was also carried out in and argon-atmosphere-protected tube furnace, and samples were heated to 300 °C, 400 °C, and 500 °C and held for 4 h. After the heat treatment, the coatings were etched to show the microstructure, and the microhardness and bonding strength of the coating were characterized.

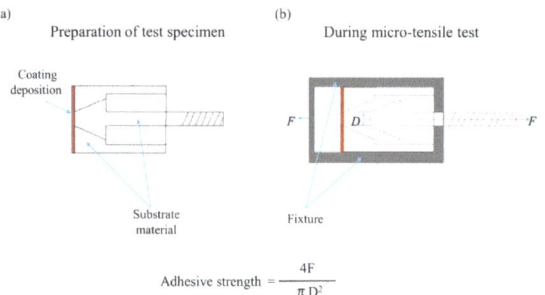

(a) Preparation of test specimen

(b) During micro-tensile test

Coating deposition

Substrate material

Fixture

$$\text{Adhesive strength} = \frac{4F}{\pi D^2}$$

Figure 2. Schematic of the micro-tensile test: (**a**) coating is deposited on the test assembly made of substrate material, (**b**) micro-tensile test is carried out to pull the test assembly apart.

3. Results and Discussion

3.1. Microstructure and Microhardness

Figure 3 shows cross-sectional microstructures of pure aluminum coatings that were cold sprayed at different parameters. At lower parameters (3 MPa 400 °C), there are a few visible pores and defects in the coating. With the increase of gas pressure and temperature, the number and size of pores and defects within the coatings are significantly reduced. The 4# coating (5 MPa 600 °C) is almost dense and free from obvious defects. It is also observed that the interfaces between pure aluminum coatings and stainless steel substrates are continuous and intimate in all scenarios. There are no obvious defects such as cracks and oxides observed, and the substrate is barely deformed at all parameters.

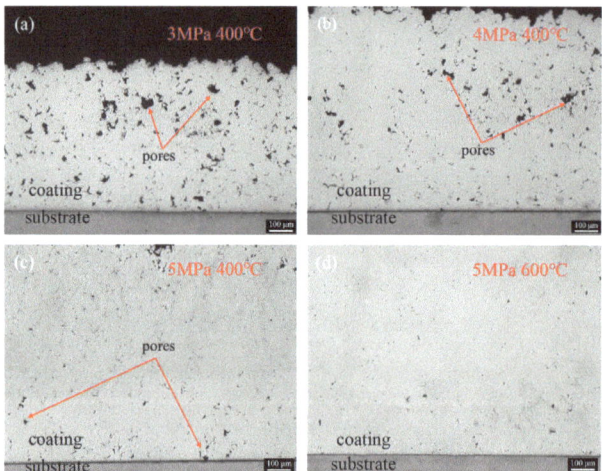

Figure 3. Cross-sectional microstructure of cold-sprayed Al coatings. (**a**) 3 MPa 400 °C; (**b**) 4 MPa 400 °C; (**c**) 5 MPa 400 °C; (**d**) 5 MPa 600 °C.

The gas temperature and pressure are two key factors that affect the in-flight speed of particles. The higher the pressure and temperature, the faster the in-flight speed of particles. Therefore, higher particle velocity and, thus, higher kinetic energy promote the plastic deformation of particles, enabling the particles to elongate into an oblate shape along the vertical direction of spraying; therefore, particle–substrate interfaces are continuously closely bonded, and the as-sprayed coatings are dense [23,24].

The coating porosity was characterized, and the results are shown in Figure 4a. The coating porosity of 1# is ~5.33%, the coating porosity of 2# is ~2.42%, the coating porosity of 3# is ~0.83%, and the coating porosity of 4# is as low as ~0.26%. The coating porosity is consistent with Figure 4a, and it monotonically decreases with the increase of the process parameters. The porosity of the coating is one of the important indicators of the performance of cold-sprayed coatings. Generally speaking, the lower the porosity, the higher the bond strength of the coating. Moreover, to obtain a better sputtering performance for the aluminum target, lower porosity of the coating is preferred.

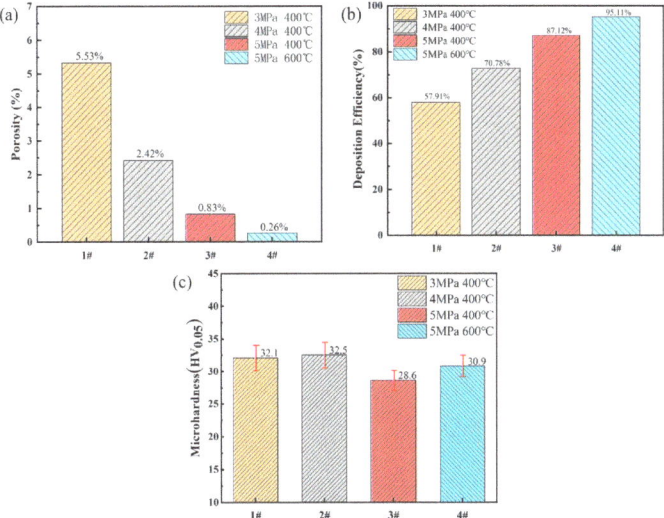

Figure 4. Key metrics of cold-sprayed Al coatings: (**a**) porosity, (**b**) deposition efficiency, and (**c**) microhardness.

The DE results were measured and are shown in Figure 4b. DE indicates the ease with which a powder can be deposited by cold spray. A higher DE for a powder would significantly increase the production efficiency and cost-effectiveness during production. The results show that the DE of Al powder increases with the increase of process parameters. The highest DE of Al powder on 304 stainless steel is 95.11% at process parameters of 5 MPa and 600 °C.

Figure 4c shows the Vickers microhardness of pure aluminum coatings that were cold sprayed under different parameters. The hardness of the 1# coating is 32.1 ± 1.96 HV$_{0.05}$, the hardness of the 2# coating is 32.5 ± 1.81 HV$_{0.05}$, the hardness of the 3# coating is 28.6 ± 1.50 HV$_{0.05}$, and the hardness of the 4# coating is 30.9 ± 1.64 HV$_{0.05}$. For reference, the Vickers microhardness of aluminum powder, as measured, is ~23 HV$_{0.01}$. Hence, the cold-sprayed aluminum coating is a more work-hardened state compared with the Al powder. It is also noted that the coating hardness at higher parameters is lower than it is at lower parameters. Generally, with the increase of gas pressure and temperature, powder particles could achieve higher in-flight velocity, leading to higher particle plastic deformation and more pronounced work-hardening effects [25]. This abnormal phenomenon should indicate that dynamic recrystallization of Al powder occurred at higher process parameters [26].

3.2. Bond Strength

As the key mechanical property index of the coating, the bonding strength determines the service performance and application range of cold-sprayed coatings. The bonding-strength results of the pure aluminum coatings on the 304 stainless steel substrate are reported in Figure 5a. The bond strength was firstly measured by ASTM C633-2013 standard, and all glue failures were observed, as shown in Figure 5b. The bonding strength of the 1# coating is 33 ± 7 MPa, that of the 2# coating is 45 ± 7 MPa, that of the 3# coating is 50 ± 8 MPa, and that of the 4# coating is 68 ± 7 MPa. Although glue failure is not the true coating bonding strength, these results still could be indicative of the relative coating adhesive strength based on our previous experience with pure aluminum coatings. Moreover, to obtain the true adhesive strength, the modified micro-tensile tests were performed to the strongest adhesive coating, #4. The micro-tensile setup is shown in Figure 5c, and the adhesive strength of the 4# coating, as measured, is as high as 98 ± 5 MPa.

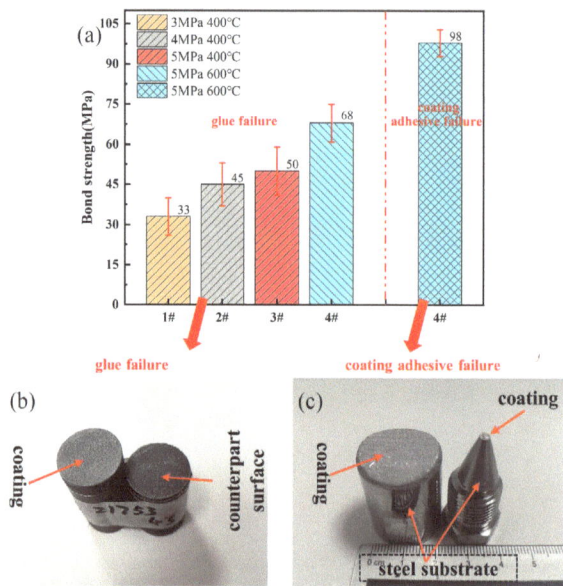

Figure 5. Bond strength results (**a**) and different testing methods of cold sprayed Al coatings: (**b**) ASTM C633 standard, (**c**) micro-tensile test.

To identify the bonding mechanisms of cold-sprayed Al powder on 304 stainless steel, the failure region after the micro-tensile tests was characterized. Figure 6 shows SEM images of the tensile failure section of the 4# coating at the substrate side, and Figure 6b,c show the EDS elemental mapping. The results show there is a certain amount of Al residuals (44.19% in area) on the failed surface, and the substrate surface seems smooth and barely deformed after deposition. Thus, it is reasonable to estimate that strong metallurgical bonding occurred to Al residuals on the stainless steel surface (44.19%), while the rest of the Al coating remained in a state of mechanical interlocking.

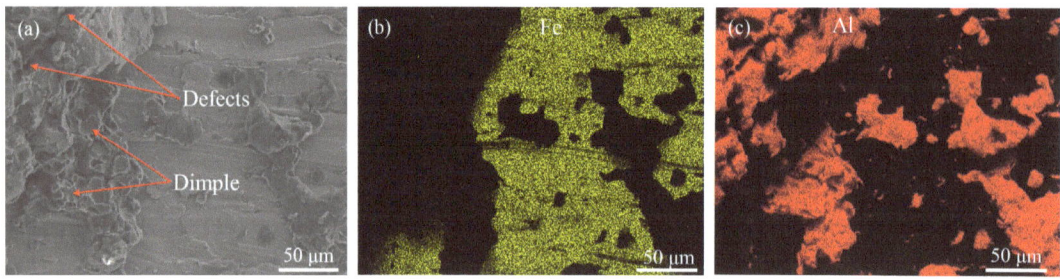

Figure 6. Fracture morphology of cold-sprayed Al coating: (**a**) SEM and (**b,c**) EDS (yellow, Fe; red, Al).

3.3. Heat Treatment

Considering that the #4 coating has the best overall performance, it was then subjected to heat-treatment studies. Figure 7 shows the etched-coating microstructure after different heat treatments. In Figure 7a, when the coating is at the as-sprayed state (RT), the particles are visibly severely deformed, and numerous particle–particle interfaces are clear and obvious. With the increase of heat treatment temperature (300 °C to 500 °C), the coating gradually becomes more homogenous, and the particle–particle interface gradually disappears. This is due to the diffusion of the Al element at evaluated temperatures to minimize the surface energy by "healing" the interfaces. However, at 500 °C (Figure 7d), there seems to be obvious phase or defect formation (in dark contrast) along the coating–substrate interface, and this is discussed in the next section.

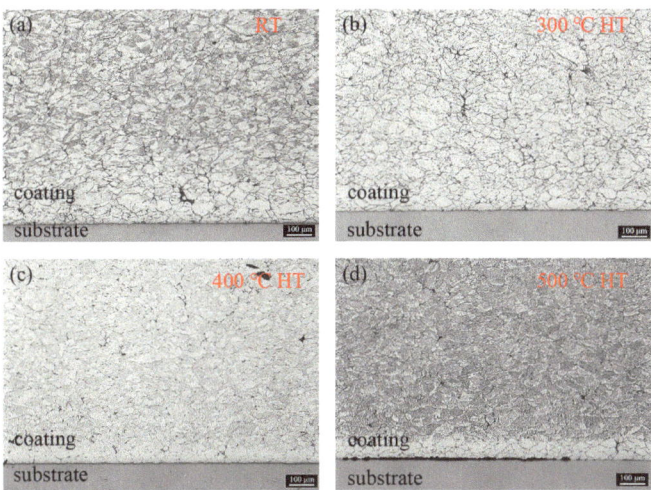

Figure 7. As-etched microstructure of Al coatings: (**a**) RT, (**b**) 300 °C HT, (**c**) 400 °C HT, and (**d**) 500 °C HT.

Figure 8a shows the microhardness of the #4 coating after heat treatment. In general, the coating hardness decreases continuously with the increasing heat-treatment temperature. The hardness of the coating after 4 h of the 500 °C heat treatment is almost identical to that of aluminum powder. The decrement of coating hardness after heat treatment is the combined effect of the release of residual stress, elimination of dislocation density by recovery and recrystallization, grain growth, etc. [27].

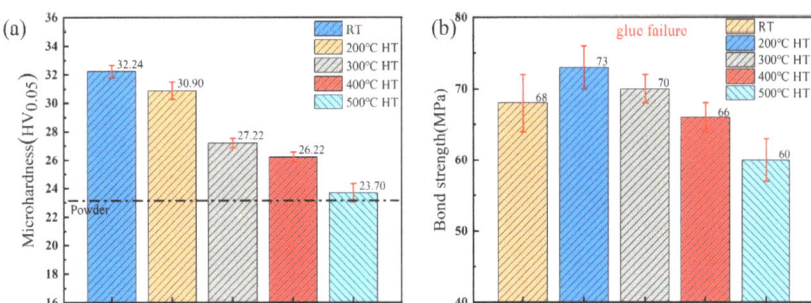

Figure 8. Mechanical properties of Al coatings after heat treatment: (**a**) microhardness and (**b**) bonding strength.

Figure 8 shows the effect of heat treatment on bonding strength, using ASTM C633-2013. Again, the failure mode for all coatings is glue failure. Glue failure means that, after tensile tests, the fracture occurs within the epoxy; thus, the actual coating strength should be higher than the measured value. However, based on our previous experience, such results are still representative of the relative coating bonding strength. The results show that the bonding strength increases slightly with a 200 °C heat treatment and then gradually decreases with the increase of the heat-treatment temperature. The slight increase of the coating bonding strength appears due to the recover phenomenon to release the internal stress. At higher temperatures, as shown in Figure 9, the interdiffusion layer between Al and stainless steel becomes obvious, as well as defects, e.g., cracks at the diffusion layer/Al coating side, and this is speculated to explain the gradual decrease of the coating's bonding strength.

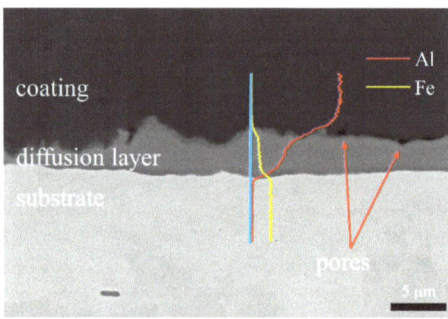

Figure 9. Diffusion layer after heat treatment at 500 °C.

4. Conclusions

In this study, cold-spray technology was used to deposit pure aluminum powder onto a stainless steel substrate as part of an overall project to prepare a large-size aluminum target. Parametric studies were carried out, and the coating microstructure and mechanical properties were characterized. The main conclusions are as follows:

(1) Using N_2 as the carrier gas, pure aluminum coatings with excellent interfacial bonding were successfully prepared on the 304 stainless steel surface. The coating has good overall performance under the process parameters of 5 MPa 600 °C, the bonding strength is ~98 MPa, the DE is ~95%, and the coating porosity is ~0.3%.

(2) With the increase of heat treatment temperature, the cold-sprayed aluminum coating becomes more homogenous in the microstructure, its microhardness is reduced, and the adhesive strength seems to be slightly reduced.

Author Contributions: Conceptualization, J.F. and H.Y.; methodology, D.C. and H.Y.; data curation, J.F. and X.Z.; writing—original draft preparation, J.F.; writing—review and editing, Y.X. and X.C.; supervision, Y.X. and X.C.; project administration, H.Y.; funding acquisition, Y.X. and G.H. All authors have read and agreed to the published version of the manuscript.

Funding: The authors gratefully acknowledge the financial support from the Guangdong Academy of Sciences Special Fund for Comprehensive Industrial Technology Innovation Center Building (No. 2022GDASZH-2022010107), International Cooperation Platform of Science and Technology, Guangdong Academy of Sciences (2022GDASZH-2022010203-003), Key-Area Research and Development Program of Guangdong Province (No. 2020B0101330001), Science Center for Gas Turbine Project (P2022-B-IV-011-001), Guangdong Provincial Key Laboratory of Modern Surface Engineering Technology (No. 2020B1212060049),Research Fund of State Key Laboratory for Marine Corrosion and Protection of Luoyang Ship Material Research Institute (LSMRI) under the contract No. 6142901200303.

Institutional Review Board Statement: Not applicable.

Informed Consent Statement: Not applicable.

Data Availability Statement: Data are contained within the article.

Acknowledgments: The authors would like to thank Liang Zeng and Yunda Nong from the Institute of New Materials, Guangdong Academy of Sciences, for cold spraying experiments. The authors also thank Xundong Sun for his assistance in the experiments and Pengyuan Diao and Yi Wu for their valuable comments.

Conflicts of Interest: The authors declare no conflict of interest.

References

1. Barajas-Valdes, U.; Suárez, O.M. Nanomechanical properties of thin films manufactured via magnetron sputtering from pure aluminum and aluminum-boron targets. *Thin Solid Film.* **2020**, *693*, 137670. [CrossRef]
2. Warren, T.L. The effect of target inertia on the penetration of aluminum targets by rigid ogive-nosed long rods. *Int. J. Impact Eng.* **2016**, *91*, 6–13. [CrossRef]
3. Tang, E.; Zhao, L.; Han, Y.; Chen, C.; Chang, M. Research on the electromagnetic propagating characteristics of hypervelocity impact on the target with aperture and different potential conditions. *Aerosp. Sci. Technol.* **2020**, *107*, 106274. [CrossRef]
4. Rajesh Kumar, B.; Hymavathi, B.; Subba Rao, T. XRD and AFM Studies on Nanostructured Zinc Aluminum Oxide Thin Films Prepared by Multi-Target Magnetron Sputtering. *Mater. Today Proc.* **2017**, *4*, 8638–8644. [CrossRef]
5. Patil, U.S.; Kadam, M.S. Microstructural analysis of SMAW process for joining stainless steel 304 with mild steel 1018 and parametric optimization by using response surface methodology. *Mater. Today Proc.* **2021**, *44*, 1811–1815. [CrossRef]
6. Li, Z.X.; Zhang, L.M.; Ma, A.L.; Hu, J.X.; Zhang, S.; Daniel, E.F.; Zheng, Y.G. Comparative study on the cavitation erosion behavior of two different rolling surfaces on 304 stainless steel. *Tribol. Int.* **2021**, *159*, 106994. [CrossRef]
7. Liao, H.; Zhang, W.; Xie, H.; Li, X.; Zhang, Q.; Wu, X.; Tian, J.; Wang, Z. Effects of welding speed on welding process stability, microstructure and mechanical performance of SUS304 welded by local dry underwater pulsed MIG. *J. Manuf. Process.* **2023**, *88*, 84–96. [CrossRef]
8. Bai, M.; Reddy, L.; Hussain, T. Experimental and thermodynamic investigations on the chlorine-induced corrosion of HVOF thermal sprayed NiAl coatings and 304 stainless steels at 700 °C. *Corros. Sci.* **2018**, *135*, 147–157. [CrossRef]
9. Gorlach, I.A. A new method for thermal spraying of Zn–Al coatings. *Thin Solid Film.* **2009**, *517*, 5270–5273. [CrossRef]
10. Gibbons, G.J.; Hansell, R.G. Thermal-sprayed coatings on aluminium for mould tool protection and upgrade. *J. Mater. Process. Technol.* **2008**, *204*, 184–191. [CrossRef]
11. Sun, B.; Fukanuma, H.; Ohno, N. Study on stainless steel 316L coatings sprayed by a novel high pressure HVOF. *Surf. Coat. Technol.* **2014**, *239*, 58–64. [CrossRef]
12. Han, M.-S.; Woo, Y.-B.; Ko, S.-C.; Jeong, Y.-J.; Jang, S.-K.; Kim, S.-J. Effects of thickness of Al thermal spray coating for STS 304. *Trans. Nonferrous Met. Soc. China* **2009**, *19*, 925–929. [CrossRef]
13. Yin, S.; Suo, X.; Guo, Z.; Liao, H.; Wang, X. Deposition features of cold sprayed copper particles on preheated substrate. *Surf. Coat. Technol.* **2015**, *268*, 252–256. [CrossRef]
14. Srikanth, A.; Mohammed Thalib Basha, G.; Venkateshwarlu, B. A Brief Review on Cold Spray Coating Process. *Mater. Today Proc.* **2020**, *22*, 1390–1397. [CrossRef]
15. Bagherifard, S.; Guagliano, M. Fatigue performance of cold spray deposits: Coating, repair and additive manufacturing cases. *Int. J. Fatigue* **2020**, *139*, 105744. [CrossRef]
16. Meng, X.-M.; Zhang, J.-B.; Han, W.; Zhao, J.; Liang, Y.-L. Influence of annealing treatment on the microstructure and mechanical performance of cold sprayed 304 stainless steel coating. *Appl. Surf. Sci.* **2011**, *258*, 700–704. [CrossRef]
17. Coddet, P.; Verdy, C.; Coddet, C.; Debray, F.; Lecouturier, F. Mechanical properties of thick 304L stainless steel deposits processed by He cold spray. *Surf. Coat. Technol.* **2015**, *277*, 74–80. [CrossRef]
18. Meng, X.; Zhang, J.; Zhao, J.; Liang, Y.; Zhang, Y. Influence of Gas Temperature on Microstructure and Properties of Cold Spray 304SS Coating. *J. Mater. Sci. Technol.* **2011**, *27*, 809–815. [CrossRef]
19. Rokni, M.R.; Widener, C.A.; Champagne, V.R. Microstructural stability of ultrafine grained cold sprayed 6061 aluminum alloy. *Appl. Surf. Sci.* **2014**, *290*, 482–489. [CrossRef]
20. Bu, H.; Yandouzi, M.; Lu, C.; Jodoin, B. Post-heat Treatment Effects on Cold-Sprayed Aluminum Coatings on AZ91D Magnesium Substrates. *J. Therm. Spray Technol.* **2012**, *21*, 731–739. [CrossRef]
21. Blochet, Q.; Delloro, F.; N'Guyen, F.; Jeulin, D.; Borit, F.; Jeandin, M. Effect of the Cold-Sprayed Aluminum Coating-Substrate Interface Morphology on Bond Strength for Aircraft Repair Application. *J. Therm. Spray Technol.* **2017**, *26*, 671–686. [CrossRef]
22. Luo, X.-T.; Li, S.-P.; Li, G.-C.; Xie, Y.-C.; Zhang, H.; Huang, R.-Z.; Li, C.-J. Cold spray (CS) deposition of a durable silver coating with high infrared reflectivity for radiation energy saving in the polysilicon CVD reactor. *Surf. Coat. Technol.* **2021**, *409*, 126841. [CrossRef]
23. Levasseur, D.; Yue, S.; Brochu, M. Pressureless sintering of cold sprayed Inconel 718 deposit. *Mater. Sci. Eng. A* **2012**, *556*, 343–350. [CrossRef]
24. Song, X.; Jin, X.-Z.; Zhai, W.; Tan, A.W.-Y.; Sun, W.; Li, F.; Marinescu, I.; Liu, E. Correlation between the macroscopic adhesion strength of cold spray coating and the microscopic single-particle bonding behaviour: Simulation, experiment and prediction. *Appl. Surf. Sci.* **2021**, *547*, 149165. [CrossRef]
25. Xie, Y.; Planche, M.-P.; Raoelison, R.; Hervé, P.; Suo, X.; He, P.; Liao, H. Investigation on the influence of particle preheating temperature on bonding of cold-sprayed nickel coatings. *Surf. Coat. Technol.* **2017**, *318*, 99–105. [CrossRef]

26.   Wong, W. *Understanding the Effects of Process Parameters on the Properties of Cold Gas Dynamic Sprayed Pure Titanium Coatings*; McGill University: Montréal, QC, Canada, 2012.

27.   Wei, Y.-K.; Luo, X.-T.; Chu, X.; Huang, G.-S.; Li, C.-J. Solid-state additive manufacturing high performance aluminum alloy 6061 enabled by an in-situ micro-forging assisted cold spray. *Mater. Sci. Eng. A* **2020**, *776*, 139024. [CrossRef]

MDPI AG
Grosspeteranlage 5
4052 Basel
Switzerland
Tel.: +41 61 683 77 34

Coatings Editorial Office
E-mail: coatings@mdpi.com
www.mdpi.com/journal/coatings

www.ingramcontent.com/pod-product-compliance
Lightning Source LLC
LaVergne TN
LVHW072352090526
838202LV00019B/2526